Lecture Notes in Computer Science 15176

Founding Editors

Gerhard Goos
Juris Hartmanis

Editorial Board Members

Elisa Bertino, *Purdue University, West Lafayette, IN, USA*
Wen Gao, *Peking University, Beijing, China*
Bernhard Steffen, *TU Dortmund University, Dortmund, Germany*
Moti Yung, *Columbia University, New York, NY, USA*

The series Lecture Notes in Computer Science (LNCS), including its subseries Lecture Notes in Artificial Intelligence (LNAI) and Lecture Notes in Bioinformatics (LNBI), has established itself as a medium for the publication of new developments in computer science and information technology research, teaching, and education.

LNCS enjoys close cooperation with the computer science R & D community, the series counts many renowned academics among its volume editors and paper authors, and collaborates with prestigious societies. Its mission is to serve this international community by providing an invaluable service, mainly focused on the publication of conference and workshop proceedings and postproceedings. LNCS commenced publication in 1973.

Svetla Petkova-Nikova · Daniel Panario
Editors

Arithmetic of Finite Fields

10th International Workshop, WAIFI 2024
Ottawa, ON, Canada, June 10–12, 2024
Revised Selected Papers

Editors
Svetla Petkova-Nikova ⓘ
KU Leuven
Leuven, Belgium

Daniel Panario ⓘ
Carleton University
Ottawa, ON, Canada

ISSN 0302-9743 ISSN 1611-3349 (electronic)
Lecture Notes in Computer Science
ISBN 978-3-031-81823-3 ISBN 978-3-031-81824-0 (eBook)
https://doi.org/10.1007/978-3-031-81824-0

© The Editor(s) (if applicable) and The Author(s), under exclusive license to Springer Nature Switzerland AG 2025

This work is subject to copyright. All rights are solely and exclusively licensed by the Publisher, whether the whole or part of the material is concerned, specifically the rights of translation, reprinting, reuse of illustrations, recitation, broadcasting, reproduction on microfilms or in any other physical way, and transmission or information storage and retrieval, electronic adaptation, computer software, or by similar or dissimilar methodology now known or hereafter developed.
The use of general descriptive names, registered names, trademarks, service marks, etc. in this publication does not imply, even in the absence of a specific statement, that such names are exempt from the relevant protective laws and regulations and therefore free for general use.
The publisher, the authors and the editors are safe to assume that the advice and information in this book are believed to be true and accurate at the date of publication. Neither the publisher nor the authors or the editors give a warranty, expressed or implied, with respect to the material contained herein or for any errors or omissions that may have been made. The publisher remains neutral with regard to jurisdictional claims in published maps and institutional affiliations.

This Springer imprint is published by the registered company Springer Nature Switzerland AG
The registered company address is: Gewerbestrasse 11, 6330 Cham, Switzerland

If disposing of this product, please recycle the paper.

Preface

These are the proceedings of WAIFI 2024, the 10th International Workshop on the Arithmetic of Finite Fields, held in Ottawa, Canada, during June 10–12, 2024. The previous editions of this workshop were held in Madrid, Spain (WAIFI 2007), Siena, Italy (WAIFI 2008), Istanbul, Turkey (WAIFI 2010), Bochum, Germany (WAIFI 2012), Gebze, Turkey (WAIFI 2014), Gent, Belgium (WAIFI 2016), Bergen, Norway (WAIFI 2018), Rennes, France (WAIFI 2020) and Chengdu, China (WAIFI 2022). Springer has published all previous volumes of the WAIFI Proceedings in the LNCS series.

Since 2008, WAIFI has been held every even year, bringing together mathematicians, computer scientists, engineers and physicists who conduct research in different areas of finite field arithmetic.

The program was composed of four invited talks and 17 presentations of contributed papers. The invited speakers were:

- Marco Baldi, Università Politecnica delle Marche, Italy;
- Koray Karabina, National Research Council, Canada;
- Chloe Martindale, University of Bristol, UK;
- Maria Montanucci, DTU, Denmark

Each of the invited speakers provided an extended abstract of their lecture for the proceedings. The contributed talks were selected from 29 submissions, each of which was assigned to at least three committee members or external sub-reviewers chosen by the PC members. Additionally, the Program Committee had a significant online discussion phase for several days.

We are very grateful to the members of the Program Committee for their dedication, professionalism and careful work during the review and selection process. We also sincerely thank the external reviewers who contributed with their expertise to review papers for this workshop.

We also would like to thank the General Co-chairs David Thomson and Qiang (Steven) Wang from Carleton University, Canada for their support of the Program Committee and their hard work in leading the overall organization of the workshop. We would also like to sincerely thank members of the Steering Committee of the workshop series for their constant support and encouragement in our efforts to create a stimulating scientific program leading to this volume. Furthermore, we thank José Luis Imaña for diligently maintaining the workshop website and to Sihem Mesnager for helping us with the LNCS proceedings. As with the previous volumes, Springer agreed to publish the revised and expanded versions of the WAIFI 2024 papers as an LNCS volume. We thank the staff from Springer for making this possible.

The submission and selection of papers were done using the EasyChair conference management system. Hence, thank you EasyChair! We would also like to acknowledge the Fields Institute, Carleton University, Tutte Institute for Mathematics and Computing and the Institute of Combinatorics and its Applications for being sponsors of the workshop.

Finally, but most importantly, we deeply thank all the authors who submitted their papers to the workshop and the participants all over the world who chose to honour us with their attendance.

September 2024

Daniel Panario
Svetla Petkova-Nikova

Organization

Steering Committee

Lilya Budaghyan	University of Bergen, Norway
Claude Carlet	Universities of Paris VIII, France and Bergen, Norway
Anwar Hasan	University of Waterloo, Canada
José Luis Imaña	Complutense University of Madrid, Spain
Çetin Kaya Koç	University of California Santa Barbara, USA
Sihem Mesnager	University of Paris VIII, France
Ferruh Özbudak	Middle East Technical University, Turkey
Svetla Petkova-Nikova	KU Leuven, Belgium
Francisco Rodríguez-Henríquez	TII, United Arab Emirates
Erkay Savaş	Sabancı University, Turkey

General Co-chairs

Daniel Panario	Carleton University, Canada
David Thomson	Carleton University, Canada
Qiang (Steven) Wang	Carleton University, Canada

Program Co-chairs

Svetla Petkova-Nikova	KU Leuven, Belgium
Daniel Panario	Carleton University, Canada

Publicity Chair

José Luis Imaña	Complutense University of Madrid, Spain

Program Committee

Claude Carlet	Universities of Paris VIII, France, and Bergen, Norway

Wouter Castryck	KU Leuven, Belgium
Thomas Decru	KU Leuven, Belgium
Sylvain Duquesne	University of Rennes, France
Guang Gong	University of Waterloo, Canada
Anna-Lena Horlemann	University of St. Gallen, Switzerland
Sophie Huczynska	University of St. Andrews, Scotland
Thais Bardini Idalino	Federal University of Santa Catarina, Brazil
José Luis Imaña	U. Complutense Madrid, Spain
Jorge Jiménez Urroz	Polytechnic University of Madrid, Spain
Angshuman Karmakar	IIT Kanpur, India
Elena Kirshanova	TII, UAE
Sihem Mesnager	University of Paris VIII, France
Lucia Moura	University of Ottawa, Canada
Alessandro Neri	University of Ghent, Belgium
Svetla Petkova-Nikova (Program Co-chair)	KU Leuven, Belgium
Daniel Panario (Program Co-chair)	Carleton University, Canada
Hilder Vitor Lima Pereira	Unicamp, Brazil
Håvard Raddum	University of Bergen, Norway
Francisco Rodríguez Henríquez	TII, UAE
Amin Sakzad	Monash University, Australia
David Thomson	Carleton University, Canada
Alev Topuzoğlu	Sabancı University, Turkey
Geertrui Van de Voorde	University of Canterbury, New Zealand
Qiang (Steven) Wang	Carleton University, Canada
Nusa Zidaric	University of Leiden, The Netherlands

Additional Reviewers

Supriya Adhikary	IIT Kanpur, India
Farzane Amirzade	Carleton University, Canada
Gustavo Biage	UFSC, Brazil
María de Los Angeles Chara	CONICET Santa Fe, Argentina
Anindya Ganguly	IIT Kanpur, India
Nikai Jagganath	Monash University, Australia
Giorgos Kapetanakis	University of Thessaly, Greece
Suparna Kundu	KU Leuven, Belgium
Reynald Lercier	University of Rennes, France
Rati Ludhani	Indian Institute of Technology Bombay, India
Kalikinkar Mandal	University of New Brunswick, Canada

Guillermo Matera	UNGS, Argentina
Wilfried Meidl	University of Klagenfurt, Austria
Puja Mondal	IIT Kanpur, India
Semyon Novoselov	Baltic Federal University, Russia
Ferruh Ozbudak	Sabanci Universtity, Turkey
Leo Perrin	Inria, France
John Proos	CSE, Canada
Paolo Santonastaso	University of Campania, Italy
Ilya Soloveychik	University of Jerusalem, Israel
Douglas Stinson	University of Waterloo, Canada
Carlos Vela	University of St. Gallen, Switzerland
Colin Weir	Canadian Centre for Cyber Security, Canada
Raymond Zhao	CSIRO, Australia
Gewei Zheng	University of Waterloo, Canada

Invited Talks

The Restricted Decoding Problem and Its Application to Post-quantum Cryptography

Marco Baldi

Università Politecnica delle Marche, Ancona, Italy
m.baldi@univpm.it

Code-based cryptographic trapdoors, along with lattice-based ones, are at the basis of some of the most interesting post-quantum cryptographic primitives. Starting from the original McEliece cryptosystem [1], the problem of decoding in the Hamming metric searching for solutions with small weight has classically been used to design code-based cryptosystems. In fact, most code-based asymmetric encryption and key exchange schemes that have been or are still competing within the NIST post-quantum cryptography standardization process follow this classical paradigm. Examples of this are Classic McEliece, which exactly follows the original McEliece scheme, LEDAcrypt and BIKE, which although replacing the classical Goppa codes with the modern families of QC-LDPC [2] and QC-MDPC [3] codes enabling more compact public keys, still rely on the difficulty of solving the decoding problem by finding solutions with small weight in the Hamming metric.

However, this approach turns out to be less promising when it comes to digital signatures. In fact, classical code-based digital signature schemes following the classical approach, such as CFS [4] and its evolutions, exhibit significant shortcomings in terms of security, efficiency, or both. For this reason, other variants of the decoding problem have begun to be considered, which appear promising for overcoming these limitations. One of these variants is the decoding problem searching for solutions with large weight. The difficulty of solving this problem over non-binary fields has been used to construct code-based digital signature schemes, such as SPANSE [5] and WAVE [6], following the hash-and-sign paradigm. Although this has produced more efficient schemes than previous ones, their public key size is still large and this may limit their practical use.

More recently, new variants of the decoding problem have been introduced that look for solutions subject to certain restrictions, such as having only entries belonging to a restricted subset of the finite field over which the code is defined. These new variants of the decoding problem are known as restricted decoding problem [7, 8]. If the restricted set does not include the null element, the restricted decoding problem also requires finding a solution having maximum Hamming weight. These new variants of the decoding problem have been successfully used in the design of new code-based digital signature schemes that no longer follow the hash-and-sign paradigm, but rather derive digital signatures from the transcription of a zero-knowledge identification protocol, by applying the Fiat-Shamir transformation [9]. A noteworthy example is the digital signature scheme named CROSS [10], which is among the candidates in the ongoing additional

NIST call for selection and standardization of post-quantum digital signature schemes [11].

References

1. McEliece, R.J.: A public-key cryptosystem based on algebraic coding theory. *DSN Progress Report*, pp. 114–116 (1978)
2. Baldi, M., Bianchi, M., Chiaraluce, F.: Security and complexity of the McEliece cryptosystem based on quasi-cyclic low-density parity-check codes. IET. Inf. Secur. **7**, 212–220 (2013)
3. Misoczki, R., Tillich, J.-P., Sendrier, N., Barreto, P. S. L.xM.: MDPC-McEliece: new McEliece variants from moderate density parity-check codes. In: 2013 IEEE International Symposium on Information Theory, pp. 2069–2073. Istanbul, Turkey (2013)
4. Courtois, N., Finiasz, M., Sendrier, N.: How to achieve a McEliece-based digital signature scheme. In: Boyd, C. (ed.) ASIACRYPT 2001. LNCS, vol. 2248, pp. 157–174. Springer, Heidelberg (2001)
5. Baldi, M., Chiaraluce, F., Santini, P.: SPANSE: Combining sparsity with density for efficient one-time code-based digital signatures. J. Algebra Appl. **23**(7), 2550099 (2024)
6. Debris-Alazard, T., Sendrier, N., Tillich, J.-P.: Wave: a new family of trapdoor one-way preimage sampleable functions based on codes. In: Galbraith, S.D., Moriai, S. (eds.) ASIACRYPT 2019. LNCS, vol. 11921, pp. 21–51. Springer, Cham (2019)
7. Baldi, M., et al.: A new path to code-based signatures via identification schemes with restricted errors. *arXiv preprint 2008.06403* (2020)
8. Baldi, M., Bitzer, S., Pavoni, A., Santini, P., Wachter-Zeh, A., Weger, V.: Zero knowledge protocols and signatures from the restricted syndrome decoding problem. In: Tang, Q., Teague, V. (eds.) PKC 2024. LNCS, vol 14602, pp. 243–274. Springer, Cham (2024)
9. Fiat, A., Shamir, A.: How to prove yourself: practical solutions to identification and signature problems. In: Odlyzko, A.M. (ed.) CRYPTO 1986. LNCS, vol. 263, pp. 186–194. Springer, Heidelberg (1987)
10. Baldi, M., et al.: CROSS: codes & restricted objects signature scheme. Submission to NIST Call for Additional Digital Signature Schemes for the Post-Quantum Cryptography Standardization Process (2023)
11. Alagic, G., et al.: Status report on the third round of the nist post-quantum cryptography standardization process. Technical Report NIST IR 8413-upd1, National Institute of Standards and Technology (NIST) (2022)

Algebraic Curves Over Finite Fields: Rational Points and Birational Invariants

Maria Montanucci [ID]

Technical University of Denmark, 2800 Kongens Lyngby, Denmark
marimo@dtu.dk
https://www.compute.dtu.dk/

Abstract. Algebraic curves over finite fields have attracted a lot of attention during the last decades, both from the purely theoretical point of view, and for their versatility in applications. In the study of algebraic curves it is important to understand the behaviour of their *birational invariants*, that is, the properties of an algebraic curve that are invariant under isomorphism. Well-known examples are the genus, the p-rank, the automorphism group and Weierstrass semigroups. Of particular relevance in applications is the family of the so-called *maximal curves*, which are the algebraic curves with the largest number of rational points among all curves of the same genus. The birational invariants of maximal curves show exceptional behaviour among algebraic curves over finite fields, and this peculiarity has been source of great fascination for many years. In this survey three main questions will be addressed: how can one costruct/classify maximal curves over finite fields using birational invariants? How do the birational invariant automorphism group behave for curves of p-rank zero (particularly maximal curves)? Since Weierstrass semigroups are of extreme important in coding theory (AG codes), how can one compute Weierstrass semigroups on maximal curves efficiently? Our contributions to answer these questions will be discussed.

Keywords: Algebraic Curves · Finite Fields · Birational Invariants

1. Birational Invariants and Maximal Curves Over Finite Fields

Algebraic curves over a finite field $\mathbb{F}_q$ have been a source of great fascination, ever since the seminal work of Hasse and Weil in the 1930s and 1940s. Many fruitful ideas have arisen out of this area, where number theory and algebraic geometry meet, and many applications of the theory of algebraic curves have been discovered during the last decades.

A very important example of such application was provided in 1977-1982 by Goppa, who found a way to use algebraic curves in coding theory (AG codes). The key point of Goppa's construction is that the code parameters are essentially expressed in terms of the features of the curve, such as the number N_q of $\mathbb{F}_q$-rational points and the genus g.

In this light, Goppa codes with good parameters are constructed from curves with large N_q with respect to their genus g.

Given a smooth projective, algebraic curve of genus g over $\mathbb{F}_q$, an upper bound for N_q is a corollary to the celebrated Hasse-Weil Theorem,

$$N_q \leq q + 1 + 2g\sqrt{q}.$$

Curves attaining this bound are called $\mathbb{F}_q$-maximal. Unless $g = 0$ (i.e. the curve is a line), maximality clearly requires that q is a square. Since isomorphic curves give rise to codes with the same parameters, it is natural to study the properties of an algebraic curve up to isomorphism, or the so-called *birational invariant*. Famous birational invariant are the genus g mentioned above, the automorphism group (over the algebraic closure of $\mathbb{F}_q$), the p-rank and the Weierstrass semigroups at points of the curve. Maximal curves over $\mathbb{F}_q$ have very structured birational invariants, for example, they always gave p-rank equal to zero and genus at most $q(q-1)/2$.

The Hermitian curve is a key example of an $\mathbb{F}_q$-maximal curve, as it is the unique curve, up to isomorphism, attaining the maximum possible genus $q(q-1)/2$ of an $\mathbb{F}_q$-maximal curve (as mentioned above). It is a result commonly attributed to Serre that any curve which is $\mathbb{F}_q$-covered by an $\mathbb{F}_q$-maximal curve is still $\mathbb{F}_q$-maximal. In particular, quotient curves of $\mathbb{F}_q$-maximal curves are $\mathbb{F}_q$-maximal. Many examples of $\mathbb{F}_q$-maximal curves have been constructed as quotient curves of the Hermitian curve by choosing a subgroup of its very large automorphism group.

The following three subsections present the three main questions of this survey, and our contribution to them.

1.1 Construction/Classification of Maximal Curves

It is a challenging problem to construct maximal curves that cannot be obtained using Serre's result, as well as to construct maximal curves with many automorphisms (in order to use the machinery described above). In fact this problem is so difficult that it was conjectured that every maximal curve is up to isomorphism a subcover of the Hermitian curve. The first known counterexample was found only in 2009 in [13] (GK-curves). This family of curves has been generalized in [8] and [2] in 2010 and 2018 respectively.

The generalization [8] emulates the equations defining the family in [13], while [2] generalizes the property of having a large automorphism group as an extension of the one of the Hermitian curve. This difference in the automorphism groups of [8] and [2] is crucial as it shows that these two families are not isomorphic in general.

Apart from constructing maximal curves, another natural question arises: given two maximal curves over the same finite field, how can one decide whether they are isomorphic or not? A way to try to give an answer to this question, as for the two examples above, is to look at the birational invariants of the two curves, that is, their properties that are invariant under isomorphism.

This idea has been used in a recent preprint to construct a large family of non-isomorphic maximal curves [5].

1.2 Large Automorphism Groups and Curves with p-rank Zero

Let $\mathcal{X}$ be an algebraic curve defined over the algebraic closure $\mathbb{K}$ of a finite field $\mathbb{F}_q$. By a classical result, $Aut(\mathcal{X})$ is finite whenever the genus g of $\mathcal{X}$ is at least two; see [22] and [11, Chapter 11; 22--24,31,33]. Several results on the interaction between the automorphism group, the genus and the p-rank of a curve can be found in the literature. A remarkable example is the work of Nakajima [20] who showed that the value of the p-rank deeply influences the order of a p-group of automorphisms of $\mathcal{X}$. He also showed that curves for which the p-rank is the largest possible, namely $\gamma = g$, have at most $84(g^2 - g)$ automorphisms.

In [12] Hurwitz showed that if $\mathcal{X}$ is defined over $\mathbb{C}$ then $|Aut(\mathcal{X})| \leq 84(g - 1)$, which is known as *Hurwitz bound*. This bound is sharp, i.e., there exist algebraic curves over $\mathbb{C}$ of arbitrarily high genus whose automorphism group has order exactly $84(g-1)$. Well-known examples are the Klein quartic and the Fricke-Macbeath curve, see [17]. Roquette [21] showed that Hurwitz bound also holds in positive characteristic p, if p does not divide $|Aut(\mathcal{X})|$.

A general bound in positive characteristic is $|Aut(\mathcal{X})| \leq 16g^4$ with one exception: the so-called Hermitian curve. This result is due to Stichtenoth [23, 24].

The quartic bound $|Aut(\mathcal{X})| \leq 16g^4$ was improved by Henn in [10]. Henn's result shows that if $|Aut(\mathcal{X})| > 8g^3$ then $\mathcal{X}$ is isomorphic to one of four explicit and well-studied algebraic curves, all of which having p-rank zero. This observation raised the following problem: is it possible to find a (optimal) function $f(g)$ such that the existence of an automorphism group G of $\mathcal{X}$ with $|G| \geq f(g)$ implies $\gamma = 0$?

Clearly from Henn's result $f(g) \leq 8g^3$. Also $f(g)$ cannot be asymptotically of order less than $g^{3/2}$ as algebraic curves of positive p-rank with approximately $g^{3/2}$ automorphisms are known; see for example [15]. The open problem above was already studied in [14] where a positive answer is given under the additional hypothesis that g is even or that the automorphism group G is solvable.

In 2022 [18] we analyzed large automorphism groups of curves of arbitrary genus $g \geq 2$ giving a partial answer to aforementioned open problem. There the following is proven for G automorphism group of an algebraic curve $\mathcal{X}$ of genus $g \geq 2$ and p-rank γ:

1. If $|G| > 24g^2$ then either G has a unique (non-tame) short orbit or it has exactly two short orbits, one tame and one non-tame.
2. If G has exactly one non-tame short orbit then $|G| \leq 336g^2$.
3. If $|G| \geq 60g^2$ and G has exactly one non-tame short orbit then γ is positive and congruent to zero modulo p.
4. If $|G| \geq 900g^2$ then G has exactly one non-tame short orbit O_1 and one tame short orbit O_2. If $\mathcal{X}/G_P^{(1)}$ is rational for $P \in O_1$ and the stabilizer $G_{P,R}$ with $R \in O_1 \setminus \{P\}$ is either a p-group or a prime-to-p group then γ is zero.

Note that this theorem implies that whenever a quadratic bound like $|Aut(\mathcal{X})| > 336g^2$ holds, then the action of the group is completely known, having one tame and one non-tame short orbits.

1.3 Weierstrass Semigroups at Points of Maximal Curves

Computing the Weierstrass semigroup at every point of a maximal curve is a well-known difficult problem, and before the investigation conducted in [3], this complete analysis has been conducted only for the Hermitian curve (and some specific subcovers) in 1986 [9]. The main idea developed in [3] (where a conjecture from 2011[7] has been solved in this way) is that the strong geometric properties of a maximal curve can be used to find function realizing non-gaps at points implicitly, that is, without the need of computing them explicitly. The geometry of a maximal curve is encoded in the so-called *Natural embedding Theorem* by Korchmáros and Torres [16]. This theorem says that a curve is maximal over $\mathbb{F}_q$ if and only if it can be embeeded in some projective space of dimension $r \geq 2$ as a curve of degree $q+1$ contained in an Hermitian variety. The method developed in [3] is versatile, and in fact opened the way for many other investigations that followed naturally for other famous families of maximal curves, see e.g. [1, 4, 6, 19] among others.

Acknowledgments. The research related to this survey was supported by a research grant (VIL"52303") from Villum Fonden. I would like to thank Ferdinando Zullo for the many hours he spent discussing about applications of algebraic curves with me during the conference; and the organisers, chairs, and community of WAIFI 2024.

References

1. Bartoli, D., Montanucci, M., Zini, G.: Weierstrass semigroups at every point of the Suzuki curve. Acta Arith. **20**, 1–20 (2021)
2. Beelen, P., Montanucci, M.: A new family of maximal curves. J. London Math. Soc. **98**, 573–592 (2018)
3. Beelen, P., Montanucci, M.: Weierstrass semigroups on the Giulietti–Korchmáros curve. Finite Fields Appl. **52**, 10–29 (2018)
4. Beelen, P., Montanucci, M., Landi, L.: Weierstrass semigroups on the Skabelund maximal curve. Finite Fields Appl. **72**, 101811 (2021)
5. Beelen, P., Montanucci, M., Niemann, J.T., Quoos, L.: Some families of non-isomorphic maximal function fields. preprint, arXiv:2404.14179 (2024)
6. Beelen, P., Montanucci, M., Vicino, L.: Weierstrass semigroups and automorphism group of a maximal curve with the third largest genus. Finite Fields Appl. **92**, 102300 (2023)
7. Duursma, I. : Two-point coordinate rings for GK-curves. IEEE Trans. Inf. Theory **57**, 593–600 (2011)
8. Garcia, A., Güneri, C., Stichtenoth, H.: A generalization of the Giulietti–Korchmáros maximal curve. Adv. Geom. **10**, 427–434 (2010)
9. Garcia, A., Viana, P.: Weierstrass points on certain non-classical curves. Arch. Math. **46**, 315–322 (1986)
10. Henn, H.W.: Funktionenkörper mit großer Automorphismengruppe. J. Reine Angew. Math. **302**, 96–115 (1978)
11. Hirschfeld, J.W.P., Korchmáros, G., Torres, F.: Algebraic Curves Over a Finite Field. Princeton Series in Applied Mathematics, Princeton (2008)

12. Hurwitz, A.: Über algebraische Gebilde mit eindeutigen Transformationen in sich. Math. Ann. **41**, 403–442 (1893)
13. Giulietti, M., Korchmáros, G.: A new family of maximal curves over a finite field. Math. Ann. **343**, 229–245 (2009)
14. Giulietti, M., Korchmáros, G.: Algebraic curves with many automorphisms. Adv. Math. **349**, 162–211 (2019)
15. Korchmáros, G., Montanucci, M., Speziali, P.: Transcendence degree one function fields over a finite field with many automorphisms. J. Pure Appl. Algebra **222**, 1810–1826 (2018)
16. Korchmáros, G., Torres, F.: Embedding of a Maximal Curve in a Hermitian Variety. Compositio Math. **128**, 95–113 (2001)
17. Macbeath, A.M. : On a theorem of Hurwitz. Proc. Glasgow Math. Assoc. **5**, 90–96 (1961)
18. Montanucci, M.: On algebraic curves with many automorphisms in characteristic p. Mathematische Zeitschrift **301**, 3695–3711 (2022)
19. Montanucci, M., Pallozzi Lavorante, V.: AG codes from the second generalization of the GK maximal curve. Discr. Math. **343**, 111810 (2020)
20. Nakajima, S.: p-ranks and automorphism groups of algebraic curves. Trans. Amer. Math. Soc. **303**, 595–607 (1987)
21. Roquette, P.: Abschätzung der Automorphismenanzahl von Funktionenkörpern bei Primzahl charakteristik. Math. Z. **117**, 157–163 (1970)
22. Schmid, H.L.: Über die Automorphismen eines algebraischen Funktionenkörpers von Primzahl charakteristik. J. Reine Angew. Math. **179**, 5–15 (1938)
23. Stichtenoth, H.: Über die Automorphismengruppe eines algebraischen Funktionenkörpers von Primzahl- charakteristik. I. Eine Abschätzung der Ordnung der Automorphismengruppe. Arch. Math. **24** (1973), 527–544
24. Stichtenoth, H.: Über die Automorphismengruppe eines algebraischen Funktionenkörpers von Primzahl- charakteristik. II. Ein spezieller Typ von Funktionenkörpern. Arch. Math. **24**, 615–631 (1973)

An Overview of Mathematical Problems, Cryptosystems, and their Interconnected Nature

Koray Karabina[1,2]

[1] National Research Council of Canada, Canada
[2] University of Waterloo, Canada

Abstract. Integer factorization, discrete logarithm, subset-sum, and lattice problems have significantly influenced cryptosystem design and motivated the development of algorithms to efficiently tackle the underlying problem instances. We present an overview of these problems and cryptosystems, focusing on their noteworthy interconnections, key applications, and security claims in the post-quantum era.

Keywords: Factorization · Discrete logarithm · Subset sum · Lattices

Could Subset Sum-Based Cryptosystems Rise Again?

Between 1976 and 1978, four public-key cryptosystems rose: Diffie-Hellman [6], RSA [12], McEliece [9], and Merkle-Hellman [10]. Their security relies on the computational intractability of solving discrete logarithm, integer factorization, decoding general linear codes, and subset sum problems. Diffie-Hellman and RSA cryptosystems have resisted cryptanalysis efforts on classical computers and seen widespread deployment. Shor's polynomial-time algorithms [14] (1994) for solving factorization and discrete logarithm problems on a quantum computer motivated the community to design new cryptosystems that are secure against quantum-powered adversaries. The original McEliece is still believed to be secure in the post-quantum era, and a significant number of post-quantum cryptosystems have emerged through code-based, hash-based, isogeny-based, lattice-based, and multivariate cryptography [1]. Interestingly, subset sum-based cryptosystems have not gained much recognition for being quantum-secure despite the underlying problem being NP-complete and the best-known quantum algorithms for solving it having exponential complexity [2]. This could mainly be because of the sudden fall of the Merkle-Hellman cryptosystem, which was broken first by Shamir [13], followed by two lattice attacks, exploiting the low-density of the system, by Brickell [3]; and by Lagarias and Odlyzko [7]. Chor and Rivest [5] (1988) avoided the low-density issue by cleverly using low weights in the subset sum to increase density. It is worth noting that implementing the Chor-Rivest cryptosystem requires solving discrete logarithms in finite fields, which poses a challenge to scale system parameters. The requirement has been removed in the powerline system designed by Lenstra [8]; and its extension fractional powerline, by Camion and Chabanne [4]. Some of the proposed Chor-Rivest instances did not withstand algebraic attacks by Vaudenay [15] (1998). Furthermore, Nguyn and Stern showed [11] (2005) that lattice attacks on subset sum-based cryptosystems can hardly be prevented due to efficient reductions from low-weight subset sum to the shortest vector problem. It remains uncertain whether efficient reductions to lattice problems lead to polynomial-time attacks on subset sum-based cryptosystems. Could subset sum-based cryptosystems rise again?

References

1. NIST Post-quantum cryptography standardization.https://csrc.nist.gov/Projects/post-quantum-cryptography/post-quantumcryptography-standardization. Accessed 10 July 2024
2. Bonnetain, X., Bricout, R., Schrottenloher, A., Shen, Y.: Improved classical and quantum algorithms for subset-sum. In: Advances in Cryptology – ASIACRYPT 2020, pp. 633–666 (2020)
3. Brickell, E.F.: Breaking iterated knapsacks. In: Advances in Cryptology, pp. 342–358. Springer, Heidelberg (1985)
4. Camion, P., Chabanne, H.: On the powerline system. In: Information and Communications Security, pp. 381–385 (1997)
5. Chor, B., Rivest, R.: A knapsack-type public key cryptosystem based on arithmetic in finite fields. IEEE Trans. Inf. Theory **34**(5), 901–909 (1988). https://doi.org/10.1109/18.21214
6. Diffie, W., Hellman, M.: New directions in cryptography. IEEE Trans. Inf. Theory **22**(6), 644–654 (1976). https://doi.org/10.1109/TIT.1976.1055638
7. Lagarias, J.C., Odlyzko, A.M.: Solving low-density subset sum problems. J. ACM **32**(1), 229–246 (1985). https://doi.org/10.1145/2455.2461
8. Lenstra, H.W.: On the chor–rivest knapsack cryptosystem. J. Cryptology **3**(3), 149–155 (1991). https://doi.org/10.1007/BF00196908
9. McEliece, R.J.: A public-key cryptosystem based on algebraic coding theory. DSN Progress Report, Jet Propulsion Laboratory, pp. 42–44 (1978)
10. Merkle, R., Hellman, M.: Hiding information and signatures in trapdoor knapsacks. IEEE Trans. Inf. Theory **24**(5), 525–530 (1978). https://doi.org/10.1109/TIT.1978.1055927
11. Nguyễn, P.Q., Stern, J.: Adapting density attacks to low-weight knapsacks. In: Advances in Cryptology - ASIACRYPT 2005, pp. 41–58 (2005)
12. Rivest, R.L., Shamir, A., Adleman, L.: A method for obtaining digital signatures and public-key cryptosystems. Commun. ACM **21**(2), 120–126 (1978). https://doi.org/10.1145/359340.359342
13. Shamir, A.: A polynomial-time algorithm for breaking the basic Merkle – Hellman cryptosystem. IEEE Trans. Inf. Theory **30**(5), 699–704 (1984). https://doi.org/10.1109/TIT.1984.1056964

14. Shor, P.: Algorithms for quantum computation: discrete logarithms and factoring. In: Proceedings 35th Annual Symposium on Foundations of Computer Science, pp. 124–134 (1994). https://doi.org/10.1109/SFCS.1994.365700
15. Vaudenay, S.: Cryptanalysis of the chor-rivest cryptosystem. In: Advances in Cryptology — CRYPTO'98, pp. 243–256 (1998)

Making and Breaking Post-quantum Cryptography from Elliptic Curves

Chloe Martidale

University of Bristol
`chloe.martindale@bristol.ac.uk`

Most of the public-key cryptography in use today relies on the hardness of either factoring or the discrete logarithm problem in a specially chosen abelian group. Here "hard" does not mean mathematically impossible but that the best known algorithm to solve the problem has complexity (sub-)exponential in the size of the input. However, once scalable quantum computers become a reality, both factoring and the discrete logarithm problem will no longer be hard problems, due to Shor's polynomial-time quantum algorithm to solve both problems. Post-quantum cryptography is about designing new cryptographic primitives based on different hard problems in mathematics for which there is no known polynomial-time classical or quantum algorithm. In this talk we will show how to design post-quantum cryptographic primitives from the hard problem of, given two elliptic curves over a large finite field, find and compute and isogeny between them (if it exists). We will then discuss recent work giving an attack on one of these primitives, Supersingular Isogeny Diffie-Hellman (SIDH). This is joint work with Luciano Maino, Lorenz Panny, Giacomo Pope, and Benjamin Wesolowski.

Contents

Coding Theory

Determining the Complete Weight Distributions of Some Families
of Cyclic Codes 3
   Gerardo Vega and Félix Hernández

Central Limit Theorem for Linear Eigenvalue Statistics of Random
Matrices from Binary Linear Codes 19
   Chin Hei Chan and Maosheng Xiong

On Decoding Hyperbolic Codes 37
   *Eduardo Camps-Moreno, Ignacio García-Marco, Hiram H. López,
   Irene Márquez-Corbella, Edgar Martínez-Moro, and Eliseo Sarmiento*

Fast Decoding of Group Testing Results from Reed-Solomon d-Disjunct
Matrices 53
   Dongxia Luo and Lucia Moura

Quantum CSS Duadic and Triadic Codes: New Insights and Properties 70
   *Reza Dastbasteh, Olatz Sanz Larrarte, Josu Etxezarreta Martinez,
   Antonio deMarti iOlius, Javier Oliva del Moral, and Pedro Crespo Bofill*

Cryptography and Boolean Functions

Prescribing Traces of Primitive Elements in Finite Fields 91
   Lucas Reis

On Cryptographic Properties of a Class of Power Permutations in Odd
Characteristic 99
   Mohit Pal

Generating Gaussian Pseudorandom Noise with Binary Sequences 117
   Francisco-Javier Soto, Ana I. Gómez, and Domingo Gómez-Pérez

An FPGA Accelerated Search Method for Maximum Period NLFSRs 127
   Amund Askeland

On Fat Linearized Polynomials 139
   Olga Polverino, Paolo Santonastaso, and Ferdinando Zullo

Counting Polynomials with Distinct Roots Using Subset Sum 154
Simon Kuttner, Zhicheng Gao, and Qiang Wang

Generalized Class Group Actions on Oriented Elliptic Curves with Level
Structure 171
*Sarah Arpin, Wouter Castryck, Jonathan Komada Eriksen,
Gioella Lorenzon, and Frederik Vercauteren*

Differential Biases, c-Differential Uniformity, and Their Relation
to Differential Attacks 191
Daniele Bartoli, Lukas Kölsch, and Giacomo Micheli

On the Walsh and Fourier-Hadamard Supports of Boolean Functions From
a Quantum Viewpoint 213
Claude Carlet, Ulises Pastor-Díaz, and José M. Tornero

Postquantum Cryptography

Efficient Batch Post-quantum Signatures with Crystals Dilithium 237
Nazlı Deniz Türe and Murat Cenk

Ursa Minor: The Implementation Framework for Polaris 258
*Mohammadtaghi Badakhshan, Guiwen Luo, Tanmayi Jandhyala,
and Guang Gong*

SMALL: Scalable Matrix OriginAted Large Integer PoLynomial
Multiplication Accelerator for Lattice-Based Post-Quantum Cryptography 274
*Jiafeng Xie, Pengzhou He, Samira Carolina Oliva Madrigal,
and Çetin Kaya Koç*

Author Index 293

Coding Theory

Determining the Complete Weight Distributions of Some Families of Cyclic Codes

Gerardo Vega[1]( ) and Félix Hernández[2]

[1] Dirección General de Cómputo y de Tecnologías de Información y Comunicación, Universidad Nacional Autónoma de México, 04510 Ciudad de México, Mexico
gerardov@unam.mx
[2] Posgrado en Ciencia e Ingeniería de la Computación, Universidad Nacional Autónoma de México, 04510 Ciudad de México, Mexico
felixhdz@ciencias.unam.mx

Abstract. Obtaining the complete weight distributions for nonbinary codes is an even harder problem than obtaining their Hamming weight distributions. In fact, obtaining these distributions is a problem that usually involves the evaluation of sophisticated exponential sums, which leaves this problem open for most of the linear codes. In this work we present a method that uses the known complete weight distribution of a given cyclic code, to determine the complete weight distributions of other cyclic codes. In addition we also obtain the complete weight distributions for a particular kind of one- and two-weight irreducible cyclic codes, and use these distributions and the method, in order to determine the complete weight distributions of infinite families of cyclic codes. As an example, and as a particular instance of our results, we determine in a simple way the complete weight distribution for one of the two families of reducible cyclic codes studied by Bae, Li and Yue [Discrete Mathematics, 338 (2015) 2275-2287].

Keywords: Complete weight enumerator · Weight distribution · One- and two-weight irreducible cyclic codes · Cyclic codes · Gauss sums

1 Introduction

The complete weight distribution of a code enumerates the codewords by the number of symbols of each kind contained in each codeword. Therefore, the complete weight distribution of a code contains much more information than the Hamming weight distribution. In fact, the complete weight distribution has a wide range of applications in many research fields as the information it contains is of vital use in practical applications. For example, as pointed out in

---

This manuscript is partially supported by PAPIIT-UNAM IN107423. The second author has also received research support from CONAHCyT, México.

© The Author(s), under exclusive license to Springer Nature Switzerland AG 2025
S. Petkova-Nikova and D. Panario (Eds.): WAIFI 2024, LNCS 15176, pp. 3–18, 2025.
https://doi.org/10.1007/978-3-031-81824-0_1

[2] the complete weight distribution of Reed-Solomon codes could be helpful in soft decision decoding. As another example, the complete weight distribution is useful in the computation of the Walsh transform of monomial functions over finite fields [6]. Unfortunately, determining the complete weight distribution is an even harder problem than obtaining the Hamming weight distribution. As a consequence, the complete weight distribution is unknown for most codes.

For this reason, determining the complete weight distributions of either linear codes or cyclic codes over finite fields has received a great deal of attention in recent years (see for example [1,3,10,11,16–19]). In this work we present a method that uses the known complete weight distribution of a given cyclic code, to determine the complete weight distribution of other cyclic codes. In addition we also obtain the complete weight distributions for a particular kind of one- and two-weight irreducible cyclic codes, and use these distributions and the method, in order to determine the complete weight distribution of infinite families of cyclic codes. As an example, and as a particular instance of our results, we determine in a simple way the complete weight distribution for one of the two families of reducible cyclic codes studied in [1]. As another example we also determine the complete weight distributions for another family of cyclic codes which, as we shall see later, can be obtained in terms of the complete weight distribution of the subclass of optimal three-weight cyclic codes recently reported in [15].

This work is organized as follows: In Sect. 2 we establish the notation, give some definitions, and recall some known results. Particularly, we recall a result that determines the Hamming weight distributions of all one- and two-weight semiprimitive irreducible cyclic codes. By using such result, the complete weight distributions for a particular kind of one- and two-weight irreducible cyclic codes is determined in Sect. 3. A method for determining new complete weight distributions, in terms of known ones, is presented in Sect. 4. In Sect. 5, we use the complete weight distributions obtained in Sect. 3, and the method in Sect. 4, in order to determine the complete weight distributions of infinite families of cyclic codes. As examples, and as particular instances of our results, two of these families are presented in this section. Finally, Sect. 6 is devoted to conclusions.

2 Notation, Definitions and Known Results

First of all we set for this section and for the rest of this work, the following:

Notation. Let p, t, q, m, and Δ, denote positive integers such that p is a prime number, $q = p^t$ and $\Delta = \frac{q^m-1}{q-1}$. From now on, γ will denote a fixed primitive element of $\mathbb{F}_{q^m}$. Let u be an integer such that $u|(q^m-1)$. For $i = 0, 1, \cdots, u-1$, we define $\mathcal{C}_i^{(u,q^m)} := \gamma^i \langle \gamma^u \rangle$, where $\langle \gamma^u \rangle$ denotes the subgroup of $\mathbb{F}_{q^m}^*$ generated by γ^u. The cosets $\mathcal{C}_i^{(u,q^m)}$ are called the *cyclotomic classes* of order u in $\mathbb{F}_{q^m}$. For an integer u, such that $\gcd(p,u) = 1$, p is said to be *semiprimitive modulo u* if there exists a positive integer d such that $u|(p^d+1)$. We will denote by "Tr", the *absolute trace mapping* from either $\mathbb{F}_{q^m}$ or $\mathbb{F}_q$ to the prime field $\mathbb{F}_p$, and by "$\mathrm{Tr}_{\mathbb{F}_{q^m}/\mathbb{F}_q}$" the *trace mapping* from $\mathbb{F}_{q^m}$ to $\mathbb{F}_q$. Let $s \in \mathbb{F}_q$, and let

$V = (v_0, v_1, \cdots, v_{n-1})$ be a vector of length n over $\mathbb{F}_q$. We define *the number of occurrences of the symbol s in V*, $\mathcal{N}(s, V)$, as the number of times that s appears as an entry in the vector V. That is:

$$\mathcal{N}(s, V = (v_0, v_1, \cdots, v_{n-1})) := |\{i | s = v_i, 0 \leq i < n\}|.$$

An $[n, l, d]$ linear code, $\mathscr{C}$, over $\mathbb{F}_q$ is an l-dimensional subspace of $\mathbb{F}_q^n$ with minimum Hamming distance d, and the vectors of $\mathscr{C}$ are called *codewords*. A code $\mathscr{C}$ is *cyclic* if it is linear and if $(c_0, c_1, \ldots, c_{n-1}) \in \mathscr{C}$ implies $(c_{n-1}, c_0, \ldots, c_{n-2}) \in \mathscr{C}$. A cyclic code is irreducible (reducible) if its *parity-check polynomial* (see for example [13, p. 194]) is irreducible (reducible). Let A_i be the number of codewords with Hamming weight i in $\mathscr{C}$ (recall that the *Hamming weight* of a codeword $\mathbf{c}$ is the number of nonzero coordinates in $\mathbf{c}$). Then, the sequence 1, A_1, ..., A_n is called the *Hamming weight distribution* of the linear code $\mathscr{C}$, and the polynomial $1 + A_1 T + \ldots + A_n T^n$ is called the *Hamming weight enumerator* of $\mathscr{C}$. If $\sharp\{1 \leq i \leq n : A_i \neq 0\} = M$, then $\mathscr{C}$ is called an M-*weight* code.

In a similar way let $\mathscr{C}$ be a code of length n over $\mathbb{F}_q$. Denote the q elements of $\mathbb{F}_q$ by $u_0 = 0, u_1, \cdots, u_{q-1}$ in some fixed order. By denoting $\mathbb{N}_0 := \mathbb{N} \cup \{0\}$, we define the *complete weight* of a vector $\mathbf{v} = (v_0, v_1, \cdots, v_{n-1}) \in \mathbb{F}_q^n$, as the vector $w_{\text{cplt}}(\mathbf{v}) := (f_1, f_2, \cdots, f_{q-1}) \in \mathbb{N}_0^{q-1}$, where f_l $(1 \leq l < q)$ is the number of components v_j $(0 \leq j < n)$ of $\mathbf{v}$ that are equal to u_l. In addition, for a vector $\boldsymbol{f} = (f_1, f_2, \cdots, f_{q-1}) \in \mathbb{N}_0^{q-1}$ we denote by $Z^{\boldsymbol{f}}$ the monomial in the $q-1$ variables $(z_1, z_2, \cdots, z_{q-1})$ given by

$$Z^{\boldsymbol{f}} := z_1^{f_1} z_2^{f_2} \cdots z_{q-1}^{f_{q-1}},$$

Now, for a linear code $\mathscr{C}$ of length n over $\mathbb{F}_q$, we define the *set of complete nonzero weights* of $\mathscr{C}$, $W_\mathscr{C}$, by the set:

$$W_\mathscr{C} := \{w_{\text{cplt}}(\mathbf{c}) \mid \mathbf{c} \text{ is a nonzero codeword in } \mathscr{C}\},$$

and for each complete nonzero weight $\boldsymbol{w} \in W_\mathscr{C}$, we define its frequency, $A_{\boldsymbol{w}}$, as:

$$A_{\boldsymbol{w}} := \sharp\{\mathbf{c} \in \mathscr{C} \mid w_{\text{cplt}}(\mathbf{c}) = \boldsymbol{w}\}.$$

The sequence $1, \{A_{\boldsymbol{w}}\}_{\boldsymbol{w} \in W_\mathscr{C}}$ is called the *complete weight distribution* of the linear code $\mathscr{C}$, whereas the polynomial

$$\text{CWE}_\mathscr{C}(Z) := 1 + \sum_{\boldsymbol{w} \in W_\mathscr{C}} A_{\boldsymbol{w}} Z^{\boldsymbol{w}}, \tag{1}$$

is called its *complete weight enumerator*.

Remark 1. Let n be as before, and let $f_0 : \mathbb{N}_0^{q-1} \to \mathbb{N}_0$ be the function given by

$$f_0(f_1, f_2, \cdots, f_{q-1}) = n - \sum_{i=1}^{q-1} f_i.$$

Thus, it is important to observe that a quite common definition for the complete weight enumerator (see for example [13, p. 141]) is:

$$\text{CWE}_{\mathscr{C}}(Z) := z_0^n + \sum_{w \in W_{\mathscr{C}}} A_w z_0^{f_0(w)} Z^w.$$

For linear codes these two definitions are equivalent and, for the convenience of this work, we are going to use (1). In addition, observe also that (1) coincides with the Hamming weight enumerator when $q = 2$ and contains much more information if $q > 2$.

The following gives an explicit description of an irreducible cyclic code of length n and dimension $\text{ord}_n(q)$ (the order of q modulo n; the smallest integer $m > 0$ for which $q^m \equiv 1 \pmod{n}$) over $\mathbb{F}_q$.

Definition 1. *Let n, N and N' be integers such that $N = \gcd(q^m - 1, N')$ and $nN = q^m - 1$. Then the set*

$$\mathcal{I}_{N'} := \{\mathbf{c}(a) | a \in \mathbb{F}_{q^m}\},$$

where

$$\mathbf{c}(a) := (\text{Tr}_{\mathbb{F}_{q^m}/\mathbb{F}_q}(a\gamma^{N'i}))_{i=0}^{n-1},$$

is an irreducible cyclic code of length n and dimension $\text{ord}_n(q)$ over $\mathbb{F}_q$.

Remark 2. Note that $\mathcal{I}_N$ and $\mathcal{I}_{N'}$ are in general two different irreducible cyclic codes, however they are equivalent in the sense that both share the same length $n = \frac{q^m - 1}{N}$, the same dimension $\text{ord}_n(q)$, and the same Hamming and complete weight distribution.

Main Assumption. From now on, we use n and N as integers in such a way that $nN = q^m - 1$, assuming that $m = \text{ord}_n(q)$. Under these circumstances, note that if $h_N(x) \in \mathbb{F}_q[x]$ is the *minimal polynomial* of γ^{-N} (see for example [13, p. 99]), then, due to Delsarte's Theorem [5], $h_N(x)$ is the parity-check polynomial of an irreducible cyclic code of length n and dimension m over $\mathbb{F}_q$.

The *canonical additive character* of $\mathbb{F}_q$ is defined as follows:

$$\chi(x) := e^{2\pi \sqrt{-1}\, \text{Tr}(x)/p}, \quad \text{for all } x \in \mathbb{F}_q.$$

Let $a \in \mathbb{F}_q$, then the orthogonality relation for χ is

$$\sum_{x \in \mathbb{F}_q} \chi(xa) = \begin{cases} q & \text{if } a = 0, \\ 0 & \text{otherwise.} \end{cases}$$

This property plays an important role in numerous applications of finite fields. Among them, this property is useful for determining the number of zero entries in a given vector; for example, if $\mathbf{v} = (a_0, a_1, \ldots, a_{n-1}) \in \mathbb{F}_q^n$, then

$$\mathcal{N}(0,\mathbf{v}) = \frac{1}{q}\sum_{i=0}^{n-1}\sum_{y\in\mathbb{F}_q}\chi(ya_i) . \qquad (2)$$

If $\langle\lambda\rangle = \mathbb{F}_q^*$, then any *multiplicative character* of $\mathbb{F}_q$ is defined by

$$\psi_j(\lambda^l) := e^{2\pi\sqrt{-1}\,jl/(q-1)} , \quad \text{for } j,l = 0,1,\cdots,q-2 .$$

If q is odd, an important multiplicative character of $\mathbb{F}_q$ is the so-called *quadratic character* which is denoted by η and defined by: $\eta(x) = 1$ if x is the square of an element of $\mathbb{F}_q^*$ and $\eta(x) = -1$ otherwise.

Let ψ be a multiplicative and χ an additive character of a finite field F. Then the *Gaussian sum*, $G_F(\psi,\chi)$, of ψ and χ over the finite field F is defined by

$$G_F(\psi,\chi) := \sum_{x\in F^*} \psi(x)\chi(x) .$$

Determining the value of a Gaussian sum is, in general, a difficult task. However, for the canonical additive character and the quadratic character of a finite field, we have the following result:

Theorem 1 *[12, Theorem 5.15, p. 199]. With our notation, let η be the quadratic character of $\mathbb{F}_q$ and let χ be the canonical additive character of $\mathbb{F}_q$. Assume that $q = p^t$ is odd. Then*

$$G_{\mathbb{F}_q}(\eta,\chi) = \begin{cases} (-1)^{t-1}q^{1/2} & \text{if } p \equiv 1 \pmod{4}, \\ (-1)^{t-1}(\sqrt{-1})^t q^{1/2} & \text{if } p \equiv 3 \pmod{4}. \end{cases}$$

The following known result gives a full description for all one-weight and semiprimitive two-weight irreducible cyclic codes over any finite field.

Theorem 2 *[14, Theorem 2]. Let n, N, and $\mathcal{I}_N$ be as in Definition 1. Fix $u = \gcd(\Delta, N)$. Assume that $u = 1$ or p is semiprimitive modulo u. Let d be the smallest positive integer such that $u|(p^d+1)$ and let $s = 1$ if $u = 1$ and $s = (mt)/(2d)$ if $u > 1$. Let $\mathbf{c}(a) \in \mathcal{I}_N$ and fix*

$$W_A = \frac{nq^{m/2-1}}{\Delta}(q^{m/2} - (-1)^{s-1}(u-1)) , \quad W_B = \frac{nq^{m/2-1}}{\Delta}(q^{m/2} - (-1)^s) ,$$

and

$$\delta := \begin{cases} 0 & \text{if } u = 1; \text{ or } p = 2; \text{ or } p > 2 \text{ and } 2|s; \\ & \text{or } p > 2, 2 \nmid s, \text{ and } 2|\frac{p^d+1}{u} , \\ \frac{u}{2} & \text{if } p > 2, 2 \nmid s \text{ and } 2 \nmid \frac{p^d+1}{u} . \end{cases}$$

Then,

$$w_H(\mathbf{c}(a)) = \begin{cases} 0 & \text{if } a = 0 , \\ W_A & \text{if } a \in C_\delta^{(u,q^m)} , \\ W_B & \text{if } a \in \mathbb{F}_{q^m}^* \setminus C_\delta^{(u,q^m)} , \end{cases} \qquad (3)$$

where $w_H(\cdot)$ stands for the usual Hamming weight function. Therefore, since $|\mathcal{C}_\delta^{(u,q^m)}| = \frac{q^m-1}{u}$, $\mathcal{I}_N$ is either a one-weight ($u = 1$) or a two-weight ($u > 1$) $[n, m]$ irreducible cyclic code whose Hamming weight enumerator is

$$1 + \frac{q^m-1}{u}T^{W_A} + \frac{(q^m-1)(u-1)}{u}T^{W_B}. \qquad (4)$$

Some desirable properties of a linear code are that it is optimal and that it has few nonzero weights (see for example [4]). The complete weight enumerator of a subclass of optimal three-weight cyclic codes was recently presented. We now recall such result by means of the following:

Theorem 3 *[15, Theorem 1]. Let e_2 and e_3 be integers. If $\gcd(e_3, q^2 - 1) = 1$ and $e_3 \equiv e_2 \pmod{q-1}$, then $h(x) := h_{(q+1)e_2}(x)h_{e_3}(x)$ is the parity-check polynomial of an optimal three-weight $[q^2 - 1, 3, q(q-1) - 1]$ cyclic code, $\mathscr{C}$, whose complete weight enumerator, $\mathrm{CWE}_{\mathscr{C}}(Z)$, is*

$$\mathrm{CWE}_{\mathscr{C}}(Z) = 1 + (q-1)\prod_{i=1}^{q-1}z_i^{q+1} + (q^2-1)\left(\prod_{i=1}^{q-1}z_i^{q} + \sum_{j=1}^{q-1}z_j\prod_{i=1,i\neq j}^{q-1}z_i^{q+1}\right).$$

3 Some Preliminary Results

Through the following result, we determine the complete weight distribution for a particular kind of one- or two-weight irreducible cyclic codes in Theorem 2.

Proposition 1. *Consider the same notation and assumption as in Theorem 2. In addition, assume also that N is a proper divisor of Δ. Then $\mathcal{I}_N$ is either a one- or two-weight irreducible cyclic code whose complete weight enumerator is*

$$\mathrm{CWE}_{\mathcal{I}_N}(Z) = 1 + \frac{q^m-1}{N}\prod_{i=1}^{q-1}z_i^{\epsilon_1} + \frac{(q^m-1)(N-1)}{N}\prod_{i=1}^{q-1}z_i^{\epsilon_2}, \qquad (5)$$

where

$$\epsilon_1 := \frac{W_A}{q-1} \quad \text{and} \quad \epsilon_2 := \frac{W_B}{q-1}.$$

Proof. In the light of Theorem 2, it is sufficient to determine the complete weight enumerator of $\mathcal{I}_N$. Since $u = \gcd(\Delta, N)$ and $N|\Delta$, $u = N$ and $n = \frac{q^m-1}{u}$. Let $\mathbf{c}(a) \in \mathcal{I}_N$, $\tau = \frac{\Delta}{u}$ and consider the $\frac{n}{\tau} = q - 1$ vectors, V_j, given by:

$$V_j := (\mathrm{Tr}_{\mathbb{F}_{q^m}/\mathbb{F}_q}(a\gamma^{u(\tau j+i)}))_{i=0}^{\tau-1}, \text{ for } j = 0, 1, \cdots, q-2.$$

Thus, note that $\mathbf{c}(a) = V_0|V_1|\cdots|V_{q-2}$, where the operator "$|$" stands for the vector concatenation. On the other hand, recall that $\langle \gamma^\Delta \rangle = \mathbb{F}_q^*$ and note that the length of the vector V_j is $\tau = \frac{\Delta}{u}$. Therefore,

$$V_j = (\mathrm{Tr}_{\mathbb{F}_{q^m}/\mathbb{F}_q}(a\gamma^{\Delta j+ui}))_{i=0}^{\tau-1} = \gamma^{\Delta j}(\mathrm{Tr}_{\mathbb{F}_{q^m}/\mathbb{F}_q}(a\gamma^{ui}))_{i=0}^{\tau-1} = \gamma^{\Delta j}V_0,$$

for $j = 0, 1, \cdots, q-2$. This means that, if $x := \mathscr{N}(1, \mathbf{c}(a))$, then $\mathscr{N}(s, \mathbf{c}(a)) = x$ for all $s \in \mathbb{F}_q^*$. In consequence, the result now follows from (3), (4), and the fact that $|\mathbb{F}_q^*| = q - 1$. □

Remark 3. In the previous proposition if q is odd and $N = 2$, then, without loss of generality, $\epsilon_1 = \frac{q^{m-1}+q^{m/2-1}}{2}$ and $\epsilon_2 = \frac{q^{m-1}-q^{m/2-1}}{2}$, and note that $\epsilon_1 + \epsilon_2 = q^{m-1}$.

When q and m are odd integers, we can also determine the complete weight distribution for some of the one-weight irreducible cyclic codes in Theorem 2.

Proposition 2. *With our current notation, suppose that q and m are odd. Let N be an integer such that $\gcd(N, q^m - 1) = 2$ and let $\mathcal{I}_N$ be as in Definition 1. Let $\lambda = \gamma^\Delta$ and fix the elements of $\mathbb{F}_q$ as $u_0 = 0$, $u_i = \lambda^{i-2\lfloor \frac{i}{2} \rfloor} \lambda^{2\lfloor \frac{i}{2} \rfloor}$, for $i = 1, 2, \cdots, q-1$ (observe that $u_{q-1} = 1$). Let $\mathcal{O}$ be the subset of odd integers in $\{1, 2, \cdots, q-1\}$, that is $\mathcal{O} := \{1, 3, \cdots, q-2\}$. Then $\mathcal{I}_N$ is a $[\frac{q^m-1}{2}, m, \frac{q^{m-1}(q-1)}{2}]$ one-weight irreducible cyclic code whose complete weight enumerator is*

$$\text{CWE}_{\mathcal{I}_N}(Z) = 1 + \frac{q^m - 1}{2} \left(\prod_{i \in \mathcal{O}} z_i^{\varepsilon_1} z_{i+1}^{\varepsilon_2} + \prod_{i \in \mathcal{O}} z_i^{\varepsilon_2} z_{i+1}^{\varepsilon_1} \right), \quad (6)$$

where

$$\varepsilon_1 := \frac{q^{m-1} + q^{\frac{m-1}{2}}}{2} \quad \text{and} \quad \varepsilon_2 := \frac{q^{m-1} - q^{\frac{m-1}{2}}}{2}.$$

Proof. Due to Remark 2 we can assume, without loss of generality, that $N = 2$. By Theorem 2, and since m is odd, $\mathcal{I}_N$ is a $[\frac{q^m-1}{2}, m, \frac{q^{m-1}(q-1)}{2}]$ one-weight irreducible cyclic code whose weight enumerator is $1 + (q^m - 1)T^{\frac{q^{m-1}(q-1)}{2}}$. We now determine the complete weight enumerator for $\mathcal{I}_N$. Let $c \in \mathbb{F}_q^*$ and $a, c' \in \mathbb{F}_{q^m}^*$ such that $\text{Tr}_{\mathbb{F}_{q^m}/\mathbb{F}_q}(c') = c$. Then

$$\mathscr{N}(c, \mathbf{c}(a)) = \sharp\{0 \leq i < \frac{q^m - 1}{2} \mid \text{Tr}_{\mathbb{F}_{q^m}/\mathbb{F}_q}(a\gamma^{2i}) - c = 0\},$$

$$= \sharp\{0 \leq i < q^m - 1 \mid \text{Tr}_{\mathbb{F}_{q^m}/\mathbb{F}_q}(a\gamma^{2i} - c') = 0\}/2.$$

If χ' and χ are the canonical additive characters of $\mathbb{F}_{q^m}$ and $\mathbb{F}_q$, respectively, then, due to (2) and since $\chi' = \chi \circ \text{Tr}_{\mathbb{F}_{q^m}/\mathbb{F}_q}$, we have

$$2\mathscr{N}(c, \mathbf{c}(a)) = \sum_{i=0}^{q^m-2} \frac{1}{q} \sum_{y \in \mathbb{F}_q} \chi(y \text{Tr}_{\mathbb{F}_{q^m}/\mathbb{F}_q}(a\gamma^{2i} - c')),$$

$$= \frac{q^m - 1}{q} + \frac{1}{q} \sum_{y \in \mathbb{F}_q^*} \sum_{x \in \mathbb{F}_{q^m}^*} \chi'(y(ax^2 - c')).$$

Therefore, by [12, Theorem 5.30, p. 217], we have

$$2\mathcal{N}(c, \mathbf{c}(a)) = \frac{q^m - 1}{q} + \frac{1}{q} \sum_{y \in \mathbb{F}_q^*} (\chi'(-yc')\eta(ya) G_{\mathbb{F}_{q^m}}(\eta, \chi') - \chi'(-yc')) ,$$

where η is the quadratic character of $\mathbb{F}_{q^m}$. Now because m is odd, η is also the quadratic character of $\mathbb{F}_q$. Thus, since $\chi'(-yc') = \chi(-yc)$ and $c \neq 0$,

$$2\mathcal{N}(c, \mathbf{c}(a)) = \frac{q^m - 1}{q} + \frac{1}{q}(G_{\mathbb{F}_{q^m}}(\eta, \chi') \sum_{y \in \mathbb{F}_q^*} \chi(-yc)\eta(ya) - \sum_{y \in \mathbb{F}_q^*} \chi(-yc)) ,$$

$$= q^{m-1} + \frac{1}{q}\eta(a) G_{\mathbb{F}_{q^m}}(\eta, \chi') \sum_{y \in \mathbb{F}_q^*} \chi(-yc)\eta(y) ,$$

$$= q^{m-1} + \frac{1}{q}\eta(a)\eta(-c) G_{\mathbb{F}_{q^m}}(\eta, \chi') \sum_{y \in \mathbb{F}_q^*} \chi(-yc)\eta(-yc) ,$$

$$= q^{m-1} + \frac{1}{q}\eta(a)\eta(-c) G_{\mathbb{F}_{q^m}}(\eta, \chi') G_{\mathbb{F}_q}(\eta, \chi) ,$$

where the second equality holds because $\sum_{y \in \mathbb{F}_q^*} \chi(-yc) = -1$. Let $l = 1$ if $p \equiv 3 \pmod 4$ and $\frac{t(m+1)}{2}$ is odd, and $l = 0$ otherwise. Then, by Theorem 1, we have

$$2\mathcal{N}(c, \mathbf{c}(a)) = q^{m-1} + \frac{1}{q}\eta(a)\eta(-c)(-1)^{t-1} q^{1/2}(-1)^{mt-1}(-1)^l q^{m/2} ,$$

$$= q^{m-1} + \eta(-a)\eta(c)(-1)^l q^{\frac{m-1}{2}} ,$$

therefore,

$$\mathcal{N}(c, \mathbf{c}(a)) = \begin{cases} \varepsilon_1 & \text{if } \eta(-a)(-1)^l = \eta(c), \\ \varepsilon_2 & \text{if } \eta(-a)(-1)^l \neq \eta(c), \end{cases}$$

and observe that if $\eta(c) = 1$ ($\eta(c) = -1$) then there must exists an even (odd) integer $1 \leq i \leq q-1$ such that $u_i = c$. Finally, since $\sharp\{a \in \mathbb{F}_{q^m}^* \mid \eta(-a)(-1)^l = 1\} = \frac{q^m-1}{2}$, both values ε_1 and ε_2 occur $\frac{q^m-1}{2}$ times. □

The Multinomial Theorem (see for example [8]) is a generalization of the Binomial Theorem, and therefore, it describes how to expand the power of a sum of more than two terms. We now recall such result by means of the following:

Theorem 4. *For a positive integer k and a non-negative integer r,*

$$(y_1 + y_2 + \cdots + y_k)^r = \sum_{e_1 + e_2 + \cdots + e_k = r} \binom{r}{e_1, e_2, \ldots, e_k} \prod_{j=1}^k y_j^{e_j} ,$$

where $e_1, \cdots, e_k$ are non-negative integers, and

$$\binom{r}{e_1, e_2, \ldots, e_k} := \frac{r!}{e_1! e_2! \cdots e_k!} .$$

As an almost direct consequence of the previous theorem we have:

Lemma 1. *Let k and r be as before. Then*

$$(1 + y_1 + y_2 + \cdots + y_k)^r = 1 + \sum_{i=1}^{r} \sum_{e_1+e_2+\cdots+e_k=i} \binom{r}{e_1, e_2, \ldots, e_k, r-i} \prod_{j=1}^{k} y_j^{e_j}.$$

Proof. Clearly, $(1+y)^r = 1 + \sum_{i=1}^{r} \binom{r}{i} y^i$, and if $y = y_1 + y_2 + \cdots + y_k$, then, by the previous theorem, we have

$$(1+y)^r = 1 + \sum_{i=1}^{r} \binom{r}{i} \sum_{e_1+e_2+\cdots+e_k=i} \binom{i}{e_1, e_2, \ldots, e_k} \prod_{j=1}^{k} y_j^{e_j}.$$

The result now follows from the fact that

$$\binom{r}{i} \binom{i}{e_1, e_2, \ldots, e_k} = \binom{r}{e_1, e_2, \ldots, e_k, r-i}.$$

□

4 Determining New Complete Weight Distributions in Terms of Known Ones

Given a cyclic code, $\mathscr{C}$, whose Hamming weight enumerator is known, it is possible to determine the Hamming weight enumerator of another cyclic code, $\mathscr{C}'$, in terms of a power of the Hamming weight enumerator of $\mathscr{C}$. A first version of this result was presented in [7,9] (see particularly Lemma 4.5 and Theorem 5.1 in [7]). An equivalent result for the complete weight enumerator is as follows:

Theorem 5. *For suitable integers n, m and d, let $\mathscr{C}$ be an $[n, m, d]$ cyclic code, over $\mathbb{F}_q$, with parity-check polynomial $h(x)$, and whose complete weight enumerator is $\mathrm{CWE}_{\mathscr{C}}(Z)$. Let also r be any positive integer, such that $\gcd(q, r) = 1$. Then, the polynomial $h(x^r)$ is the parity-check polynomial of an $[nr, mr, d]$ cyclic code, $\mathscr{C}'$, whose complete weight enumerator, $\mathrm{CWE}_{\mathscr{C}'}(Z)$, is $\mathrm{CWE}_{\mathscr{C}'}(Z) = \mathrm{CWE}_{\mathscr{C}}(Z)^r$.*

Proof. Clearly, $h(x^r) | (x^{nr} - 1)$ and $\deg(h(x)) = m$. Therefore, since $\gcd(q, nr) = 1$, we have that $h'(x) := h(x^r)$ is the parity-check polynomial of an $[nr, mr]$ cyclic code, $\mathscr{C}'$, over $\mathbb{F}_q$. Suppose that $W_{\mathscr{C}} = \{\boldsymbol{w}_1, \boldsymbol{w}_2, \cdots, \boldsymbol{w}_k\}$ is the set of complete nonzero weights of $\mathscr{C}$, and, for $1 \leq j \leq k$, let $A_{\boldsymbol{w}_j}$ be the number of codewords in $\mathscr{C}$ whose complete weight is equal to $\boldsymbol{w}_j$. In a similar way, suppose that $W_{\mathscr{C}'} = \{\boldsymbol{w}'_1, \boldsymbol{w}'_2, \cdots, \boldsymbol{w}'_{k'}\}$ is the set of complete nonzero weights of $\mathscr{C}'$, and, for $1 \leq j \leq k'$, let $A'_{\boldsymbol{w}'_j}$ be the number of codewords in $\mathscr{C}'$ whose complete weight is equal to $\boldsymbol{w}'_j$. Then $\mathrm{CWE}_{\mathscr{C}}(Z) = 1 + \sum_{j=1}^{k} A_{\boldsymbol{w}_j} z^{\boldsymbol{w}_j}$ and $\mathrm{CWE}_{\mathscr{C}'}(Z) = 1 + \sum_{j=1}^{k'} A'_{\boldsymbol{w}'_j} z^{\boldsymbol{w}'_j}$ are the complete weight enumerators of $\mathscr{C}$ and $\mathscr{C}'$, respectively.

Through the correspondence

$$\pi : \mathbb{F}_q^{nr} \to \mathcal{R}_{nr} := \mathbb{F}_q[x]/(x^{nr} - 1),$$

with

$$\pi(a_0, a_1, ..., a_{nr-1}) := a_0 + a_1 x + \cdots + a_{nr-1} x^{nr-1},$$

we can view the cyclic code $\mathscr{C}'$ as an ideal in the ring $\mathcal{R}_{nr}$, whose generator polynomial is $g'(x) = (x^{nr} - 1)/h'(x)$. By considering this, let us define, for $i = 1, \cdots, r$,

$$o_i := \{\pi^{-1}(x^{i-1} g'(x)), \pi^{-1}(x^{i-1+r} g'(x)), \cdots, \pi^{-1}(x^{i-1+(m-1)r} g'(x))\} \subset \mathbb{F}_q^{nr}.$$

Observe that if $\mathcal{S}_i \subset \mathbb{F}_q^{nr}$ ($i = 1, \cdots, r$) is the linear span of o_i (that is $\mathcal{S}_i = \langle o_i \rangle$), then we have for sure the following three facts:

(i) $\mathscr{C}' = \bigoplus_{i=1}^{r} \mathcal{S}_i$ (where $\bigoplus$ denotes direct sum of subspaces), and $\mathcal{S}_i \cap \mathcal{S}_l = \{\mathbf{0}\}$ (the zero codeword in $\mathscr{C}'$) if and only if $1 \leq i \neq l \leq r$. That is, any subspace $\mathcal{S}_i$ ($i = 1, \cdots, r$) is independent of all other subspaces $\mathcal{S}_l$ in the sense that there does not exist any nonzero codeword in $\mathcal{S}_i$ which is a linear combination of codewords in the other subspaces. Therefore, for each $c' \in \mathscr{C}'$ there must exist unique codewords $c_1, \cdots, c_r$, with $c_i \in \mathcal{S}_i$ and $1 \leq i \leq r$, such that $c' = c_1 + \cdots + c_r$.

(ii) For each pair of codewords a and b, such that $a \in \mathcal{S}_i$ and $b \in \mathcal{S}_l$, with $1 \leq i \neq l \leq r$, we have that $w_{\text{cplt}}(a + b) = w_{\text{cplt}}(a) + w_{\text{cplt}}(b)$.

(iii) Each $\mathcal{S}_i$ is an $[nr, m, d]$ linear code (not necessarily cyclic), whose complete weight enumerator is given by $\text{CWE}_{\mathscr{C}}(Z)$.

With the idea of clarifying the previous facts we briefly interrupt this proof in order to present the following:

Example 1. Let $\mathbb{F}_4 = \mathbb{F}_2(\alpha)$, with $\alpha^2 + \alpha + 1 = 0$, and we denote the elements of $\mathbb{F}_4$ as: $u_0 = 0$, $u_1 = 1$, $u_2 = \alpha$ and $u_3 = \alpha + 1$. Let $n = 5$, $h(x) = 1 + \alpha x + x^2$, and $r = 3$. Then $g(x) := \frac{x^5 - 1}{h(x)} = 1 + \alpha x + \alpha x^2 + x^3$, and $g'(x) := g(x^3)$. Under these conditions, it is not difficult to see that the cyclic code $\mathscr{C}$ is a $[5, 2, 4]$ one-weight irreducible cyclic code over $\mathbb{F}_4$, whose complete weight enumerator is $\text{CWE}_{\mathscr{C}}(Z) = 1 + 5z_1^2 z_2^2 z_3^0 + 5z_1^2 z_2^0 z_3^2 + 5z_1^0 z_2^2 z_3^2$. Therefore note that $W_{\mathscr{C}} = \{(2, 2, 0), (2, 0, 2), (0, 2, 2)\}$. The generator matrices for the cyclic codes $\mathscr{C}$ and $\mathscr{C}'$ are, respectively,

$$G = \begin{bmatrix} 1 & \alpha & \alpha & 1 & 0 \\ 0 & 1 & \alpha & \alpha & 1 \end{bmatrix},$$

and
$$G' = \begin{bmatrix} 1\,0\,0\,\alpha\,0\,0\,\alpha\,0\,0\,1\,0\,0\,0\,0\,0 \\ 0\,1\,0\,0\,\alpha\,0\,0\,\alpha\,0\,0\,1\,0\,0\,0\,0 \\ 0\,0\,1\,0\,0\,\alpha\,0\,0\,\alpha\,0\,0\,1\,0\,0\,0 \\ 0\,0\,0\,1\,0\,0\,\alpha\,0\,0\,\alpha\,0\,0\,1\,0\,0 \\ 0\,0\,0\,0\,1\,0\,0\,\alpha\,0\,0\,\alpha\,0\,0\,1\,0 \\ 0\,0\,0\,0\,0\,1\,0\,0\,\alpha\,0\,0\,\alpha\,0\,0\,1 \end{bmatrix}.$$

Therefore

$$o_1 = \left\{ \begin{array}{l} (1\,0\,0\,\alpha\,0\,0\,\alpha\,0\,0\,1\,0\,0\,0\,0\,0), \\ (0\,0\,0\,1\,0\,0\,\alpha\,0\,0\,\alpha\,0\,0\,1\,0\,0) \end{array} \right\},$$

$$o_2 = \left\{ \begin{array}{l} (0\,1\,0\,0\,\alpha\,0\,0\,\alpha\,0\,0\,1\,0\,0\,0\,0), \\ (0\,0\,0\,0\,1\,0\,0\,\alpha\,0\,0\,\alpha\,0\,0\,1\,0) \end{array} \right\},$$

and

$$o_3 = \left\{ \begin{array}{l} (0\,0\,1\,0\,0\,\alpha\,0\,0\,\alpha\,0\,0\,1\,0\,0\,0), \\ (0\,0\,0\,0\,0\,1\,0\,0\,\alpha\,0\,0\,\alpha\,0\,0\,1) \end{array} \right\}.$$

In this way, note that $\mathcal{S}_1$, $\mathcal{S}_2$ and $\mathcal{S}_3$ are $[15, 2, 4]$ linear codes (they are not cyclic), whose complete weight enumerator is $\mathrm{CWE}_{\mathscr{C}}(Z)$, and clearly $\mathscr{C}' = \mathcal{S}_1 \oplus \mathcal{S}_2 \oplus \mathcal{S}_3$ and $\mathcal{S}_i \cap \mathcal{S}_j = \{\mathbf{0}\}$, $1 \leq i \neq j \leq 3$.

Continuing with the proof, we now define $\mathcal{U} := \{\mathcal{S}_1, \cdots, \mathcal{S}_r\}$ and recall that $W_{\mathscr{C}} = \{\boldsymbol{w}_1, \boldsymbol{w}_2, \cdots, \boldsymbol{w}_k\} \subset \mathbb{N}_0^{q-1}$. Thus, as a consequence of the previous facts, note that for each codeword $c' \in \mathscr{C}'$, with $w_{\mathrm{cplt}}(c') = \boldsymbol{w}' \in W_{\mathscr{C}'}$, there must exist an integer i; k non-negative integers, $e_1, \cdots, e_k$; k disjoint subsets, $\mathcal{V}_1, \cdots, \mathcal{V}_k$, of the set $\mathcal{U}$; and k codewords, $a_1, \cdots, a_k$, of $\mathscr{C}'$, in such a way that the following conditions are met:

(1) $1 \leq i \leq r$, $i = e_1 + \cdots + e_k$ and $c' = a_1 + \cdots + a_k$.
(2) $|\mathcal{V}_j| = e_j$, $a_j \in \bigoplus_{S \in \mathcal{V}_j} S$ and $w_{\mathrm{cplt}}(a_j) = e_j \boldsymbol{w}_j$, for $j = 1, \cdots, k$.
(3) The complete nonzero weight $\boldsymbol{w}'$, can be expressed as $\boldsymbol{w}' = w_{\mathrm{cplt}}(a_1 + \cdots + a_k) = w_{\mathrm{cplt}}(a_1) + \cdots + w_{\mathrm{cplt}}(a_k) = e_1 \boldsymbol{w}_1 + \cdots + e_k \boldsymbol{w}_k$, where the penultimate equality holds because the subsets $\mathcal{V}_j$ are disjoint.

Then, we can summarize all the above by saying that for each $\boldsymbol{w}' \in W_{\mathscr{C}'}$, there must exist at least one $(k+1)$-tuple of non-negative integers $(i, e_1, \cdots, e_k)$, that satisfies $1 \leq i \leq r$, $i = e_1 + \cdots + e_k$ and $\boldsymbol{w}' = e_1 \boldsymbol{w}_1 + \cdots + e_k \boldsymbol{w}_k$. In consequence, if we construct, for each $\boldsymbol{w}' \in W_{\mathscr{C}'}$, the set

$$\mathcal{T}_{\boldsymbol{w}'} := \{\, \tau \mid \tau \text{ is an } (k+1)\text{-tuple of non-negative integers of the form}$$
$$(i, e_1, \cdots, e_k) \text{ that satisfies, } 1 \leq i \leq r, i = e_1 + \cdots + e_k \text{ and}$$
$$\boldsymbol{w}' = e_1 \boldsymbol{w}_1 + \cdots + e_k \boldsymbol{w}_k \,\},$$

then it is clear that $|\mathcal{T}_{w'}| \neq 0$. Now, by taking a fixed $(i, e_1, \cdots, e_k) \in \mathcal{T}_{w'}$, we have that there are

$$\binom{r}{e_1}\binom{r-e_1}{e_2}\cdots\binom{r-(e_1+\cdots+e_{k-1})}{e_k} = \binom{r}{e_1, e_2, \ldots, e_k, r-i}$$

possible choices for the construction of the disjoint subsets: $\mathcal{V}_1, \cdots, \mathcal{V}_k$. But recall that all the $\mathbb{F}_q$-linear subspaces $\mathcal{S}_i$, have the same complete weight enumerator $\text{CWE}_{\mathscr{C}}(Z)$. Therefore, for all these $\mathbb{F}_q$-linear subspaces $\mathcal{S}_i$, the integer value A_{w_j} (for $j = 1, \cdots, k$) is the frequency of occurrence of the complete nonzero weight w_j. Consequently we have

$$A'_{w'} = \sum_{(i, e_1, +\cdots+e_k) \in \mathcal{T}_{w'}} \binom{r}{e_1, e_2, \ldots, e_k, r-i} \prod_{j=1}^{k} A_{w_j}^{e_j},$$

but, since $w' = e_1 w_1 + \cdots + e_k w_k$ and $A_{w_j}^{e_j} Z^{e_j w_j} = (A_{w_j} Z^{w_j})^{e_j}$,

$$A'_{w'} Z^{w'} = \sum_{(i, e_1, +\cdots+e_k) \in \mathcal{T}_{w'}} \binom{r}{e_1, e_2, \ldots, e_k, r-i} \prod_{j=1}^{k} (A_{w_j} Z^{w_j})^{e_j}.$$

Conversely, note that for each $(k+1)$-tuple of non-negative integers of the form $(i, e_1, \cdots, e_k)$, that satisfies $1 \leq i \leq r$ and $i = e_1 + \cdots + e_k$, there must exist a unique $w' \in W_{\mathscr{C}'}$ such that $w' = e_1 w_1 + \cdots + e_k w_k$. Therefore,

$$\text{CWE}_{\mathscr{C}'}(Z) = 1 + \sum_{i=1}^{r} \sum_{e_1+e_2+\cdots+e_k=i} \binom{r}{e_1, e_2, \ldots, e_k, r-i} \prod_{j=1}^{k} (A_{w_j} Z^{w_j})^{e_j},$$

and, by Lemma 1, we conclude that $\text{CWE}_{\mathscr{C}'}(Z) = \text{CWE}_{\mathscr{C}}(Z)^r$. Finally, both $\mathscr{C}$ and $\mathscr{C}'$ have the same minimum Hamming distance, d, because $W_{\mathscr{C}} \subseteq W_{\mathscr{C}'}$. □

Example 2. Let $\mathscr{C}$ and $\mathscr{C}'$ be as in Example 1. Thus, due to Theorem 5, $\mathscr{C}'$ is a $[15, 6, 4]$ cyclic code over $\mathbb{F}_4$, whose complete weight enumerator is:

$$\begin{aligned}
\text{CWE}_{\mathscr{C}'}(Z) &= \text{CWE}_{\mathscr{C}}(Z)^3 = (1 + 5z_1^2 z_2^2 z_3^0 + 5z_1^2 z_2^0 z_3^2 + 5z_1^0 z_2^2 z_3^2)^3 \\
&= 1 + 15(z_1^2 z_2^2 z_3^0 + z_1^2 z_2^0 z_3^2 + z_1^0 z_2^2 z_3^2) + 75(z_1^4 z_2^4 z_3^0 + z_1^4 z_2^0 z_3^4 + z_1^0 z_2^4 z_3^4) \\
&\quad + 125(z_1^6 z_2^6 z_3^0 + z_1^6 z_2^0 z_3^6 + z_1^0 z_2^6 z_3^6) + 150(z_1^2 z_2^2 z_3^4 + z_1^2 z_2^4 z_3^2 + z_1^4 z_2^2 z_3^2) \\
&\quad + 375(z_1^2 z_2^4 z_3^6 + z_1^2 z_2^6 z_3^4 + z_1^4 z_2^2 z_3^6 + z_1^4 z_2^6 z_3^2 + z_1^6 z_2^2 z_3^4 + z_1^6 z_2^4 z_3^2) \\
&\quad + 750 z_1^4 z_2^4 z_3^4.
\end{aligned}$$

By using directly the cyclic code $\mathscr{C}'$, the previous numerical result was verified by a computer program.

5 Complete Weight Distribution of Families of Cyclic Codes

In order to observe the usefulness of Theorem 5, we now determine in a simple way the complete weight distribution for one of the two families of reducible cyclic codes studied in [1].

Theorem 6 *[1, Theorem 3.1]. With the notation of Propositions 1 and 2, suppose that q is odd and let $\mathscr{C}'$ be the $[q^m - 1, 2m]$ reducible cyclic code with parity-check polynomial $h'(x) := h_1(x)h_{\frac{q^m-1}{2}+1}(x)$. Then the complete weight enumerator of $\mathscr{C}'$ is*

$$\mathrm{CWE}_{\mathscr{C}'}(Z) = \left[1 + \frac{q^m - 1}{2}\left(\prod_{i=1}^{q-1} z_i^{\epsilon_1} + \prod_{i=1}^{q-1} z_i^{\epsilon_2}\right)\right]^2$$

$$= 1 + (q^m - 1)\left(\prod_{i=1}^{q-1} z_i^{\epsilon_1} + \prod_{i=1}^{q-1} z_i^{\epsilon_2}\right) + \frac{(q^m - 1)^2}{2}\prod_{i=1}^{q-1} z_i^{q^{m-1}}$$

$$+ \frac{(q^m - 1)^2}{4}\left(\prod_{i=1}^{q-1} z_i^{2\epsilon_1} + \prod_{i=1}^{q-1} z_i^{2\epsilon_2}\right), \tag{7}$$

if m is even and

$$\mathrm{CWE}_{\mathscr{C}'}(Z) = \left[1 + \frac{q^m - 1}{2}\left(\prod_{i \in \mathcal{O}} z_i^{\epsilon_1} z_{i+1}^{\epsilon_2} + \prod_{i \in \mathcal{O}} z_i^{\epsilon_2} z_{i+1}^{\epsilon_1}\right)\right]^2$$

$$= 1 + (q^m - 1)\left(\prod_{i \in \mathcal{O}} z_i^{\epsilon_1} z_{i+1}^{\epsilon_2} + \prod_{i \in \mathcal{O}} z_i^{\epsilon_2} z_{i+1}^{\epsilon_1}\right) + \frac{(q^m - 1)^2}{2}\prod_{i=1}^{q-1} z_i^{q^{m-1}}$$

$$+ \frac{(q^m - 1)^2}{4}\left(\prod_{i \in \mathcal{O}} z_i^{2\varepsilon_1} z_{i+1}^{2\varepsilon_2} + \prod_{i \in \mathcal{O}} z_i^{2\varepsilon_2} z_{i+1}^{2\varepsilon_1}\right), \tag{8}$$

if m is odd.

Proof. Let $h_2(x)$ be the minimal polynomial of γ^{-2}. Since q is odd, $\deg(h_2(x)) = \deg(h_1(x)) = \deg(h_{\frac{q^m-1}{2}+1}(x)) = m$. Additionally, since γ^{-2} is a root of $h_2(x)$, and because $\gamma^{-\frac{q^m-1}{2}-1} = -\gamma^{-1}$, we see that γ^{-1} and $\gamma^{-\frac{q^m-1}{2}-1}$ are both roots of $h_2(x^2)$. Thus, $h_2(x^2) = h_1(x)h_{\frac{q^m-1}{2}+1}(x)$ and, by Definition 1 and Propositions 1 and 2, $h(x) := h_2(x)$ is the parity-check polynomial of a $[\frac{q^m-1}{2}, m]$ irreducible cyclic code, $\mathcal{I}_N$, whose complete weight enumerator is given by (5) if m is even and (6) if m is odd, where $N = 2$ for these two equations. Clearly $\gcd(q, 2) = 1$, thus, by Theorem 5 and Remark 3, $\mathscr{C}'$ is a $[q^m - 1, 2m]$ reducible cyclic code whose complete weight enumerator is given by (7) if m is even and by (8) if m is odd. Finally, note that (7) and (8) coincide with Tables 1 and 2 in [1], respectively. □

Example 3. (A) Let $q = 3$ and $m = 2$. Then by Theorem 6 $\epsilon_1 = 2$, $\epsilon_2 = 1$, and $h'(x) = h_1(x)h_5(x)$ is the parity-check polynomial of an $[8, 4]$ reducible cyclic code, $\mathscr{C}'$, whose complete weight enumerator is

$$\mathrm{CWE}_{\mathscr{C}'}(Z) = (1 + 4(z_1^2 z_2^2 + z_1 z_2))^2 = 1 + 32 z_1^3 z_2^3 + 16 z_1^4 z_2^4 + 24 z_1^2 z_2^2 + 8 z_1 z_2 ,$$

which coincides with Example 3.2(1) in [1] (take into consideration Remark 1).

(B) Let $q = 5$ and $m = 2$. Then by Theorem 6 $\epsilon_1 = 2$, $\epsilon_2 = 3$, and $h'(x) = h_1(x)h_{13}(x)$ is the parity-check polynomial of a $[24, 4]$ reducible cyclic code, $\mathscr{C}'$, whose complete weight enumerator is

$$\begin{aligned}\mathrm{CWE}_{\mathscr{C}'}(Z) &= (1 + 12(z_1^2 z_2^2 z_3^2 z_4^2 + z_1^3 z_2^3 z_3^3 z_4^3))^2 \\ &= 1 + 288 z_1^5 z_2^5 z_3^5 z_4^5 + 144(z_1^4 z_2^4 z_3^4 z_4^4 + z_1^6 z_2^6 z_3^6 z_4^6) \\ &\quad + 24(z_1^2 z_2^2 z_3^2 z_4^2 + z_1^3 z_2^3 z_3^3 z_4^3) ,\end{aligned}$$

which coincides with Example 3.2(2) in [1].

(C) Let $q = 3$ and $m = 3$. Then by Theorem 6 $\varepsilon_1 = 3$, $\varepsilon_2 = 6$, and $h'(x) = h_1(x)h_{14}(x)$ is the parity-check polynomial of a $[26, 6]$ reducible cyclic code, $\mathscr{C}'$, whose complete weight enumerator is

$$\begin{aligned}\mathrm{CWE}_{\mathscr{C}'}(Z) &= (1 + 13(z_1^3 z_2^6 + z_1^6 z_2^3))^2 \\ &= 1 + 338 z_1^9 z_2^9 + 169(z_1^{12} z_2^6 + z_1^6 z_2^{12}) + 26(z_1^3 z_2^6 + z_1^6 z_2^3) ,\end{aligned}$$

which coincides with Example 3.2(3) in [1].

(D) Let $q = 5$ and $m = 3$. Then by Theorem 6 $\varepsilon_1 = 10$, $\varepsilon_2 = 15$, and $h'(x) = h_1(x)h_{63}(x)$ is the parity-check polynomial of a $[124, 6]$ reducible cyclic code, $\mathscr{C}'$, whose complete weight enumerator is

$$\begin{aligned}\mathrm{CWE}_{\mathscr{C}'}(Z) &= (1 + 62(z_1^{10} z_2^{15} z_3^{10} z_4^{15} + z_1^{15} z_2^{10} z_3^{15} z_4^{10}))^2 \\ &= 1 + 7688 z_1^{25} z_2^{25} z_3^{25} z_4^{25} + 3844(z_1^{20} z_2^{30} z_3^{20} z_4^{30} + z_1^{30} z_2^{20} z_3^{30} z_4^{20}) \\ &\quad + 124(z_1^{10} z_2^{15} z_3^{10} z_4^{15} + z_1^{15} z_2^{10} z_3^{15} z_4^{10}) ,\end{aligned}$$

which coincides with Example 3.2(4) in [1] (be careful, for this last example the authors of [1] choose a different order for the elements in $\mathbb{F}_5$).

As another instance of Theorem 5, we can now determine the complete weight distributions for another family of cyclic codes which, as we shall see below, can be obtained in terms of the complete weight distribution of the subclass of optimal three-weight cyclic codes given in Theorem 3:

Theorem 7. *Consider the same notation and assumption as in Theorem 3. Let r be any positive integer, such that $\gcd(q, r) = 1$. Then $h(x^r)$ is the parity-check polynomial of a $[(q^2 - 1)r, 3r, q(q - 1) - 1]$ cyclic code, $\mathscr{C}'$, whose complete weight enumerator, $\mathrm{CWE}_{\mathscr{C}'}(Z)$, is*

$$\mathrm{CWE}_{\mathscr{C}'}(Z) = \left[1 + (q - 1) \prod_{i=1}^{q-1} z_i^{q+1} + (q^2 - 1) \left(\prod_{i=1}^{q-1} z_i^q + \sum_{j=1}^{q-1} z_j \prod_{i=1, i \neq j}^{q-1} z_i^{q+1} \right) \right]^r .$$

Proof. Direct from Theorems 5 and 3. □

Example 4. Let $(q, e_2, e_3, r) = (4, 1, 1, 3)$. Thus, due to Theorem 3 and [15, Example 1], $h(x) = h_5(x)h_1(x)$ is the parity-check polynomial of an optimal three-weight [15, 3, 11] cyclic code, $\mathscr{C}$, over $\mathbb{F}_4$, whose complete weight enumerator is $\text{CWE}_{\mathscr{C}}(Z) = 1 + 3A + 15(B + C + D + E)$, where $A = (z_1 z_2 z_3)^5$, $B = (z_1 z_2 z_3)^4$, $C = z_1 z_2^5 z_3^5$, $D = z_1^5 z_2 z_3^5$, and $E = z_1^5 z_2^5 z_3$. On the other hand, by Theorem 7, $h(x^3)$ is the parity-check polynomial of a [45, 9, 11] cyclic code, $\mathscr{C}'$, whose complete weight enumerator, $\text{CWE}_{\mathscr{C}'}(Z)$, is

$$\begin{aligned}\text{CWE}_{\mathscr{C}'}(Z) &= \text{CWE}_{\mathscr{C}}(Z)^3 \\ &= 1 + 9A + 27(A^2 + A^3) + (45 + 270A + 405A^2)(B + C + D + E) \\ &\quad + (4050AB + 1350B + 10125B^2)(C + D + E) \\ &\quad + (675 + 2025A)(B^2 + C^2 + D^2 + E^2) \\ &\quad + 3375(B^3 + C^3 + D^3 + E^3) \\ &\quad + (1350 + 4050A + 20250B)(CD + CE + DE) + 20250CDE \\ &\quad + 10125(B(C^2 + D^2 + E^2) + C^2(D + E) \\ &\quad + D^2(C + E) + E^2(C + D)) \,.\end{aligned}$$

By using directly the cyclic code $\mathscr{C}'$, the previous numerical result was verified by a computer program.

6 Conclusions

In this work we determined the complete weight distributions for a particular kind of one- and two-weight irreducible cyclic codes (Propositions 1 and 2). After this, a method that determines new complete weight distributions in terms of known ones was presented (Theorem 5). Then, we used such method in order to determine the complete weight distribution of infinite families of cyclic codes (Sect. 5). As an example of such families, the complete weight distribution for one of the two families of reducible cyclic codes studied in [1] was determined in a simple way (Theorem 6). As another example, the complete weight distribution for another family of cyclic codes was also determined which, as shown earlier, can be obtained in terms of the complete weight distribution of the subclass of optimal three-weight cyclic codes presented recently in [15] (Theorem 7).

As it is known, Theorem 2 gives us the Hamming weight distributions for all one-weight and semiprimitive two-weight irreducible cyclic codes over any finite field. On the other hand, by means of Propositions 1 and 2, the complete weight distributions for a particular kind of one- and two-weight irreducible cyclic codes were determined. Thus, as a complement of this work, we believe that it would be interesting to determine the complete weight distributions of the remaining part of the family of one- and semiprimitive two-weight irreducible cyclic codes in Theorem 2.

References

1. Bae, S., Li, C., Yue, Q.: On the complete weight enumerator of some reducible cyclic codes. Discret. Math. **338**, 2275–2287 (2015)
2. Blake, I.F., Kith, K.: On the complete weight enumerator of Reed-Solomon codes. SIAM J. Discret. Math. **4**(2), 164–171 (1991)
3. Chan, C.H., Xiong, M.: On the complete weight distribution of subfield subcodes of algebraic-geometric codes. IEEE Trans. Inf. Theory **65**(11), 7079–7086 (2019)
4. Cheng, Y., Cao, X., Luo, G.: Three new constructions of optimal linear codes with few weights. Comput. Appl. Math. **42**(321) (2023)
5. Delsarte, P.: On subfield subcodes of Reed-Solomon codes. IEEE Trans. Inf. Theory **21**(5), 575–576 (1975)
6. Helleseth, T., Kholosha, A.: Monomial and quadratic bent functions over the finite fields of odd characteristic. IEEE Trans. Inf. Theory **52**, 2018–2032 (2006)
7. Helleseth, T., Kløve, T., Mykkeltveit, J.: The weight distribution of irreducible cyclic codes with block lengths $n_1((q^l - 1)n)$. Discret. Math. **18**(2), 179–211 (1977)
8. Hilliker, D.L.: On the multinomial theorem. Fibonacci Q. **15**(1), 22–24 (1977)
9. Kløve, T.: The weight distribution for a class of irreducible cyclic codes. Discret. Math. **20**, 87–90 (1977)
10. Li, C., Bae, S., Ahn, J., Yang, S., Yao, Z.A.: Complete weight enumerators of some linear codes and their applications. Designs Codes Cryptogr. **81**, 153–168 (2016)
11. Li, C., Yue, Q., Fu, F.W.: Complete weight enumerators of some cyclic codes. Designs Codes Cryptogr. **80**, 295–315 (2015)
12. Lidl, R., Niederreiter, H.: Finite Fields. Cambridge University Press, Cambridge (1983)
13. MacWilliams, F.J., Sloane, N.J.A.: The Theory of Error-Correcting Codes. North-Holland, Amsterdam (1977)
14. Vega, G.: The b-symbol weight distributions of all semiprimitive irreducible cyclic codes. Des. Codes Cryptogr. **91**, 2213–2221 (2023)
15. Vega, G., Hernández, F.: The complete weight distribution of a subclass of optimal three-weight cyclic codes. Cryptogr. Commun. **15**(2), 317–330 (2023)
16. Yang, S.: Complete weight enumerators of a class of linear codes from weil sums. IEEE Access **8**, 194631–194639 (2020)
17. Yang, S., Yao, Z.: Complete weight enumerators of a family of three-weight linear codes. Des. Codes Cryptogr. **82**, 663–674 (2017)
18. Yang, S., Yao, Z.A.: Complete weight enumerators of a class of linear codes. Discret. Math. **340**(4), 729–739 (2017)
19. Yang, S., Yao, Z.A., Zhao, C.A.: A class of three-weight linear codes and their complete weight enumerators. Cryptogr. Commun. **9**, 133–149 (2017)

Central Limit Theorem for Linear Eigenvalue Statistics of Random Matrices from Binary Linear Codes

Chin Hei Chan[(✉)] and Maosheng Xiong[(✉)]

Department Mathematics, Hong Kong University of Science and Technology,
Hong Kong, People's Republic of China
chchanam@connect.ust.hk, mamsxiong@ust.hk

Abstract. It was known that the empirical spectral distribution of random matrices constructed from binary linear codes of increasing length converges to the Marchenko-Pastur law as long as the dual distance of the codes is at least 5, and the condition of the dual distance 5 is optimal because there are binary linear codes of dual distance 4 that do not satisfy this property. In this article, we push this result a little further: we show that a Gaussian central limit theorem holds for the linear spectral statistics associated with such random matrices from binary linear codes of increasing length when the dual distance is at least 7. We also show that the condition of dual distance 7 is optimal as there are binary linear codes of dual distance 6 that do not satisfy this property. This result can be interpreted as that pseudorandom sequences constructed from long binary linear codes of dual distance 7 in general satisfy a more stringent pseudorandom test than those from binary linear codes of dual distance 5.

Keywords: Binary linear codes · Linear spectral statistics · Central limit theorem · Random matrix

1 Introduction

The theory of random matrices has been actively studied in the past two decades. A particular major topic is the statistical behaviour of the eigenvalues of large-dimensional random matrix models. There is a universality phenomenon that the limiting spectral behaviour does not depend on the exact distribution of the matrix entries. Therefore it has been applied to a wide variety of areas including statistics [16], theoretical physics [15], economics [12], communication theory [13] and number theory [13].

In a series of papers (initiated by [4], followed by [3,5,17]), the authors studied the eigenvalue behaviour of a random matrix model based on linear codes over finite fields and proved that the empirical spectral distribution (ESD) of the matrix converges to the Marchenko-Pastur (MP) law in probability as long as the minimum Hamming distance of the dual code is at least 5. As this is the property satisfied by truly random matrices (i.e., matrices with i.i.d. entries), this was called a "group randomness" property [3] and may be considered as a joint randomness test for pseudorandom sequences generated from linear codes, which may have many potential applications. It is also interesting to note that the condition of dual distance 5 is optimal in the sense that there are linear codes of dual distance 4 that do not satisfy this "group randomness" property. Recently it was established further that the convergence to the Marchenko-Pastur law of such matrices from linear codes of dual distance 5 is quite fast, with the rate of convergence at least $n^{-1/4}$ in probability where n is the length of the code [7].

The proof of [17] was based on the moment method. To describe the result, we need some notation. For the sake of simplicity, let us focus on binary linear codes.

Consider a family of binary linear codes $\mathcal{C}$ of length n with dual distance $d^\perp$ with $n \to \infty$. The standard additive character $\psi : 0 \mapsto 1, 1 \mapsto -1$ extends component-wisely to $\psi : \mathbb{F}_2^n \to \{\pm 1\}^n$. We pick p (not necessarily distinct) sequences from $\psi(\mathcal{C})$ uniformly and independently. This makes $\psi(\mathcal{C})^p$ a uniform probability space. These p sequences form the rows of a $p \times n$ random matrix $\Phi_\mathcal{C}$ with (± 1)-entries. Let $\mathcal{G}_\mathcal{C}$ be the *Gram matrix* of $\frac{1}{\sqrt{n}}\Phi_\mathcal{C}$ given by

$$\mathcal{G}_\mathcal{C} = \frac{1}{n}\Phi_\mathcal{C}\Phi_\mathcal{C}^T. \tag{1}$$

Since $\mathcal{G}_\mathcal{C}$ is a symmetric random matrix with respect to the probability space $\psi(\mathcal{C})^p$, its eigenvalues $\lambda_1, \lambda_2, \cdots, \lambda_p$ are real random variables. What was proved in [17] is that if the dual distance $d^\perp$ of $\mathcal{C}$ always satisfies $d^\perp \geq 5$ as $n \to \infty$, then for any fixed positive integer k, one has

$$\lim_{n \to \infty} \mathbb{E}\left(\frac{1}{p}\sum_{j=1}^{p}\lambda_j^k\right) = \beta_{k,\text{MP}}. \tag{2}$$

Here $y = p/n \in (0,1)$ is fixed and $\beta_{k,\text{MP}}$ is the well-known k-th moment of the MP law. The main result of [17] then follows immediately from (2) by the moment method (see [17, Theorems 1 and 3]).

The purpose of this paper is to investigate the behaviour of the linear statistic $\frac{1}{p}\sum_{j=1}^{p}\lambda_j^k$ in (2) a little further. We prove the following:

Theorem 1. *Assume $d^\perp \geq 7$, then for any fixed integer $k \geq 2$, under the above setting with $y = p/n \in (0,1)$ fixed and $n \to \infty$, we have*

$$\sum_{j=1}^{p}\lambda_j^k - \mathbb{E}\left(\sum_{j=1}^{p}\lambda_j^k\right) \xrightarrow{d} N(0,\sigma_k^2). \tag{3}$$

Here $\xrightarrow{d}$ denotes the weak convergence, σ_k is a positive constant depending only on k and $N(0,\sigma_k^2)$ is the Gaussian distribution with mean zero and variance σ_k^2.

Remark 1. Theorem 1 was proved only for binary linear codes, but the proof could be easily adopted for general linear codes. For the sake of simplicity, we do not do it here.

Remark 2. When $k = 1$, we always have

$$\sum_{j=1}^{p} \lambda_j^k = p,$$

hence the left side of (3) is always zero, i.e., the (limiting) distribution is degenerate.

Remark 3. The result (3) shows that the linear statistics $\sum_{j=1}^{p} \lambda_j^k$ of the random matrix formed from binary linear codes of length n and dual distance $d^\perp \geq 7$ satisfies a central limit theorem (CLT) as $n \to \infty$. Since (3) is the property satisfied by large truly random matrices (i.e. matrices with i.i.d entries), it is a new "group randomness" property (see [3]) and can be interpreted as a joint randomness test for sequences constructed from binary linear codes, showing that pseudorandom sequences arising from long binary linear codes of dual distance 7 in general satisfy a more stringent pseudorandom test than those from binary linear codes of dual distance 5.

Remark 4. A simple example of a family of binary linear codes of dual distance $d^\perp = 7$ is $\{\mathcal{C}_i\}_{i=3}^{\infty}$ where the binary linear code $\mathcal{C}_i$ is defined as

$$\mathcal{C}_i = \{(\text{Tr}(a\beta^j + b\beta^{3j} + c\beta^{5j}))_{j=0}^{n_i-1} : a, b, c \in \mathbb{F}_{2^i}\},$$

and the length $n_i = 2^i - 1$, Tr is the absolute trace map from $\mathbb{F}_{2^i}$ to $\mathbb{F}_2$ and β is a primitive element of $\mathbb{F}_{2^i}$. They are the dual codes of the primitive narrow-sense binary BCH codes with designed distance 7.

Remark 5. We also remark that the condition $d^\perp \geq 7$ is optimal in Theorem 1. Actually, the augmented codes from binary Gold codes of length $n_i = 2^i - 1$ have dual distance either 6 or 4, depending on whether or not the Gold function is almost perfect nonlinear (APN) [9]. In both cases, it will be seen that (3) does not hold. This is similar to the condition $d^\perp \geq 5$ that was required in [17].

The paper is organized as follows. In Sect. 2 we recall some basic facts about the Gaussian law, as well as a useful result related to binary linear codes. In Sects. 3 and 4 we present a detailed proof of Theorem 1 by using the moment method. The computation of the exact value of the variance σ_k^2 is quite technical, since it does not affect the shape of the limiting law, we provide all the details of computation in Sect. 6 **Appendix**. Finally in Sect. 5 we study augmented binary Gold codes and show that Theorem 1 does not hold for these codes.

2 Preliminaries

2.1 The Gaussian Law

The density function of the centralized Gaussian distribution $\mathcal{N}(0, \sigma^2)$ is given by

$$\phi_\sigma(\mathrm{d}x) = \frac{1}{\sqrt{2\pi}\sigma} e^{-\frac{x^2}{2\sigma^2}} \, \mathrm{d}x.$$

The corresponding cumulative distribution function (cdf) is given by

$$\Phi_\sigma(z) = \int_{-\infty}^{z} \phi_\sigma(\mathrm{d}x).$$

In particular, if $\sigma = 1$, we commonly drop the subscript and simply write the cdf as $\Phi(z)$, and the distribution is called the standard Gaussian (or normal) distribution.

The moments of $\mathcal{N}(0, \sigma^2)$ are given by

$$\beta_{\ell, \phi_\sigma} = \int_{-\infty}^{\infty} x^\ell \phi_\sigma(\mathrm{d}x) = \begin{cases} 0 & (\ell \text{ odd}) \\ (\ell - 1)!! \sigma^\ell & (\ell \text{ even}), \end{cases} \quad (4)$$

where $(\ell - 1)!!$ is the product of all the positive integers not larger than $\ell - 1$ and having the same parity as $\ell - 1$.

It is known that the quantities $\beta_{\ell, \phi_\sigma}$ satisfy the Carleman's condition [6, Lemma B.3]

$$\sum_{\ell=1}^{\infty} \beta_{2\ell, \phi_\sigma}^{-1/2\ell} = \infty,$$

which guarantees that Gaussian law is the unique probability distribution with all the ℓ-th moments given by (4). This is essential for the moment method to work.

2.2 A Bound on Weight Distribution of Binary Linear Codes

The following is a bound about the weight distribution of binary linear codes introduced by Levy (see [10,11]). It is essentially a sphere-packing bound for the number of codewords with a given fixed weight.

Lemma 1 *[Levy's bound]. Let $\mathcal{C}$ be a binary linear code of length n with minimum distance d and let $A_j(\mathcal{C})$ be the number of weight j codewords in $\mathcal{C}$. Then for any $d \leq j \leq n$ we have*

$$A_j(\mathcal{C}) \leq \min_{\nu} \frac{\binom{n}{j+\nu}}{N_n(d, j, \nu)}.$$

Here the minimum is taken over all integers ν in the range $-\lfloor \frac{d-1}{2} \rfloor \leq \nu \leq \lfloor \frac{d-1}{2} \rfloor$, and the quantity $N_n(d, j, \nu)$ is given by

$$N_n(d, j, \nu) = \sum_i \binom{j}{i} \binom{n-j}{\nu+i}, \quad (5)$$

where the summation is over all integers i satisfying $2i + \nu \leq \lfloor \frac{d-1}{2} \rfloor$.

3 Moment Estimation

We use notations from Sect. 1. Let $\mathcal{C}$ be a binary linear code of length n with dual distance $d^\perp \geq 7$. Let $\psi : \mathbb{F}_2 \to \{1, -1\}$ denote the standard additive character defined by $\psi(a) = (-1)^a$ for $a \in \mathbb{F}_2$ and extend coordinate-wisely to $\mathbb{F}_2^n$.

In order to choose p elements at random from $\psi(\mathcal{C})$, we define Ω_p to be the set of all maps $s : [1..p] \to \mathcal{D}$ endowed with the uniform probability, here for any integers $a < b$, $[a..b]$ denotes the set of integers from a to b. Hence Ω_p is a probability space with cardinality $|\Omega_p| = |\mathcal{C}|^p$. For each $s \in \Omega_p$, the $p \times n$ matrix Φ_s corresponding to s is given by

$$\Phi_s^T = \left[s(1)^T, s(2)^T, \cdots, s(p)^T\right]_{n \times p},$$

here we have written $s(i) \in \psi(\mathcal{C})$ as $1 \times n$-row vectors.

Let $\mathcal{G}_s := \frac{1}{n}\Phi_s\Phi_s^T$, where $y := \frac{p}{n} \in (0,1)$ is fixed. Let $k \geq 1$ be a fixed positive integer. We define

$$A_{n,k} := \sum_{i=1}^{p} \lambda_i^k - \mathbb{E}\left(\sum_{i=1}^{p} \lambda_i^k\right) = \mathbf{Tr}(\mathcal{G}_s^k) - \mathbb{E}\mathbf{Tr}(\mathcal{G}_s^k).$$

Here λ_i's are the real eigenvalues of the matrix $\mathcal{G}_s$, $\mathbf{Tr}(\mathcal{G}^k)$ denotes the trace of the matrix $\mathcal{G}^k$, and the expectation $\mathbb{E}$ is taken over the with respect to the probability space Ω_p.

Then the ℓ-th moment $\beta_{k,\ell}$ of $A_{n,k}$ is given by

$$\beta_{k,\ell} = \mathbb{E}A_{n,k}^\ell = \mathbb{E}\left(\mathbf{Tr}(\mathcal{G}^k) - \mathbb{E}\mathbf{Tr}(\mathcal{G}^k)\right)^\ell. \tag{6}$$

It is easy to see that when $\ell = 1$, trivially we have $\beta_{k,1} = 0$ for all k. For $\ell \geq 2$ we have the following result.

Theorem 2. *Let $k, \ell \geq 2$ be fixed. Then*

$$\beta_{k,\ell} = \begin{cases} O_{k,\ell}(n^{-1}) & (\ell \text{ odd}), \\ (\ell - 1)!!\sigma_k^\ell + O_{k,\ell}(n^{-1}) & (\ell \text{ even}), \end{cases}$$

where $\sigma_k^2 = \lim_{n \to \infty} \beta_{k,2} > 0$ which depends only on k and will be determined later, and the notation $O_{k,\ell}$ means that the absolute constant derived from the Big-O notation depends only on k and ℓ.

By the moment method, Theorem 1 can be obtained immediately from Theorem 2. For the remaining part of the paper, we give a proof of Theorem 2.

3.1 Problem Set-up

Let $\Pi_{k,p}$ denote the set of all maps $\gamma : [0..k] \to [1..p]$ such that $\gamma(0) = \gamma(k)$. This γ is called a k-closed path in short. Expanding the k-th power of the matrix $\mathcal{G}_s$, we obtain

$$\mathbf{Tr}(\mathcal{G}_s^k) = \frac{1}{n^k} \sum_{\gamma \in \Pi_{k,p}} \omega_\gamma(s),$$

where
$$\omega_\gamma(s) := \prod_{j=0}^{k-1} \langle s \circ \gamma(j), s \circ \gamma(j+1) \rangle.$$

Here for two vectors $\mathbf{u} = (u_1, \ldots, u_n), \mathbf{v} = (v_1, \ldots, v_n) \in \mathbb{R}^n$, the inner product $\langle \mathbf{u}, \mathbf{v} \rangle$ is defined as
$$\langle \mathbf{u}, \mathbf{v} \rangle = u_1 v_1 + \cdots + u_n v_n.$$

Using the above setting and expanding the right side of (6), we have
$$\beta_{k,\ell} = \frac{1}{n^{k\ell}} \sum_{\gamma_1, \gamma_2, \cdots, \gamma_\ell \in \Pi_{k,p}} \mathbb{E} W^{\gamma_1, \gamma_2, \cdots, \gamma_\ell}, \tag{7}$$

where for any $\gamma_1, \cdots, \gamma_\ell \in \Pi_{k,p}$, we have defined
$$W^{\gamma_1, \gamma_2, \cdots, \gamma_\ell} = \prod_{i=1}^{\ell} \left(\omega_{\gamma_i}(s) - \mathbb{E}\omega_{\gamma_i}(s) \right). \tag{8}$$

Let Σ_p denote the set of permutations on $[1..p]$. It is easy to see that for any $\tau \in \Sigma_p$, we have
$$\mathbb{E} W^{\tau \circ \gamma_1, \tau \circ \gamma_2, \cdots, \tau \circ \gamma_\ell} = \mathbb{E} W^{\gamma_1, \gamma_2, \cdots, \gamma_\ell}.$$

We define $(\gamma_1, \gamma_2, \cdots, \gamma_\ell) \sim (\gamma'_1, \gamma'_2, \cdots, \gamma'_\ell)$ if there is $\tau \in \Sigma_p$ such that $\gamma'_i = \tau \circ \gamma_i$ for all i. It is easy to see that this is an equivalence relation on $\Pi^\ell_{k,p}$.

Now denote by $\Pi^\ell_{k,p}/\Sigma_p$ a set of representatives of the equivalence classes of $\Pi^\ell_{k,p}$ under the above equivalence relation. Note that for any $(\gamma_1, \gamma_2, \cdots, \gamma_\ell) \in \Pi^\ell_{k,p}$, we have
$$\#[(\gamma_1, \gamma_2, \cdots, \gamma_\ell)] = \frac{p!}{(p - v_{\gamma_1, \gamma_2, \cdots, \gamma_\ell})!},$$

where we denote
$$v_{\gamma_1, \gamma_2, \cdots, \gamma_\ell} = \#V_{\gamma_1, \gamma_2, \cdots, \gamma_\ell}, \quad V_{\gamma_1, \gamma_2, \cdots, \gamma_\ell} = \bigcup_{i=1}^{\ell} \gamma_i([0..k]).$$

Hence (7) can be rewritten as
$$\beta_{k,\ell} = \frac{1}{n^{k\ell}} \sum_{(\gamma_1, \gamma_2, \cdots, \gamma_\ell) \in \Pi^\ell_{k,p}/\Sigma_p} \frac{p!}{(p - v_{\gamma_1, \gamma_2, \cdots, \gamma_\ell})!} \mathbb{E} W^{\gamma_1, \gamma_2, \cdots, \gamma_\ell}. \tag{9}$$

It is clear from Equation (9) that to prove Theorem 2, we need a detailed study of the quantity $\mathbb{E} W^{\gamma_1, \gamma_2, \cdots, \gamma_\ell}$ for any $(\gamma_1, \gamma_2, \cdots, \gamma_\ell) \in \Pi^\ell_{k,p}/\Sigma_p$.

3.2 Proof of Theorem 2

For a fixed element $\gamma^{(\ell)} := (\gamma_1, \gamma_2, \cdots, \gamma_\ell) \in \Pi_{k,p}^\ell / \Sigma_p$, we may consider each γ_i as an undirected graph with on the vertex set $[1..\ell]$ with undirected edges given by $\overline{\gamma_i(j)\gamma_i(j+1)}$ for $0 \leq j \leq k-1$. This is also a closed path of length k on the vertex set $\{\gamma_i(j) : 0 \leq j \leq k-1\}$.

From this section onwards, for any subset $A \subset [1..\ell]$, denote by γ_A the union $\bigcup_{i \in A} \gamma_i$, that is, γ_A is the union of the paths γ_i for $i \in A$. We define

$$V_{\gamma_A} = \bigcup_{i \in A} \gamma_i([0..k]), \quad v_{\gamma_A} = \#V_{\gamma_A},$$

and

$$W^{\gamma_A} = \mathbb{E} \prod_{i \in A} \left(\omega_{\gamma_i}(s) - \mathbb{E}\omega_{\gamma_i}(s) \right).$$

Since the rows $s(i)$ of Φ_s are mutually independent, it is clear that for any $A, B \subset [1..\ell]$, if $V_{\gamma_A} \cap V_{\gamma_B} = \emptyset$, then $\mathbb{E}W^{\gamma_{A \cup B}} = \mathbb{E}W^{\gamma_A}\mathbb{E}W^{\gamma_B}$. This means that we may decompose the vertex set $[1..\ell]$ into a disjoint union of connected components of the graph $\gamma^{(\ell)}$. To be more precise, let

$$[1..\ell] = \bigsqcup_{i \in \mathfrak{C}} A_i, \quad \gamma_{A_i} \cap \gamma_{A_j} = \emptyset \quad \forall i \neq j \in \mathfrak{C}, \tag{10}$$

and each $A_i \subset [1..\ell]$ is connected in $\gamma^{(\ell)}$ and maximal in the sense of (10). Then we have

$$\mathbb{E}W^{\gamma_1, \gamma_2, \cdots, \gamma_\ell} = \mathbb{E}W^{\gamma_{[1..\ell]}} = \prod_{i \in \mathfrak{C}} \mathbb{E}W^{\gamma_{A_i}}. \tag{11}$$

Therefore the analysis can be reduced to the case $\mathbb{E}W^{\gamma_A}$, where the union graph γ_A is connect on the vertex set A. The computation of $\mathbb{E}W^{\gamma_A}$ in this case involves quite technical arguments. To streamline the idea of the proof of Theorem 2, let us assume the following result the proof of which will be postponed in Sect. 4:

Lemma 2. *Let $k \geq 2$ and $\gamma_A = (\gamma_1, \gamma_2, \cdots, \gamma_t) \in \Pi_{k,p}^t$ such that the graph γ_A on the vertex set $A := [1..t]$ is connected. Then there is a non-empty subset $\Gamma_2 \subset \Pi_{k,p}^2$ such that*

$$\mathbb{E}W^{\gamma_A} = \begin{cases} 0 & (t=1) \\ C_{\gamma_1,\gamma_2} n^{2k-v_{\gamma_1,\gamma_2}} + O(n^{2k-1-v_{\gamma_1,\gamma_2}}) & (t=2 \text{ and } (\gamma_1,\gamma_2) \in \Gamma_2) \\ O(n^{kt-1-v_{\gamma_1,\gamma_2,\cdots,\gamma_t}}) & (otherwise). \end{cases}$$

Here C_{γ_1,γ_2} is a positive constant depending only on $(\gamma_1, \gamma_2) \in \Gamma_2$.

Armed with Lemma 2, we can give a proof of Theorem 2 now.

Proof of Theorem 2

We start from $\ell = 2$. First, if $V_{\gamma_1} \cap V_{\gamma_2} = \emptyset$, then we have $W^{\gamma_1,\gamma_2} = W^{\gamma_1}W^{\gamma_2} = 0$ by Lemma 2.

Now assume that $V_{\gamma_1} \cap V_{\gamma_2} \neq \emptyset$. Then the union path $\gamma_{1,2} = \gamma_1 \cup \gamma_2$ is connected. By using Lemma 2 and the fact that

$$\frac{p!}{(p-v)!} = p^v \left(1 + O_k\left(\frac{1}{p}\right)\right), \tag{12}$$

we obtain from Eq. (9) that

$$\beta_{k,2} = \frac{1}{n^{2k}} \sum_{(\gamma_1,\gamma_2) \in \Gamma_2} \frac{p!}{(p-v_{\gamma_1,\gamma_2})!} C_{\gamma_1,\gamma_2} n^{2k-v_{\gamma_1,\gamma_2}}$$

$$+ \frac{1}{n^{2k}} \sum_{\substack{(\gamma_1,\gamma_2) \in \Pi^2_{k,p}/\Sigma_p \\ V_{\gamma_1} \cap V_{\gamma_2} \neq \emptyset}} \frac{p!}{(p-v_{\gamma_1,\gamma_2})!} O(n^{2k-v_{\gamma_1,\gamma_2}-1})$$

$$= \sum_{(\gamma_1,\gamma_2) \in \Gamma_2} C_{\gamma_1,\gamma_2} \left(\frac{p}{n}\right)^{v_{\gamma_1,\gamma_2}} \left(1 + O_k\left(\frac{1}{p}\right)\right) + \sum_{\substack{(\gamma_1,\gamma_2) \in \Pi^2_{k,p}/\Sigma_p \\ V_{\gamma_1} \cap V_{\gamma_2} \neq \emptyset}} O_k\left(\left(\frac{p}{n}\right)^{v_{\gamma_1,\gamma_2}} \frac{1}{n}\right).$$

Using the facts that $\frac{p}{n} = y \in (0,1)$ and

$$\sum_{\substack{(\gamma_1,\gamma_2,\cdots,\gamma_\ell) \in \Pi^\ell_{k,p}/\Sigma_p \\ v_{\gamma_1,\gamma_2,\cdots,\gamma_\ell} = v}} 1 \leq v^{k\ell} \leq (k\ell)^{k\ell} \quad \forall 1 \leq v \leq k\ell, \tag{13}$$

we finally get

$$\beta_{k,2} = \sum_{(\gamma_1,\gamma_2) \in \Gamma_2} C_{\gamma_1,\gamma_2} y^{v_{\gamma_1,\gamma_2}} + O_k\left(\frac{1}{n}\right).$$

Denote

$$\sigma_k^2 = \sum_{(\gamma_1,\gamma_2) \in \Gamma_2} C_{\gamma_1,\gamma_2} y^{v_{\gamma_1,\gamma_2}}. \tag{14}$$

Then we have

$$\beta_{k,2} = \sigma_k^2 + O_k\left(n^{-1}\right) = (2-1)!!\sigma_k^2 + O_k\left(n^{-1}\right). \tag{15}$$

The explicit evaluations of C_{γ_1,γ_2} and hence σ_k^2 in (14) are quite complicated in general. They are closely related to a standard computation in random matrix theory and do not affect the shape of the limiting distribution. We postpone the computation of C_{γ_1,γ_2} and σ_k^2 to Sect. 6 **Appendix**. It is easy to see from (15) that we have proved Theorem 2 for $\ell = 2$.

Now let us assume $\ell \geq 3$. Consider the decomposition of $[1..\ell]$ into disjoint union of connected components as in (10). By (11) and Lemma 2, if there is an $i \in \mathfrak{C}$ such that $\#A_i = 1$, then $W^{\gamma_1,\gamma_2,\cdots,\gamma_\ell} = 0$. Now suppose $\#A_i \geq 2$ for any $i \in \mathfrak{C}$. If there is a $i_0 \in \mathfrak{C}$ with $\#A_{i_0} \geq 3$, then we have

$$\mathbb{E}W^{\gamma_1,\gamma_2,\cdots,\gamma_\ell} \ll n^{k\#A_{i_0}-1-v_{\gamma_{A_{i_0}}}} \prod_{i \in \mathfrak{C}\setminus\{i_0\}} n^{k\#A_i - v_{\gamma_A}} \ll_{k,\ell} n^{k\ell-1-v_{\gamma_1,\gamma_2,\cdots,\gamma_\ell}}.$$

The total contribution for these cases is bounded by

$$I_1 = \frac{1}{n^{k\ell}} \sum_{\substack{(\gamma_1,\gamma_2,\cdots,\gamma_\ell) \in \Pi^\ell_{k,p}/\Sigma_p \\ \#A_i \geq 2 \,\forall i \in \mathfrak{C} \\ \#A_{i_0} \geq 3 \,\exists i_0 \in \mathfrak{C}}} \frac{p!}{(p-v_{\gamma_1,\gamma_2,\cdots,\gamma_\ell})!} \mathbb{E}W^{\gamma_1,\gamma_2,\cdots,\gamma_\ell}$$

$$\ll_{k,\ell} \sum_{(\gamma_1,\gamma_2,\cdots,\gamma_\ell) \in \Pi^\ell_{k,p}/\Sigma_p} \left(\frac{p}{n}\right)^{v_{\gamma_1,\gamma_2,\cdots,\gamma_\ell}} \frac{1}{n} \ll_{k,\ell} \frac{1}{n}, \qquad (16)$$

which is negligible as $n \to \infty$. Now we consider the last case, that is, $\#A_i = 2$ for any $i \in \mathfrak{C}$. Note that this can happen only when ℓ is even. Hence for any odd $\ell \geq 3$, we have $\beta_{k,\ell} = O_{k,\ell}(n^{-1})$, confirming Theorem 2 for $\ell \geq 3$ and ℓ is odd.

Now we consider $\ell \geq 4$ even. Denote by $P = \{\{p_{1,1}, p_{1,2}\}, \{p_{2,1}, p_{2,2}\}, \cdots, \{p_{\ell/2,1}, p_{\ell/2,2}\}\}$ a partition of $[1\mathinner{..}\ell]$ into $\ell/2$ pairs of indices. Then for any $\gamma^{(\ell)} = (\gamma_1, \gamma_2, \cdots, \gamma_\ell)$ that corresponds to this partition P, which we denote as condition $P(*)$, we have

$$\mathbb{E}W^{\gamma_1,\gamma_2,\cdots,\gamma_\ell} = \prod_{j=1}^{\ell/2} \mathbb{E}W^{\gamma_{p_{j,1}},p_{j,2}} = \left(\prod_{j=1}^{\ell/2} C_{\gamma_{p_{j,1}},\gamma_{p_{j,2}}}\right) n^{k\ell - v_{\gamma_1,\gamma_2,\cdots,\gamma_\ell}} + O(n^{k\ell - v_{\gamma_1,\gamma_2,\cdots,\gamma_\ell} - 1}),$$

where $C_{\gamma_{p_{j,1}},\gamma_{p_{j,2}}} = 0$ if $(\gamma_{p_{j,1}}, \gamma_{p_{j,2}}) \notin \Gamma_2$.

Hence by (13), for any such P,

$$\frac{1}{n^{k\ell}} \sum_{\substack{(\gamma_1,\gamma_2,\cdots,\gamma_\ell) \in \Pi^\ell_{k,p}/\Sigma_p \\ P(*)}} \frac{p!}{(p-v_{\gamma_1,\gamma_2,\cdots,\gamma_\ell})!} \mathbb{E}W^{\gamma_1,\gamma_2,\cdots,\gamma_\ell}$$

$$= \sum_{\substack{(\gamma_1,\gamma_2,\cdots,\gamma_\ell) \in \Pi^\ell_{k,p}/\Sigma_p \\ P(*)}} \left(\prod_{j=1}^{\ell/2} C_{\gamma_{p_{j,1}},\gamma_{p_{j,2}}}\right) y^{v_{\gamma_1,\gamma_2,\cdots,\gamma_\ell}} + O_{k,\ell}\left(\frac{1}{n}\right)$$

$$= \prod_{j=1}^{\ell/2} \left(\sum_{(\gamma_{p_{j,1}},\gamma_{p_{j,2}}) \in \Gamma_2} C_{\gamma_{p_{j,1}},\gamma_{p_{j,2}}} y^{v_{\gamma_{p_{j,1}},\gamma_{p_{j,2}}}}\right) + O_{k,\ell}\left(\frac{1}{n}\right)$$

$$= \sigma_k^\ell + O_{k,\ell}\left(\frac{1}{n}\right).$$

Since the number of such partitions P of $[1\mathinner{..}\ell]$ is obviously $(\ell-1)!!$, the total contribution to $\beta_{k,\ell}$ by these cases is

$$I_2 = (\ell-1)!!\sigma_k^\ell + O_{k,\ell}\left(\frac{1}{n}\right). \qquad (17)$$

Combining the estimate on I_1 in (16) and I_2 above, we obtain for all even $\ell \geq 4$,

$$\beta_{k,\ell} = I_1 + I_2 = (\ell-1)!!\sigma_k^\ell + O_{k,\ell}\left(\frac{1}{n}\right).$$

This completes the proof of Theorem 2. $\square$

4 Analysis of $\mathbb{E}W^{\gamma_A}$

In this section we provide a proof of Lemma 2 by giving a deep analysis of the quantity W^{γ_A} where we may assume $\gamma_A = (\gamma_1, \gamma_2, \cdots, \gamma_t) \in \Pi_{k,p}^t$ and the union graph γ_A is connected on the vertex set $[1..t]$. Under a suitable renaming of the domain indices, γ_A can be considered as a closed path of length kt.

First, we observe that

$$\mathbb{E}W^{\gamma_1, \gamma_2, \cdots, \gamma_t} = \sum_{j=0}^{t} (-1)^{t-j} \sum_{\substack{B \subset [1..t] \\ \#B = j}} \left(\mathbb{E}W_{\gamma_B} \prod_{i \in [1..t] \setminus B} \mathbb{E}W_{\gamma_i} \right),$$

where $W_{\gamma_B} := \prod_{i \in B} \omega_{\gamma_i}(s)$.

When $\ell = 1$, it is trivial from the above formula that $\mathbb{E}W^{\gamma_1} = 0$. Now assume $\ell \geq 2$. We shall remark that a term very similar to $\mathbb{E}W_{\gamma_B}$ has been studied in [17, Section II] by using a graph method. Following the same ideas from [17], we see that the quantity $\mathbb{E}W_{\gamma_B}$ counts the number of solutions for the variables $(t_{i,j})_{i \in B, j \in [0..k-1]}$ to the system of equations given by

$$\sum_{i \in B} \sum_{j \in \gamma_i^{-1}(a)} \mathbf{g}_{t_{i,j}} = \sum_{i \in B} \sum_{j \in \gamma_i^{-1}(a)} \mathbf{g}_{t_{i,j-1}} \quad \forall a \in V_{\gamma_B} := \bigcup_{i \in B} \gamma_i([0..k]). \tag{18}$$

Here the index j is treated modulo k, and $\mathbf{g}_t$ is the t-th column of a fixed generator matrix of $\mathcal{C}$. Similar to [17], one can see that in the system of equations (18), exactly one equation is redundant, and the quantity $\mathbb{E}W^{\gamma_1, \gamma_2, \cdots, \gamma_t}$ is always non-negative and not larger than the leading term $\mathbb{E}W_{\gamma_{[1..t]}}$. In particular, it suffices to consider (18) for $B = [1..t]$.

Before proceeding, we first state and prove an important result that is directly related to the code condition $d^\perp \geq 7$.

Lemma 3. *Assume $d^\perp \geq 7$, and $7 \leq m \leq \sqrt{n}$. Then the number of solutions of $(t_1, t_2, \cdots, t_m) \in [1..n]^m$ to the equation*

$$\sum_{i=1}^{m} \mathbf{g}_{t_i} = \mathbf{0} \tag{19}$$

is $O(n^{m-3})$.

Proof (Proof of Lemma 3). Denote by W_m the number of solutions to (19). In addition, for $m \in \mathbb{N}$, define W'_m as the number of such solutions with t_i's pairwise distinct. For $m = 0$, define $W'_0 := 1$. Then we can write

$$W_m \ll \sum_{a=0}^{\lfloor m/2 \rfloor} \binom{m}{2a} (2a-1)!! n^a W'_{m-2a}, \tag{20}$$

where the binomial and double-factorial factors correspond to the number of ways to choose $2a$ coordinates for pairs and pairing these $2a$ coordinates respectively.

We first claim that $W'_a = O(n^{a-3})$ for all fixed $a \in \mathbb{N}$.
Note that $d^\perp \geq 7$ implies that $W'_a = 0 = O(n^{a-3})$ for any $1 \leq a \leq 6$ trivially.
Now assume $a \geq 7$. Then (5) with $d = 7$ and $\nu = -3$ yields

$$N_n(7, a, -3) = \binom{a}{3}.$$

Hence by Lemma 1 on the dual code $\mathcal{C}^\perp$, we get

$$A_a(\mathcal{C}^\perp) \leq \frac{\binom{n}{a-3}}{\binom{a}{3}} \ll \frac{n^{a-3}}{a!}.$$

Hence $W'_a = a! A_a(\mathcal{C}^\perp) \ll n^{a-3}$ as we want.
Inequality (20) then yields

$$W_m \ll \sum_{a=0}^{\lfloor m/2 \rfloor} \binom{m}{2a} (2a-1)!! n^{m-a-3} + n^{m/2}$$

$$\ll \sum_{a=0}^{\lfloor m/2 \rfloor} \frac{m^{2a}}{2^a a!} n^{m-a-3} + n^{m/2}$$

$$\ll n^{m-3} \left(\sum_{a=0}^{\lfloor m/2 \rfloor} \frac{\left(\frac{m^2}{2n}\right)^a}{a!} + 1 \right) \ll n^{m-3} \left(e^{\frac{m^2}{2n}} + 1 \right) \ll n^{m-3},$$

where in the third inequality we have used the assumption $m \geq 7$, and in the last inequality we have use the assumption $m \leq \sqrt{n}$. $\square$

Proof (Proof of Lemma 2). For any connected γ, we can apply the reduction processes successively as mentioned in [17, Section IV-C] until we get a reduced path γ'. Denote by $k_{\gamma'}$ the length of γ'. Here we recall that a path γ' is reduced if either of the following holds:

1. $k_{\gamma'} = v_{\gamma'} = 1$ OR
2. $v_{\gamma'} \geq 2$, each vertex in $V_{\gamma'}$ is traversed at least twice by γ', and the preimages of γ' do not contain consecutive elements.

We have two cases:

Case I. The reduced path γ' of γ has a vertex traversed at least four times.

In this case, the corresponding equation has $m \geq 8$ variables. After fixing each of the $O(n^{m-3})$ solutions for this equation from the system (18) by Lemma 3, we have $k_{\gamma'} - m$ variables left for $v_{\gamma'} - 2$ independent (not necessarily homogeneous) linear equations, and so can have at most $O(n^{k_{\gamma'}-m-v_{\gamma'}+2})$ solutions in average. Altogether we have

$$\mathbb{E} W_{\gamma'} = O(n^{k_{\gamma'}-m-v_{\gamma'}+2+m-3}) = O(n^{k_{\gamma'}-v_{\gamma'}-1}).$$

Using the argument in [17, Section IV-D], we get $\mathbb{E}W_\gamma = O(n^{k\ell - v_\gamma - 1})$. This implies $\mathbb{E}W^{\gamma_1, \gamma_2, \cdots, \gamma_\ell} = O(n^{k\ell - v_\gamma - 1})$ too.

Case II. Each vertex in the reduced path γ' is traversed no more than three times.

In this case, each equation has at most 6 variables. As $d^\perp \geq 7$, each equation can be satisfied only if the indices hold values in pairs. This is the same as in the full binary code case, that is, when the entries in X are i.i.d. Rademacher variables. A more general combinatorial argument is true for the entries of X being i.i.d. centralized and normalized real random variables. This is described by the following lemma, whose proof can be found in [1,2].

Lemma 4. *Let a be the union of ℓ closed walks $w_1, w_2, \cdots, w_\ell$ each of length k. Define G_a to be the graph of non-coincident edges of a. Assume that G_a is connected and each edge in G_a is traversed at least twice in a, and that for any i, there is $j \neq i$ such that w_i and w_j share at least one common edge. Then the number of vertices in G_a is at most $1 + \left\lfloor \frac{(k-1)\ell}{2} \right\rfloor$.*

In the setting of $\mathcal{G}$, the closed walks are of length $2k$. As we know that v_γ of the vertices are for γ, Lemma 4 then implies that $\mathbb{E}W^{\gamma_1, \gamma_2, \cdots, \gamma_\ell} = O(n^{k\ell + 1 - \lceil \frac{\ell}{2} \rceil - v_\gamma})$.

Hence if $\ell \geq 3$, then we get $\mathbb{E}W^{\gamma_1, \gamma_2, \cdots, \gamma_\ell} = O(n^{k\ell - v_\gamma - 1})$.

Now let $\ell = 2$. Denote by $\Gamma_2 \subset \Pi_{k,p}^2 / \Sigma_p$ the set of all k-closed path pairs (γ_1, γ_2) such that $W^{\gamma_1, \gamma_2} \asymp n^{2k - v_{\gamma_1, \gamma_2}}$. From the above argument it is easy to see that any $(\gamma_1, \gamma_2) \in \Gamma_2$ must correspond to γ in **Case II**, so that Γ_2 is independent of $\mathcal{C}$. To see that $\Gamma_2 \neq \emptyset$, we may consider the pair $\gamma_1 = \gamma_2 = \tilde{\gamma}$ where $\tilde{\gamma}(0) = 1$ and $\tilde{\gamma}(1) = \tilde{\gamma}(2) = \cdots = \tilde{\gamma}(k-1) = 2$. Then we have $v_{\gamma_1, \gamma_2} = 2$. For a single path γ_j ($j = 1, 2$), we see that it corresponds to a single equation in the form $\mathbf{g}_{t_0} = \mathbf{g}_{t_{k-1}}$ after removing $k-2$ identical variables in two sides (this is equivalent to applying a total of $k-2$ reductions of consecutive elements as defined in [17, Section IV-C]), so we have $\mathbb{E}W_{\gamma_i} = n^{k-2+1} = n^{k-1}$. For the union path $\gamma = \gamma_1 \cup \gamma_2$, we have a single equation in the form

$$\mathbf{g}_{t_0} + \mathbf{g}_{u_0} = \mathbf{g}_{t_{k-1}} + \mathbf{g}_{u_{k-1}} \tag{21}$$

after applying a total of $2(k-2)$ reductions. Hence γ indeed belongs to **Case II**. The only solutions are when the values of the four indices come in pairs. Considering all combinations of pairs, we get $3n^2 + O(n)$ solutions for $(t_0, u_0, t_{k-1}, u_{k-1})$. Together with the $2(k-2)$ removed (free) variables, we get $W_\gamma = 3n^{2k-2} + O(n^{2k-3})$. This implies that $\mathbb{E}W^{\gamma_1, \gamma_2} = \mathbb{E}W_\gamma - \mathbb{E}W_{\gamma_1} W_{\gamma_2} = 2n^{2k-2} + O(n^{2k-3})$. Hence $(\gamma_1, \gamma_2) \in \Gamma_2$.

Finally, as the involved equations are linear and each variable can take on at most n distinct values, the number of solutions can always be written as a polynomial in n. Therefore there exists some positive constant C_{γ_1, γ_2} such that $W^{\gamma_1, \gamma_2} = C_{\gamma_1, \gamma_2} n^{2k - v_{\gamma_1, \gamma_2}} + O(n^{2k - v_{\gamma_1, \gamma_2} - 1})$ for $(\gamma_1, \gamma_2) \in \Gamma_2$.

This completes the proof of Lemma 2. □

5 Augmented Codes from Binary Gold Codes

In this section we show that the Gaussian central limit theorem does not hold for the augmented codes from binary Gold code $\mathcal{C}$ of length $n = 2^i - 1$ [9] under any normalization scale factor α_n. The Gold code is the binary linear code with dimension $2i$ given by

$$\mathcal{C} = \left\{ (\text{Tr}(a\beta^j + b\beta^{(2^r+1)j}))_{j=0}^{n-1} : a, b \in \mathbb{F}_{2^i} \right\}.$$

The augmented Gold code is simply the binary linear code of dimension $2i + 1$ generated by the Gold code and the all-one codeword of the same length.

The Gold codes have dual distance 5 or 3 according to whether or not $\gcd(i, r) = 1$ (which corresponds to whether or not the function $f(x) = x^{2^r+1}$ is APN). The corresponding augmented codes $\mathcal{C}$ have dual distance 6 or 4. All the non-zero weights w of $\mathcal{C}$ with $w \neq n$ satisfy the inequality

$$2^{i-1} - 2^{\lfloor i/2 \rfloor} \leq w \leq 2^{i-1} + 2^{\lfloor i/2 \rfloor}.$$

Hence we can apply [14, Corollary 3.3] for Gold codes or [8, Theorem 5] for augmented Gold codes to obtain

$$A_m(\mathcal{C}^\perp) = \frac{1}{2^{2i}} \binom{n}{m} + O(n^{m/2}) = \frac{1}{m!} n^{m-2} + O(n^{m-3}) \qquad (22)$$

for all fixed $m \geq 6$. For $m = 4$, we have

$$A_4(\mathcal{C}^\perp) = O(n^2) \text{ (zero if } \gcd(i,r) = 1). \qquad (23)$$

First, we can use the estimation (22) and the arguments in Sect. 4 to show that $W^{\gamma_1, \gamma_2, \cdots, \gamma_\ell} = O(n^{k\ell - v_{\gamma_1, \gamma_2, \cdots, \gamma_\ell}})$, which implies that $\beta_{k,\ell}$ is bounded in n for all fixed k, ℓ, and also that $\beta_{k,2} \asymp 1$. Therefore $\alpha_n = p$ is the correct normalizing scale.

In the following we show that $\beta_{k,3}$ converges to a finite non-zero number as n tends to infinity. This implies that the limiting distribution is not symmetric and hence cannot be Gaussian.

First, we note that for any triple $(\gamma_1, \gamma_2, \gamma_3)$ of k-closed paths, we always have $W^{\gamma_1, \gamma_2, \gamma_3} \geq 0$. Therefore it suffices to prove the existence of such a triple that contributes to the constant term.

In fact, if $\gamma_1 = \gamma_2 = \gamma_3 = \tilde{\gamma}$ is defined by $\tilde{\gamma}(0) = 1$ and $\tilde{\gamma}(1) = \tilde{\gamma}(2) = \cdots = \tilde{\gamma}(k-1) = 2$, then as in Sect. 4, we have $W_{\gamma_j} = n^{k-1}$ for $j = 1, 2, 3$. For $W_{\gamma_j, j'}$ with $j \neq j' \in \{1, 2, 3\}$, it counts the number of solutions to an equation of the form (21). By (23), we also have $W_{\gamma_{j,j'}} = O(n^{2k-2})$.

Finally the quantity $W_{\gamma_{1,2,3}}$ counts the number of solutions to the equation

$$\mathbf{g}_{t_{11}} + \mathbf{g}_{t_{21}} + \mathbf{g}_{t_{31}} = \mathbf{g}_{t_{10}} + \mathbf{g}_{t_{20}} + \mathbf{g}_{t_{30}}$$

after performing a total of $3(k-2)$ reductions of consecutive elements. Hence (22) with $m = 6$ yields

$$\mathbb{E} W_{\gamma_{1,2,3}} = n^{3(k-2)} \left(6! A_6(\mathcal{C}^\perp) + O(n^3) \right) = n^{3k-2} + O(n^{3k-3}).$$

These imply that

$$\mathbb{E} W^{\gamma_1, \gamma_2, \gamma_3} = W_{\gamma_{1,2,3}} - W_{\gamma_{1,2}} W_{\gamma_3} - W_{\gamma_{2,3}} W_{\gamma_1} - W_{\gamma_{1,3}} W_{\gamma_2} + 2 W_{\gamma_1} W_{\gamma_2} W_{\gamma_3} = n^{3k-2} + O(n^{3k-3}).$$

As $v_{\gamma_1, \gamma_2, \gamma_3} = 2$, we see that this choice of $\gamma_1, \gamma_2, \gamma_3$ indeed contributes non-negligibly positively to the limit $\lim_{n \to \infty} \beta_{k,3}$.

6 Appendix: Explicit Formula for σ_k

In this section we provide a technical formula for the quantity σ_k in Theorem 1. A similar formula was known for Wigner matrices in [1,2]. However we cannot find a reference for the analogue for sample covariance matrices, which is the i.i.d. version of our model. Hence here we also give a proof of it.

Theorem 3.

$$\sigma_k^2 = \sum_{u=2}^{2k-2} y^u \sum_{s=2}^{k} \frac{2k^2}{s} \sum_{u'=s}^{u} D(k, s, u') D(k, s, u+s-u'),$$

where

$$D(k, s, u') = \sum_{(\mathbf{k}, \mathbf{u}) \in \mathfrak{U}(u')} N(\mathbf{k}, \mathbf{u}), \tag{24}$$

$$\mathbf{k} = (k_{I,1}, k_{T,1}, k_{I,2}, k_{T,2}, \cdots, k_{I,s}, k_{T,s}),$$
$$\mathbf{u} = (u_{I,1}, u_{T,1}, u_{I,2}, u_{T,2}, \cdots, u_{I,s}, u_{T,s}),$$
$$\mathfrak{U}(u') = \{(\mathbf{k}, \mathbf{u}) : \sum_{i=1}^{s}(k_{I,i} + k_{T,i}) = k - s, \sum_{i=1}^{s}(u_{I,i} + u_{T,i}) = u',$$
$$k_{I,i}, k_{T,i} \geq 0, u_{I,i} \geq 1, u_{T,i} \geq 0, (k_{T,i}, u_{T,i}) \neq (0,1)\},$$

$$N(\mathbf{k}, \mathbf{u}) = \prod_{i=1}^{s} \left(N(k_{I,i}, u_{I,i}) N(k_{T,i}, u_{T,i}) \right)$$

and

$$N(k, u) = \begin{cases} \frac{1}{u}\binom{k}{u-1}\binom{k-1}{u-1} & (u > 0) \\ \delta_{k,0} & (u = 0). \end{cases}$$

Here the numbers $N(k, u)$ for $k > 0$ and $u > 0$ are called the Narayana numbers. They also appear in the formulas for the moments of the MP law.

Before giving a proof, here we need to define several concepts and notations, partially borrowed from [1].

Given two disjoint sets of alphabets I and T, we define an (I, T)-*word* to be a finite sequence of alphabets (with at least one element) starting from an I-alphabet so that no two adjacent alphabets come from the same set. It is called *closed* if the first and last alphabets are the same. In particular any closed (I, T)-words have odd length. Two (I, T)-words w_1 and w_2 are said to be *equivalent* if there is a bijection on (I, T) that maps one word into another.

Given an (I,T)-word $w = (s_1, s_2, \cdots, s_k)$, we denote by $\ell(w) = k$ the length of w, $\mathrm{supp}(w)$ the *support* of w, that is, the set of alphabets defined in w, and $\mathrm{wt}(w)$ the *weight* of w, which is the size of the support of w. We also denote $\mathrm{supp}_I(w)$ and $\mathrm{supp}_T(w)$ to be the set of I and T-alphabets defined in w respectively. $\mathrm{wt}_I(w)$ and $\mathrm{wt}_T(w)$ (the I-weight and T-weight of w) are defined similarly.

Moreover, we associate with w a graph $G_w = (V_w, E_w)$ whose set of vertices V_w and edges E_w are given by $V_w = \mathrm{supp}(w)$ and $E_w = \{\{s_i, s_{i+1}\} : i = 1, 2, \cdots, k-1\}$ respectively. Note that by the definition of (I,T)-word G_w is actually (I,T)-bipartite. In particular G_w contains no self-loops. Also, G_w is connected. For $e \in E_w$, we denote by N_e^w the number of times e is traversed in w (in any direction). Then equivalent words generate the same G_w up to isomorphism, and the multiset $\{N_e^w\}$ derived from them is also the same.

An (I,T)-pair a is a pair of (I,T)-words (w_1, w_2). Similar to words, we define two (I,T)-pairs a_1 and a_2 to be equivalent if there is a bijection on (I,T) mapping one to another.

Then the support and weight of a are defined by $\mathrm{supp}(a) := \mathrm{supp}(w_1) \cup \mathrm{supp}(w_2)$ and $\mathrm{wt}(a) := \#\mathrm{supp}(a)$ respectively. $\mathrm{supp}_I(a), \mathrm{supp}_T(a), \mathrm{wt}_I(a)$ and $\mathrm{wt}_T(a)$ are also defined analogously with single words.

Also, the graph $G_a = (V_a, E_a)$ associated with a is defined simply as $V_a = V_{w_1} \cup V_{w_2}$ and $E_a = E_{w_1} \cup E_{w_2}$. And G_a is connected if and only if $V_{w_1} \cap V_{w_2} \neq \emptyset$. N_e^a is again defined similarly to N_e^w.

Proof (Proof of Theorem 3). Note that we can write

$$\sigma_k^2 = \lim_{n \to \infty} \frac{1}{n^{2k}} \sum_a X^a. \tag{25}$$

Here the summation is over all (I,T)-pairs $a = (w_1, w_2)$ with

$$w_j = (i_0^j, t_0^j, i_1^j, t_1^j, \cdots, i_{k-1}^j, t_{k-1}^j, i_0^j), \quad j = 1, 2$$

both being (I,T)-closed words of length $2k+1$,

$$X^a := \mathbb{E}(X_{w_1} X_{w_2}) - \mathbb{E}X_{w_1} \mathbb{E}X_{w_2}$$

and

$$X_{w_j} := x_{i_0^j, t_0^j} x_{i_1^j, t_0^j} x_{i_1^j, t_1^j} x_{i_2^j, t_1^j} \cdots x_{i_{k-1}^j, t_{k-1}^j} x_{i_0^j, t_{k-1}^j}.$$

Noting that σ_k is independent of $\mathcal{C}$, we may assume $\mathcal{C} = \mathbb{F}_2^n$. Then the entries of X are i.i.d. Rademacher variables, so $\mathbb{E}\prod_{i,j} x_{ij}^{m_{ij}} = 1$ if all m_{ij} are even and 0 otherwise. This implies, in the following situations, we have $X^a = 0$:

(1) N_e^a is odd for some $e \in E_a$;
(2) $N_e^{w_1}$ and $N_e^{w_2}$ are even for all $e \in E_a$;
(3) G_{w_1} and G_{w_2} have no common edges.

On the other hand, if N_e^a is even for all $e \in E_a$ but there exists $e \in E_{w_j}$ so that $N_e^{w_j}$ is odd (j can be 1 or 2), then $E_{w_1} \cap E_{w_2} \neq \emptyset$ (in fact any such e satisfying above condition is in the intersection, and also if $N_e^{w_1}$ is odd, then so is $N_e^{w_2}$ and vice versa) and $X^a = 1$.

In addition, equivalent (I,T)-pairs give rise to same X^a. Therefore we may further write the sum (25) as

$$\sigma_k^2 = \lim_{n \to \infty} \frac{1}{n^{2k}} \sum_{u=1}^{2k} \frac{p!}{(p-u)!} \sum_{v=1}^{2k} \frac{n!}{(n-v)!} \# \mathscr{W}_{2k,(u,v)}^{(2)} \quad (26)$$

where $\mathscr{W}_{2k,(u,v)}^{(2)}$ is the set of equivalence classes of (I,T)-pairs a consisting of two closed (I,T)-words (w_1, w_2) each of length $2k+1$, so that $\mathrm{wt}_I(a) = u$ and $\mathrm{wt}_T(a) = v$, N_e^a is even for all $e \in E_a$ but for $j = 1$ or 2, $N_e^{w_j}$ is odd for some $e \in E_{w_j}$.

By Lemma 4, we always have $u + v = \mathrm{wt}(a) \le 2k$. Also noting that $p = yn$ and $\frac{p!}{(p-u)!} = p^u(1 + o(1))$, (26) yields

$$\sigma_k^2 = \sum_{u=1}^{2k-1} y^u \# \mathscr{W}_{2k,(u,2k-u)}^{(2)}. \quad (27)$$

Note that if $a \in \mathscr{W}_{2k,(u,2k-u)}^{(2)}$, then G_a must be connected. Hence $\#E_a \ge \#V_a - 1 = 2k - 1$. On the other hand, since $N_e^a \ge 2$ for all $e \in E_a$, we have $\#E_a \le \frac{4k}{2} = 2k$. Hence we have $\#E_a = 2k - 1$ or $2k$.

Case I. $\#E_a = 2k - 1$. Then G_a is a tree, and so are the subgraphs G_{w_1} and G_{w_2}. However then we must have $N_e^{w_j}$ all even so that w_j can be closed. Therefore $a \notin \mathscr{W}_{2k,(u,2k-u)}^{(2)}$.

Case II. $\#E_a = 2k$. Then in this case G_a is a unicyclic graph, that is, a graph with a unique cycle. Since G_a is bipartite, this unique cycle must be of even length. Say its length is $2s$ where $2 \le s \le k$. Also, $N_e^a = 2$ for all $e \in E_a$.

First note that if $u = 1$ or $2k-1$, then G_a has only one I-vertex or T-vertex, so that it cannot contain any cycle and $\#\mathscr{W}_{2k,(u,2k-u)}^{(2)} = 0$. Therefore in the following we assume that $2 \le u \le 2k - 2$.

By [1, Lemma 2.1.33], it can be shown that $N_e^{w_j} = 1$ for $j = 1, 2$ if and only if e is an edge of the unique cycle. This implies that indeed $a \in \mathscr{W}_{2k,(u,2k-u)}^{(2)}$. In addition, the other vertices are in either V_{w_1} or V_{w_2} but not both. In each w_j, after removing the $2s$ edges in the cycle, there are $2k - 2s$ traversed "edges", and each edge is traversed twice, so that the total number of edges in G_{w_j} outside the cycle is $k - s$. Furthermore, assuming $\mathrm{wt}_I(w_1) = u'$, we have $\mathrm{wt}_I(w_2) = u + s - u'$ as G_{w_1} and G_{w_2} share precisely the s I-vertices in the cycle. This also implies that $u' \ge s$ and $u + s - u' \ge s$, which is equivalent to $s \le u' \le u$.

Counting the number of such words up to equivalence is the same as counting the number of non-equivalent ways to attach trees (some or all may be empty) to the vertices of a $2s$-cycle so that the total number of edges is $k - s$. Here if nothing is to be attached to a given vertex of the cycle, then we treat it as attaching an empty graph with one vertex to that vertex (so that the equality $\#V = \#E + 1$ holds). As there are a total of $N(k,u)$ non-equivalent trees with k edges and u I-vertices, the total number of non-equivalent trees attached to the $2s$ vertices with a total of $k-s$ edges and u' I-vertices is given by $D(k,s,u')$ as shown in Eq. (24).

Considering possible choices for the starting points of w_1 and w_2, the fact that the cycle may be traversed in same or opposite orientations in w_1 and w_2, and also the cyclic equivalence, the number of $a \in \mathscr{W}_{2k,(u,2k-u)}^{(2)}$ in which G_a contains a unique cycle of length $2s$ is given by

$$\frac{2k^2}{s} \sum_{u'=s}^{u} D(k,s,u')D(k,s,u+s-u').$$

Putting this into (27) completes the proof of Theorem 3. □

7 Conclusion

In this paper, we proved a central limit theorem for linear eigenvalue statistics of matrices formed by picking codewords at random from long binary linear codes under the condition that the dual distance of the code is at least 7. We also provided a family of binary linear codes with dual distance 6 which do not satisfy this central limit theorem, showing that the condition of the dual distance being at least 7 is optimal in this respect. Since the central limit theorem holds for linear eigenvalue statistics of large truly random matrices, i.e., matrices with i.i.d entries, this gives a new "group randomness" property for binary linear codes of dual distance 7.

It was known that the empirical spectral distribution of random matrices constructed the same way from binary linear codes of increasing length converges to the Marchenko-Pastur law as long as the dual distance of the code is at least 5, and the rate of convergence is of the magnitude $n^{-1/4}$ where n is the length of the code. It might be interesting to see if this rate of convergence result can be improved under the condition that the dual distance of the code is at least 7.

Acknowledgments. The authors are grateful to Zhigang Bao for raising this question and for many valuable suggestions. The first author would like to thank the Dept. of Mathematics at Hong Kong University of Science and Technology for financial support. Both authors also thank the anonymous referees for raising interesting questions and for valuable suggestions that improve the quality of the paper.

Disclosure of Interests. The authors have no competing interests to declare that are relevant to the content of this article.

References

1. Anderson, G.W., Guionnet, A., Zeitouni, O.: An Introduction to Random Matrices. Cambridge University Press, Cambridge (2005)
2. Anderson, G.W., Zeitonui, O.: A CLT for a band matrix model. Probab. Theory Relat. Fields **134**(2), 283–338 (2006)
3. Babadi, B., Ghassemzadeh, S.S., Tarokh, V.: Group randomness properties of pseudo-noise and gold sequences. In: Proceedings of 12th Canadian Workshop on Information Theory (CWIT), British Columbia, Canada, pp. 42–46 (2011)
4. Babadi, B., Tarokh, V.: Random frames from binary linear block codes. In: 2010 44th Annual Conference on Information Sciences and Systems (CISS), Princeton, NJ, USA, pp. 1–3 (2010)
5. Babadi, B., Tarokh, V.: Spectral distribution of random matrices from binary linear block codes. IEEE Trans. Inform. Theory **57**(6), 3955–3962 (2011)
6. Bai, Z., Silverstein, J.W.: Spectral Analysis of Large Dimensional Random Matrices. Springer Series in Statistics, 2nd edn. New York (2010)
7. Chan, C., Tarokh, V., Xiong, M.: Convergence rate of empirical spectral distribution of random matrices from linear codes. IEEE Trans. Inform. Theory **67**(2), 1080–1087 (2019)
8. Chan, C., Xiong, M.: On the complete weight distribution of subfield subcodes of algebraic-geometric codes. IEEE. Trans. Inform. Theory **65**(11), 7079–7086 (2019)
9. Gold, R.: Maximal recursive sequences with 3-valued recursive cross-correlation functions. IEEE Trans. Inform. Theory **14**(1), 154–156 (1968)
10. Kløve, T., Luo, J.: Upper bounds on the weight distribution function for some classes of linear codes. IEEE Trans. Inform. Theory **58**(8), 5512–5521 (2012)
11. Levy, J.E.: A weight distribution bound for linear codes. IEEE Trans. Inform. Theory **14**(3), 487–490 (1968)
12. Pafka, S., Potters, M., Kondor, I.: Exponential weighting and random-matrix-theory-based filtering of financial covariance matrices for portfolio optimization (2004). http://arxiv.org/abs/cond-mat/0402573
13. Tulino, A.M., Verdú, S.: Random matrix theory and wireless communications. Commun. Inf. Theory **1**, 1–182 (2004)
14. Vlăduts, S.G., Skorobogatov, A.N.: Weight distributions of subfield subcodes of algebraic-geometric codes. Probl. Inform. Transm. **27**(1), 19–29 (1991)
15. Wigner, E.P.: Characteristic vectors of bordered matrices with infinite dimensions. Ann. Math. **62**(2), 548–564 (1955)
16. Wishart, J.: The generalised product moment distribution in samples from a normal multivariate population. Biometrika **20A**(1/2), 32–52 (1928)
17. Xia, J., Xiong, M.: On a question of Babadi and Tarokh. IEEE Trans. Inform. Theory **60**(11), 7355–7367 (2014)

On Decoding Hyperbolic Codes

Eduardo Camps-Moreno[1], Ignacio García-Marco[2], Hiram H. López[1],
Irene Márquez-Corbella[2], Edgar Martínez-Moro[3(✉)], and Eliseo Sarmiento[4]

[1] Department of Mathematics, Virginia Tech, Blacksburg, VA, USA
{eduardoc,hhlopez}@vt.edu
[2] Facultad de Ciencias and Instituto de Matemáticas y Aplicaciones (IMAULL),
Universidad de La Laguna, San Cristóbal de La Laguna, Spain
{iggarcia,imarquec}@ull.es
[3] Institute of Mathematics, University of Valladolid, Valladolid, Castilla, Spain
edgar.martinez@uva.es
[4] Instituto Politécnico Nacional, Mexico City, Mexico
esarmiento@ipn.mx

Abstract. This work studies several decoding algorithms for hyperbolic codes. We use some previous ideas to describe how to decode a hyperbolic code using the largest Reed-Muller code contained in it or using the smallest Reed-Muller code that contains it. A combination of these two algorithms is proposed when hyperbolic codes are defined by polynomials in two variables. Then, we compare hyperbolic codes and Cube codes (tensor product of Reed-Solomon codes) and propose decoding algorithms of hyperbolic codes based on their closest Cube codes. Finally, we adapt to hyperbolic codes the Geil and Matsumoto's generalization of Sudan's list decoding algorithm.

Keywords: Reed-Muller codes · evaluation codes · hyperbolic codes · decoding algorithms

1 Introduction

Let $\mathbb{F}_q$ be a finite field with q elements. Given two vectors $\mathbf{x} = (x_1, \ldots, x_n)$ and $\mathbf{y} = (y_1, \ldots, y_n) \in \mathbb{F}_q^n$, the *Hamming distance* between $\mathbf{x}$ and $\mathbf{y}$ is defined as $d_H(\mathbf{x}, \mathbf{y}) = |\{i \mid x_i \neq y_i\}|$, where $|\cdot|$ denotes the cardinality of the set. The *Hamming weight* of $\mathbf{x}$ is given by $w_H(\mathbf{x}) = d_H(\mathbf{x}, \mathbf{0})$, where $\mathbf{0}$ denotes de zero vector in $\mathbb{F}_q^n$. The *support* of $\mathbf{x}$ is the set $\mathrm{supp}(\mathbf{x}) = \{i \mid x_i \neq 0\}$. An $[n(C), k(C), \delta(C)]_q$ *linear code* C over $\mathbb{F}_q$ is an $\mathbb{F}_q$-vector space of $\mathbb{F}_q^{n(C)}$

E. Camps-Moreno, H. H. López, E. Martínez-Moro and I. Márquez-Corbella were partially supported by Grant TED2021-130358B-I00 funded by MCIU/AEI/10.13039/501100011033 and by the "European Union NextGenerationEU/PRTR".
H. H. López was partially supported by the NSF grants DMS-2201094 and DMS-2401558.
I. García-Marco and I. Márquez-Corbella were partially supported by the Spanish MICINN PID2019-105896GB-I00.

© The Author(s), under exclusive license to Springer Nature Switzerland AG 2025
S. Petkova-Nikova and D. Panario (Eds.): WAIFI 2024, LNCS 15176, pp. 37–52, 2025.
https://doi.org/10.1007/978-3-031-81824-0_3

with dimension $k(C)$, and minimum distance $\delta(C) = \min\{d_H(\mathbf{c}, \mathbf{c}') : \mathbf{c}, \mathbf{c}' \in C, \mathbf{c} \neq \mathbf{c}'\}$, thus its error correction capability is $t_C = \left\lfloor \frac{\delta(C)-1}{2} \right\rfloor$. When there is no ambiguity, we write n, k, δ, and t instead of $n(C), k(C), \delta(C)$, and t_C, respectively.

Let $\mathbb{N}$ be the set of non-negative integers. For $A \subseteq \mathbb{N}^m$, $\mathbb{F}_q[A]$ is the $\mathbb{F}_q$-vector subspace of polynomials in $\mathbb{F}_q[\mathbf{X}] = \mathbb{F}_q[X_1, \ldots, X_m]$ with basis given by the set of monomials $\{\mathbf{X}^{\mathbf{i}} = X_1^{i_1} \cdots X_m^{i_m} \mid \mathbf{i} = (i_1, \ldots, i_m) \in A\}$. Let $\mathcal{P} = \mathbb{F}_q^m$, where $\mathcal{P} = \{P_1, \ldots, P_n\}$ and $n = |\mathcal{P}| = q^m$. Define the following evaluation map

$$\mathrm{ev}_{\mathcal{P}} : \mathbb{F}_q[X_1, \ldots, X_m] \longrightarrow \mathbb{F}_q^n$$
$$f \longmapsto (f(P_1), \ldots, f(P_n)).$$

The *monomial code associated* to A, denoted by $\mathcal{C}_A$, is defined as

$$\mathcal{C}_A = \mathrm{ev}_{\mathcal{P}}(\mathbb{F}_q[A]) = \{\mathrm{ev}_{\mathcal{P}}(f) \mid f \in \mathbb{F}_q[A]\}.$$

For $a, b \in \mathbb{R}$ and $a \leq b$, we denote by $[\![a, b]\!]$ the integer interval $[a, b] \cap \mathbb{Z}$.

Definition 1 (Reed-Muller and Hyperbolic codes)

- Let $s \geq 0, m \geq 1$ be integers. The monomial code $\mathcal{C}_R$ where the set R is given by $R = \left\{\mathbf{i} = (i_1, \ldots, i_m) \in [\![0, q-1]\!]^m \mid \sum_{j=1}^m i_j \leq s\right\}$, is called the Reed-Muller code *over $\mathbb{F}_q$ of degree s with m variables*. This code is denoted by $\mathrm{RM}_q(s, m)$.
- Let $d, m \geq 1$ be integers. The monomial code $\mathcal{C}_H$ where the set H is given by $H = \left\{\mathbf{i} = (i_1, \ldots, i_m) \in [\![0, q-1]\!]^m \mid \prod_{j=1}^m (q - i_j) \geq d\right\}$, is called the hyperbolic code *over $\mathbb{F}_q$ of order d with m variables*. This code is denoted by $\mathrm{Hyp}_q(d, m)$.

In our previous work [2] we proved there are two optimal Reed-Muller codes such that $\mathrm{RM}_q(s, m) \subseteq \mathrm{Hyp}_q(d, m) \subseteq \mathrm{RM}_q(s', m)$. In other words, the largest Reed-Muller code $\mathrm{RM}_q(s, m)$ contained in $\mathrm{Hyp}_q(d, m)$, and the smallest Reed-Muller code $\mathrm{RM}_q(s', m)$ that contains $\mathrm{Hyp}_q(d, m)$. We will use that result to propose several decoding procedures.

The paper is organized as follows. In Sect. 2, we describe two different algorithms to decode a hyperbolic code $\mathrm{Hyp}_q(d, m)$. These algorithms are based on the optimal Reed-Muller code that approximates to our hyperbolic code, that is, the largest Reed-Muller code $\mathrm{RM}_q(s, m)$ contained in $\mathrm{Hyp}_q(d, m)$, or the smallest Reed-Muller code $\mathrm{RM}_q(s', m)$ that contains $\mathrm{Hyp}_q(d, m)$. We will study the advantages and disadvantages in terms of efficiency and correction capability of these proposed algorithms. The choice of the algorithm to be used depends on which Reed-Muller code is closest to the hyperbolic code $\mathrm{Hyp}_q(d, m)$ as well as the efficiency or effectiveness that we need. At the end of that section, a third algorithm, which is a combination of the previous two, is adapted for $m = 2$, the case of two variables. In Sect. 3 a decoding algorithm for a hyperbolic code $\mathrm{Hyp}_q(d, m)$ based on the tensor product of Reed-Solomon codes is presented. In

Sect. 4 we recover Geil and Matsumoto's generalization of Sudan's List Decoding for order domain codes [3] but focus explicitly on hyperbolic codes. The novel idea that we present here is that we explicitly describe each of the sets involved in the algorithm or at least we give a subset of such sets. Finally in Sect. 5, we compare the performance of the five decoding algorithms proposed in this paper.

2 Decoding Based on Known Reed-Muller Decoders

2.1 Decoding Algorithm Based on the Smallest Reed-Muller Code

The main idea that we present in this section is well-known and works for any pair of nested linear codes. Let $\mathcal{C}_1 \subseteq \mathcal{C}_2 \subseteq \mathbb{F}_q^n$ be two linear codes with parameters $[n, k_1, d_1]_q$ and $[n, k_2, d_2]_q$, respectively. Observe that a decoding algorithm for $\mathcal{C}_2$ that corrects up to t_2 errors is also a decoding algorithm for $\mathcal{C}_1$ that requires the same number of operations as in $\mathcal{C}_2$, but corrects up to t_2 errors. Note that $d_2 \leq d_1$ and the difference between these two values might be huge. That is, a decoding algorithm for $\mathcal{C}_2$ is also a decoding algorithm for $\mathcal{C}_1$, but there is a loss in the number of errors that one might expect to correct.

Given a hyperbolic code $\mathrm{Hyp}_q(d, m)$, by [2, Theorem 3.6] we know the smallest integer s such that $\mathrm{Hyp}_q(d, m) \subseteq \mathrm{RM}_q(s, m)$. Thus, for each decoding algorithm for $\mathrm{RM}_q(s, m)$ we have one for $\mathrm{Hyp}_q(d, m)$. For example, if we use the already mentioned Pellikaan-Wu's list-decoding algorithm, one can correct up to $q^m(1 - \sqrt{(q^m - \delta(\mathrm{RM}_q(s, m)))/q^m})$ errors.

Example 1. We have $\mathcal{C}_1 = \mathrm{Hyp}_9(9, 2) \subseteq \mathcal{C}_2 = \mathrm{RM}_9(s, 2)$, where $s \geq 12$. Note that $\delta(\mathcal{C}_2) = 5$, while $\delta(\mathcal{C}_1) = 9$. Using the algorithm explained in this section and Pellikkan-Wu's decoder for $\mathcal{C}_2$, we can correct up to 2 errors in $\mathcal{C}_1$ (which coincides with the error correcting capability of $\mathcal{C}_2$), while the error-correcting capability of the code $\mathcal{C}_1$ is $t_{\mathcal{C}_1} = 4$.

2.2 Decoding Algorithm Based on the Largest Reed-Muller Code

Take $A \subseteq B \subseteq \mathbb{N}^m$. We consider the set $\mathcal{Q} = \{f \in \mathbb{F}_q[\mathbf{X}] \mid \mathrm{supp}(f) \subseteq B \setminus A\}$, where the support of a polynomial $f \in \mathbb{F}_q[\mathbf{X}]$ is defined as

$$\mathrm{supp}(f) = \left\{(i_1, \ldots, i_m) \mid f = \sum_{j=1}^m \alpha_{i_j} \mathbf{X}^{i_j}, \alpha_{i_j} \in \mathbb{F}_q \setminus \{0\}\right\}.$$

Observe that $|\mathcal{Q}| = q^{|B \setminus A|}$.

Let $\mathbf{y} \in \mathbb{F}_q^n$ be a received word. The following proposition tells us that, if the number of errors with respect to $\mathcal{C}_B$ is at most its error-correcting capability, i.e. $d_H(\mathbf{y}, \mathcal{C}_B) \leq t_{\mathcal{C}_B}$, then, there exists a unique polynomial $f \in \mathcal{Q}$ such that the nearest codeword to $\mathbf{y} - \mathrm{ev}_\mathcal{P}(f)$ is unique in $\mathcal{C}_A$.

Proposition 1. *Let $\mathbf{y} \in \mathbb{F}_q^n$ be a received word. Then there exists a polynomial $f \in \mathcal{Q}$ such that $d_H(\mathbf{y} - \mathrm{ev}_\mathcal{P}(f), \mathcal{C}_A) = d_H(\mathbf{y}, \mathcal{C}_B)$. Moreover, if $d_H(\mathbf{y}, \mathcal{C}_B) \leq t_{\mathcal{C}_B}$, the polynomial f is unique.*

Proof. We first prove the existence of such polynomial. Set $t := d_H(\mathbf{y}, \mathcal{C}_B)$, then there exists a codeword $\mathbf{z} = \text{ev}_\mathcal{P}(g) \in \mathcal{C}_B$, with $\text{supp}(g) \subseteq B$, such that $d_H(\mathbf{y}, \mathbf{z}) = t$. Writing $g = f + \tilde{f}$, with $\text{supp}(f) \subseteq B \setminus A$ and $\text{supp}(\tilde{f}) \subseteq A$, we have $t = d_H(\mathbf{y}, \mathbf{z}) = d_H(\mathbf{y} - \text{ev}_\mathcal{P}(f), \text{ev}_\mathcal{P}(\tilde{f}))$, with $f \in \mathcal{Q}$ and $\text{ev}_\mathcal{P}(\tilde{f}) \in \mathcal{C}_A$.

Now we will prove the uniqueness when $t \leq t_{\mathcal{C}_B}$. Let $h \in \mathcal{Q}$ be another possible option. There exist two error-vectors $\mathbf{e}_1, \mathbf{e}_2 \in \mathbb{F}_q^n$ with weight smaller or equal to $t_{\mathcal{C}_B}$, such that $\text{ev}_\mathcal{P}(f) + \mathbf{e}_1 = \text{ev}_\mathcal{P}(h) + \mathbf{e}_2 = \mathbf{y}$. Therefore $\text{ev}_\mathcal{P}(f) - \text{ev}_\mathcal{P}(h) = \mathbf{e}_2 - \mathbf{e}_1$ is an element of $\mathcal{C}_B$, with weight at most $2t_{\mathcal{C}_B} < \delta(\mathcal{C}_B) - 1$. Thus, the difference $\text{ev}_\mathcal{P}(f) - \text{ev}_\mathcal{P}(h)$ must be zero and $f = h$. □

Let $\mathcal{C}_A \subseteq \mathcal{C}_B$ monomial codes such that $A \subseteq B \subseteq \mathbb{N}^m$. Let $\text{Dec}_{\mathcal{C}_A}$ be a decoding algorithm for $\mathcal{C}_A$, which corrects up to E errors. By Proposition 1, we can define the following decoding algorithm for $\mathcal{C}_B$ that corrects also up to E errors.

Initialization: Let $\mathbf{y} \in \mathbb{F}_q^n$ be the received word. For each $f \in \mathcal{Q}$ we follow these steps

Step 1. Compute $\mathbf{y} - \text{ev}_\mathcal{P}(f)$.

Step 2. Decode using $\text{Dec}_{\mathcal{C}_A}$ the word $\mathbf{y} - \text{ev}_\mathcal{P}(f)$.

Step 3. Denote by L the output list of **Step 2**, that is $L = \{\mathbf{c} \in \mathcal{C}_A \mid d_H(\mathbf{c}, \mathbf{y} - \text{ev}_\mathcal{P}(f)) \leq E\}$. If L is not empty, then for each $\mathbf{c}_A \in L$, we add to the output list $\mathbf{c}_A + \text{ev}_\mathcal{P}(f)$.

The previous list-decoding algorithm for $\mathcal{C}_B$ corrects up to E errors and requires $q^{|B \setminus A|}$ calls to $\text{Dec}_{\mathcal{C}_A}$. Moreover, if $E \leq t_{\mathcal{C}_B}$, one can easily transform the previous list-decoding algorithm into the following unique decoding one:

Initialization: Let $\mathbf{y} \in \mathbb{F}_q^n$ be the received word. Assume that the number of errors is at most $E \leq t_{\mathcal{C}_B}$.

Step 1 Compute $\mathbf{y} - \text{ev}_\mathcal{P}(f)$ for some $f \in \mathcal{Q}$.

Step 2 Decode using the decoder $\text{Dec}_{\mathcal{C}_A}$ the word $\mathbf{y} - \text{ev}_\mathcal{P}(f)$.

Step 3 If the result of **Step 2** is codeword $\mathbf{c}_A$ such that $w_H(y - ev_P(f), \mathbf{c}_A) \leq t_{\mathcal{C}_B}$, then return $\mathbf{c}_A + \text{ev}_\mathcal{P}(f)$.

Step 4 Otherwise, go back to **Step 1**.

The correctness of this algorithm is justified by Proposition 1 and the fact that $E \leq t_{\mathcal{C}_B} \leq t_{\mathcal{C}_A}$ and, hence, the output in **Step 2** has at most one element.

The algorithms proposed involve at most $|\mathcal{Q}| = q^{|B \setminus A|}$ calls to $\text{Dec}_{\mathcal{C}_A}$, which could be interpreted as an inefficient algorithm from a theoretical point of view. Nevertheless, for practical purposes, if the difference between the sets B and A is small, this algorithm defines an efficient algorithm for $\mathcal{C}_B$ that corrects up to E errors, as long as an efficient algorithm for $\mathcal{C}_A$ exists and corrects the same number of errors.

2.3 Intermediate Case

In this section we are going to study an intermediate proposal between the two previous options. We will do our study for the case of two variables.

Proposition 2. *Let s be the smallest integer such that $\mathrm{Hyp}_q(d,2) \subseteq \mathrm{RM}_q(s,2)$. Let $H, R \subseteq \mathbb{N}^2$ such that $\mathcal{C}_H = \mathrm{Hyp}_q(d,2)$ and $\mathcal{C}_R = \mathrm{RM}_q(s-1,2)$. Then*

$$|H - R| \leq 2(\sqrt[4]{d} + 1).$$

Proof. We have that $\mathcal{C}_H = \mathrm{Hyp}_q(d,2) \subseteq \mathrm{RM}_q(s,2)$. An easy observation is that whenever $(i_1, i_2) \in H - R$, then $i_1 + i_2 = s$. Thus,

$$H - R = \{(t, s-t) \mid t \in [\![0, s]\!] \text{ and } (q-t)(q-s+t) \geq d\}. \tag{1}$$

Hence we are looking for those $t \in [\![0, s]\!]$ such that $(q-t)(q-s+t) \geq d$, or equivalently,

$$\left(q - \frac{s}{2}\right)^2 - \left(\frac{s}{2} - t\right)^2 \geq d \Rightarrow \left|\frac{s}{2} - t\right| \leq \underbrace{\sqrt{\left(q - \frac{s}{2}\right)^2 - d}}_{\Delta}.$$

If we compute Δ, then: $|H - R| \leq 2\Delta + 1$. We separate the proof in two cases depending on the parity of s. Recall that, by [2, Proposition 3.2], we know that $s = \left\lfloor 2(q - \sqrt{d}) \right\rfloor$.

1. If $s + 1 = 2r$, then $(r, r) \notin H$. As a consequence, $(q - \frac{s+1}{2})^2 < d$. Or equivalently, $\left(q - \frac{s}{2}\right)^2 - q + \frac{2s+1}{4} < d$. Thus,

$$\Delta \leq \sqrt{q - \frac{2q+1}{4}} < \sqrt{q - \frac{4(q-\sqrt{d}) - 1}{4}} = \sqrt{\sqrt{d} + \frac{1}{4}} < \sqrt[4]{d} + \frac{1}{2}, \text{ for } d \geq 1.$$

2. If $s = 2r$, then $(r, r+1) \notin H$. As a consequence, $\left(q - \frac{s}{2}\right)\left(q - \frac{s}{2} - 1\right) < d$. Or equivalently, $\left(q - \frac{s}{2}\right)^2 - \left(q - \frac{s}{2}\right) < d$. Thus,

$$\Delta \leq \sqrt{q - \frac{s}{2}} < \sqrt{q - \frac{2(q - \sqrt{d}) - q}{2}} = \sqrt{\sqrt{d} + \frac{1}{2}} < \sqrt[4]{d} + \frac{1}{2} \text{ for } d \geq 1. \quad \square$$

Now we have the ingredients to define a new decoding algorithm for the code $\mathcal{C}_H = \mathrm{Hyp}_q(d,2)$. Let s be the smallest integer such that $\mathrm{Hyp}_q(d,2) \subseteq \mathrm{RM}_q(s,2)$. See [2, Proposition 3.2] for a precise description of such parameter s. We define $R \subseteq \mathbb{N}^2$ such that $\mathcal{C}_R = \mathrm{RM}_q(s-1,2)$. By Proposition 2 we know that $|H - R| \leq c\sqrt[4]{d}$. Let Dec_R be a decoding algorithm for $\mathcal{C}_R$ that corrects up to $t_R = \left\lfloor \frac{d(\mathcal{C}_R) - 1}{2} \right\rfloor$ errors. Unifying the ideas of the above decoding algorithms (see Sects. 2.1 and 2.2), we have a decoding algorithm for $\mathcal{C}_H$ that corrects up to E errors and requires $q^{c\sqrt[4]{d}}$ calls to Dec_R. If we compare it with the algorithm proposed in Sect. 2.1, we have a poorer complexity but we can correct more errors. This approach is particulary well suited when $d' = \delta(\mathrm{RM}_q(s,2)) \geq q$ and $|H - R|$ is small. In this case, $\delta(\mathcal{C}_R) = d' - q + 1$ and, as a consequence, the correction capability of the auxiliary Reed-Muller code we are using to decode is increased by around $q/2$.

Example 2. Consider $\mathcal{C} = \mathrm{Hyp}_q(d,2)$ with $q = 11$, and $d = 32$. See Fig. 1 for a representation of this example. Throughout this example, we use Pellikaan-Wu's list-decoder (PW) for Reed-Muller codes. By [2, Proposition 3.2, Proposition 4.2] we have that $\mathrm{RM}_q(s',2) \subseteq \mathcal{C} \subseteq \mathrm{RM}_q(s,2)$ for $s' \leq 8$ and $s \geq 10$.

1. Thus, $\mathcal{C} \subseteq \mathcal{C}_{R_1} = \mathrm{RM}_q(10,2)$. Using Sect. 2.1 and applying PW algorithm on $\mathcal{C}_{R_1}$ we have an efficient decoding algorithm for $\mathcal{C}$ that **corrects up to 5 errors** and requires just one call to $\mathrm{Dec}_{\mathcal{C}_{R_1}}$.
2. Thus, $\mathcal{C}_{R_2} = \mathrm{RM}_q(8,2) \subseteq \mathcal{C}$. Using Sect. 2.2 to decode $\mathcal{C}$ and applying PW algorithm on $\mathcal{C}_{R_2}$ we have a decoding algorithm for $\mathcal{C}$ that **corrects up to 16 errors**. This algorithm uses the decoder $\mathrm{Dec}_{\mathcal{C}_{R_2}}$ plus some brute force. In particular, it requires q^{11} calls to $\mathrm{Dec}_{\mathcal{C}_{R_2}}$, where $11 = |H - R_2|$, and H and R_2 are the sets in $\mathbb{N}^2$ that define $\mathcal{C}$ and $\mathcal{C}_{R_2}$, respectively.
3. The intermediate proposal between the two previous options is described in Sect. 2.3. More precisely, we consider $\mathcal{C}_{R_3} = \mathrm{RM}_q(9,2)$ and we perform $q^{|H-R_3|} = q^5$ calls to the PW decoder for $\mathcal{C}_{R_3}$ to **correct up to 10 errors** in $\mathcal{C}$, where R_3 is the set in $\mathbb{N}^2$ that defines $\mathcal{C}_{R_3}$.

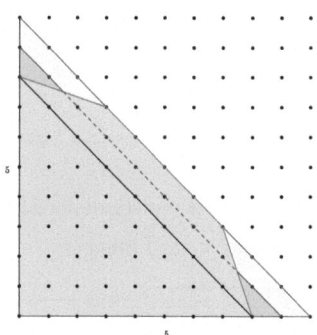

Fig. 1. Let $q = 11$ and $m = 2$. In this Figure the code $\mathrm{Hyp}_q(32,2)$ is equal to $\mathcal{C}_H$, $\mathrm{RM}_q(10,2)$ is equal to $\mathcal{C}_{R_1}$, $\mathrm{RM}_q(8,2)$ is equal to $\mathcal{C}_{R_2}$ and the code $\mathrm{RM}_q(9,2)$ equals to $\mathcal{C}_R$, where H, R_1, R_2 and R are the sets of lattice points below the red, the blue, the black and the green curve, respectively. (Color figure online)

3 Decoding Based on Tensor Products of RS Codes

Let $f \in \mathbb{F}_q[X_1, \ldots, X_m]$ be a polynomial. The maximum degree of f with respect to X_i is denoted by $\deg_{X_i}(f)$. Now we define a family of monomial codes that we call cube codes of order s, which consist of the evaluation of the polynomials $f \in \mathbb{F}_q[X_1, \ldots, X_m]$ that satisfy that $\deg_{X_i}(f) \leq s$ for each $1 \leq i \leq m$, on the q^m points of $\mathbb{F}_q^m$. The formal definition is as follows.

Definition 2 (Cube codes). Take $s \in \mathbb{N}$ and define $A = [\![0, s]\!]^m$. The monomial code $\mathcal{C}_A$, denoted by $\mathrm{Cube}_q(s, m)$, is called the cube code over $\mathbb{F}_q$ of order $s \geq 0$ with $m \geq 1$ variables.

A Reed-Solomon code can be seen as a Reed-Muller code $\mathrm{RM}_q(d, m)$ of order d and $m = 1$ variables. That is, the Reed-Solomon code of order s is defined as

$$\mathrm{RS}_q(s) = \{\mathrm{ev}_\mathcal{P}(f) \mid f \in \mathbb{F}_q[X] \text{ and } \deg(f) \leq s\} = \mathrm{RM}_q(s, 1).$$

Reed-Solomon codes are one of the most popular and important families of codes. They are maximum distance separable (MDS) codes, thus a $\mathrm{RS}_q(s)$ is a code with parameters $[q, s+1, q-s]_q$. Reed-Solomon codes have efficient decoding algorithms. In the literature, the two primary decoding algorithms for Reed-Solomon codes are the Berlekamp-Massey algorithm [1], and the Sugiyama et al. adaptation of the Euclidean algorithm [4], both designed to solve a key equation.

Remark 1. Asume that the polynomials that define the Reed-Solomon code $\mathrm{RS}_q^i(s)$ belong to $\mathbb{F}_q[X_i]$. It is easy to see that the cube code $\mathrm{Cube}_q(s, m)$ is the tensor product of the m Reed-Solomon codes $\mathrm{RS}_q^1(s), \ldots, \mathrm{RS}_q^m(s)$. In other words, we have that

$$\mathrm{Cube}_q(s, m) = \mathrm{RS}_q^1(s) \otimes \cdots \otimes \mathrm{RS}_q^m(s).$$

Proposition 3. *The minimum distance of the cube code* $\mathrm{Cube}_q(s, m)$ *coincides with its footprint bound for the deglex monomial ordering. Therefore, the code* $\mathcal{C} = \mathrm{Cube}_q(s, m)$ *has length* $n(\mathcal{C}) = q^m$, *dimension* $k(\mathcal{C}) = (s+1)^m$, *and minimum distance* $\delta(\mathcal{C}) = (q-s)^m$.

Proof. This is a consequence of the fact that the cube code is the tensor product of Reed-Solomon codes. □

Proposition 4. *Take* $d \in \mathbb{N}$. *Then,*

(a) $\mathrm{Cube}_q(s, m) \subseteq \mathrm{Hyp}_q(d, m)$ *if and only if* $s \leq q - \sqrt[m]{d}$.

(b) $\mathrm{Hyp}_q(d, m) \subseteq \mathrm{Cube}_q(s', m)$ *if and only if* $s' \geq q - \left\lceil \frac{d}{q^{m-1}} \right\rceil$.

Proof. Let $H \subset [\![0, q-1]\!]^m$ such that $\mathcal{C}_H = \mathrm{Hyp}_q(d, m)$. If $\mathbf{i} = (i_1, \ldots, i_m)$ satisfies that $0 \leq i_j \leq q - \sqrt[m]{d}$ for all $1 \leq j \leq m$, then $\prod_{j=1}^m (q - i_j) \geq d$. Hence, $\mathrm{Cube}_q(s, m) \subseteq \mathrm{Hyp}_q(d, m)$ for all $s \leq q - \sqrt[m]{d}$. If $s > q - \sqrt[m]{d}$, then $(q-s)^m < d$. Therefore $(s, \ldots, s) \notin H$ and $\mathrm{Cube}_q(s, m) \not\subseteq \mathrm{Hyp}_q(d, m)$.

We now check that $\mathrm{Hyp}_q(d, m) \subseteq \mathrm{Cube}_q(s', m)$ if and only if $s' \geq r := q - \left\lceil \frac{d}{q^{m-1}} \right\rceil$. For $s' \geq r$, it suffices to observe that for all $\mathbf{i} = (i_1, \ldots, i_m) \in H$, we have that $\max\{i_j\} \leq r$. Otherwise $\prod_{j=1}^m (q - i_j) \leq q^{m-1}(q - \max\{i_j\}) < q^{m-1}(q - r) \leq d$, a contradiction. Moreover, we have that $(r, 0, \ldots, 0) \in H$, so if $s' < r$ then $\mathrm{Hyp}_q(d, m) \not\subseteq \mathrm{Cube}_q(s', m)$. □

Theorem 1. *Let* Dec_{RS} *be a decoding algorithm for* $\text{RS}_q(s)$ *that corrects up to* t_{RS} *errors. Then, there exists a decoding algorithm* $\text{Dec}_{\text{Cube}_q(s,m)}$ *for* $\text{Cube}_q(s,m)$ *that corrects up to* $(t_{\text{RS}}+1)^m - 1$ *errors and requires calling* $f(m)$ *times the decoding algorithm* Dec_{RS}, *where*

$$f(m) = \sum_{i=0}^{m-1} (s+1)^{m-1-i} q^i \leq m q^{m-1} = \frac{mn}{q}, \text{ for all } m \geq 1.$$

Proof. We denote by $\alpha_1, \ldots, \alpha_q$ all the elements of $\mathbb{F}_q$. We proceed by induction on $m \in \mathbb{N}$. For $m = 1$, since $\text{Cube}_q(s,1) = \text{RS}_q(s)$ then, there exists Dec_{RS} that corrects up to t_{RS} errors.

Now assume that there exists a decoding algorithm for $\text{Cube}_q(s, m-1)$ that corrects up to $(t_{\text{RS}}+1)^{m-1} - 1$ errors. Without loss of generality we reorder the points $\mathcal{P} = \{P_1, \ldots, P_n\} = \mathbb{F}_q^m$ with $n = q^m$ in such a way that the first q^{m-1} points of $\mathcal{P}$ are those that have α_1 in their first coordinate, then those that have α_2, and so on. Let $\mathbf{v} = (\mathbf{v}_1, \ldots, \mathbf{v}_q) \in \mathbb{F}_q^n$ where $\mathbf{v}_i \in \mathbb{F}_q^{q^{m-1}}$ be such that there exists $\mathbf{u} = (\mathbf{u}_1, \ldots, \mathbf{u}_q) \in \text{Cube}_q(s, m)$ where $\mathbf{u}_i \in \mathbb{F}_q^{q^{m-1}}$ with $d(\mathbf{v}, \mathbf{u}) < (t_{RS}+1)^m$. As $\mathbf{u} \in \text{Cube}_q(s,m)$, there exists

$$f(X_1, \ldots, X_m) = \sum_{i_1, \ldots, i_m = 0}^{s} \beta_{i_1, \ldots, i_m} X_1^{i_1} \cdots X_m^{i_m} \in \mathbb{F}_q[X_1, \ldots, X_m],$$

such that $\mathbf{u} = \text{ev}_{\mathcal{P}}(f)$ and $\mathbf{u}_i = \text{ev}_{\mathcal{P}'}(f(\alpha_i, X_2, \ldots, X_m))$ with $\mathcal{P}' = \mathbb{F}_q^{m-1}$. We are going to show how to recover $f(X_1, \ldots, X_m)$ from $\mathbf{v}$ by calling q times the decoder $\text{Dec}_{\text{Cube}_q(s,m-1)}$ and $(s+1)^{m-1}$ times the decoder Dec_{RS}.

For all $i \in \{1, \ldots, q\}$, we define $d_i = d(\mathbf{u}_i, \mathbf{v}_i)$. We say that $i \in \{1, \ldots, q\}$ is GOOD if $d_i < (t_{RS}+1)^{m-1}$; otherwise we say that i is BAD. Let d_{bad} be the number of BAD values $i \in \{1, \ldots, q\}$. Since

$$(t_{RS}+1)^m > \sum_{i=1}^{q} d_i \geq \sum_{i \text{ is BAD}} d_i \geq (t_{RS}+1)^{m-1} d_{bad},$$

we have that $d_{bad} < (t_{RS}+1)$.

Write $f(X_1, \ldots, X_m) = \sum_{j_2, \ldots, j_m = 0}^{s} h_{j_2, \ldots, j_m}(X_1) X_2^{j_2} \cdots X_m^{j_m}$ where the univariate polynomial $h_{j_2, \ldots, j_m}(X_1) \in \mathbb{F}_q[X_1]$ has degree at most s, for all $j_2, \ldots, j_m \in \{0, \ldots, s\}$. For all $\ell \in \{1, \ldots, q\}$, consider

$$g_\ell(X_2, \ldots, X_m) := f(\alpha_\ell, X_2, \ldots, X_m) = \sum_{j_1, \ldots, j_m = 0}^{s} \beta_{j_1, \ldots, j_m} \alpha_\ell^{j_1} X_2^{j_2} \cdots X_m^{j_m}$$

$$= \sum_{j_2, \ldots, j_m = 0}^{s} h_{j_2, \ldots, j_m}(\alpha_\ell) X_2^{j_2} \cdots X_m^{j_m} \in \mathbb{F}_q[X_2, \ldots, X_m],$$

which satisfies that $\deg_{X_i}(g_\ell) \leq s$ for all $i \in \{2, \ldots, m\}$. Moreover, whenever ℓ is GOOD, we can recover g_ℓ by means of $\text{Dec}_{\text{Cube}_q(s,m-1)}$ and the values $\mathbf{v}_\ell = (v_{\ell_1}, \ldots, v_{\ell_{q^{m-1}}}) \in \mathbb{F}_q^{q^{m-1}}$.

Thus, if ℓ is GOOD, we have recovered the value $h_{j_2,\ldots,j_m}(\alpha_\ell, X_2, \ldots, X_m)$ for all $j_2, \ldots, j_m \in \{0, \ldots, s\}$. As there are at least $(q - t_{RS})$ GOOD values, then for each $j_2, \ldots, j_m \in \{0, \ldots, s\}$ we have at least $(q - t_{RS})$ correct evaluations of $h_{j_2,\ldots,j_m}(X_1)$. Thus, by induction we can recover $\beta_{j_1,\ldots,j_m}$ using $(s+1)^{m-1}$ times the decoder Dec_{RS}.

Now, let $f(m)$ be the number of times that algorithm $\text{Dec}_{\text{Cube}_q(s,m)}$ calls algorithm Dec_{RS}. We will deduce a formula for $f(m)$ by induction on $m \in \mathbb{N}$. First notice that for $m = 1$, since $\text{Cube}_q(s,1) = \text{RS}_q(s)$, we have that $f(1) = 1$. Moreover, from the above paragraphs, we can deduce that $f(m) = qf(m-1) + (s+1)^{m-1}f(1)$. Now we assume that $f(r) = qf(r-1) + (s+1)^{r-1}f(1)$ for all $r \leq m$ and we try to show the result for $f(m)$. Indeed,

$$\begin{aligned}
f(m) &= qf(m-1) + (s+1)^{m-1}f(1) \\
&= q((s+1)^{m-2} + qf(m-2)) + (s+1)^{m-1} \\
&= (s+1)^{m-1} + q(s+1)^{m-2} + q^2 f(m-2) \\
&= \cdots = (s+1)^{m-1} + q(s+1)^{m-2} + q^2(s+1)^{m-3} + \cdots + q^{m-1}f(1) \\
&= \sum_{i=0}^{m-1}(s+1)^{m-1-i}q^i.
\end{aligned}$$

This completes the proof. $\square$

Example 3. Consider $\mathcal{C} = \text{Hyp}_q(d,2)$ with $q = 32$, and $d = 225$. See Fig. 2 for a representation of this example. We will give different decoding algorithms for $\mathcal{C}$ using all ideas proposed in this article so far.

1. By [2, Proposition 3.2], $\mathcal{C} \subseteq \text{RM}_q(s,2)$, for $s \geq 34$. Therefore, using Sect. 2.1, and applying PW algorithm on $\mathcal{C}_{R_1} = \text{RM}_q(34,2)$, we have an efficient decoding algorithm for $\mathcal{C}$ that **corrects up to 11 errors** and requires just one call to $\text{Dec}_{\mathcal{C}_{R_1}}$.
2. By [2, Theorem 4.3], $\text{RM}_q(s,2) \subseteq \mathcal{C}$, for $s \leq 24$. Therefore, using Sect. 2.2 and applying PW algorithm on $\mathcal{C}_{R_2} = \text{RM}_q(24,2)$, we have a decoding algorithm for $\mathcal{C}$ that **corrects up to** 127 **errors** because the minimum distance of $RM_{32}(24,2)$ is 256 and so the error-correcting capability will be 127, which is beyond the error correction capability of $\mathcal{C}$. The algorithm uses the decoder $\text{Dec}_{\mathcal{C}_{R_2}}$ plus brute force. In particular it requires q^{156} calls to $\text{Dec}_{\mathcal{C}_{R_2}}$ where $156 = |H - R_2|$, and H and R_2 are the sets in $\mathbb{N}^2$ defining $\mathcal{C}$ and $\mathcal{C}_{R_2}$, respectively.
3. The intermediate proposal between the two previous options described in Sect. 2.3 has a slight advantage in this example with respect to the first option. More precisely, one should consider $\mathcal{C}_{R'} = \text{RM}_q(33,2)$ and perform $q^{|H-R'|} = q$ calls to the PW decoder for $\mathcal{C}_{R'}$ to **correct up to 14 errors** because the minimum distance of $RM_{32}(33,2)$ is 30 and so the error-correcting capability will be 14.
4. By Proposition 4(b), we have $\mathcal{C} \subseteq \text{Cube}_q(s,2)$ for $s \geq 24$. We know an efficient decoding algorithm for $\mathcal{C}_3 = \text{Cube}_q(24,2)$ that corrects up to $t_3 =$

$(t_{RS}+1)^m - 1 = 15$, where $t_{RS} = \lfloor \frac{q-s-1}{2} \rfloor = 3$ denotes the error correction capability of $\mathrm{RS}_q(s)$ with $s = 24$. Therefore, using Theorem 1, we have an efficient decoding algorithm for $\mathcal{C}$ that **corrects up to** 15 **errors** and requires calling $q + s + 1 = 57$ times the decoder Dec_{RS}.

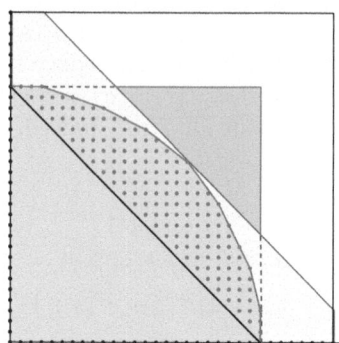

Fig. 2. Let $q = 32$ and $m = 2$. In this Figure the code $\mathrm{Hyp}_q(225, 2)$ is equal to $\mathcal{C}_H$, $\mathrm{RM}_q(24, 2)$ equals $\mathcal{C}_{R_1}$, the code $\mathrm{RM}_q(34, 2)$ is equal to $\mathcal{C}_{R_2}$ and the code $\mathrm{Cube}_q(24, 2)$ equals $\mathcal{C}_A$, where H, R_1, R_2 and A are the sets of lattice points below the red, the black, the blue and the green curve, respectively. (Color figure online)

Remark 2. All the ideas proposed in Sect. 2 can be adapted to decode a hyperbolic code using a cube code. That is, by Proposition 4, given a hyperbolic code Hyp we can find the largest (respectively smallest) cube code contained in (respectively that contains) Hyp. And this result allow us to give decoding algorithms for hyperbolic codes in terms of the decoding algorithms of cube codes. Furthermore, an intermediate proposal between the two options can be used (as in Sect. 2.3).

4 Decoding Based on a Generalized Sudan's List Decoding

Definition 3. *Let $\mathcal{C}_H = \mathrm{Hyp}_q(d, m)$ be a hyperbolic code with $H \subset [\![q-1]\!]^m$. For $r \geq 1$, take $H_r \subset [\![q-1]\!]^m$ such that $\mathcal{C}_{H_r} = \mathrm{Hyp}_q(r+1, m)$. For $i \geq 0$, we define*

$$L(d, r, i) = \{\mathbf{a} \in [\![q-1]\!]^m \mid \mathbf{a} + iH \subseteq H_r\}.$$

Observe that $L(d, r, 0) = H_r$ for any d. We also have that $L(d, r, i+1) \subseteq L(d, r, i)$. Indeed, as $L(d, r, i+1) + (i+1)H \subset H_r$, then $L(d, r, i+1) \subset L(d, r, i+1) + H \subset L(d, r, i)$.

For the following algorithm, we assume that the numbers r and t satisfy the following conditions:

$$\sum_{i=0}^{\infty} \#L(d,r,i) > n \quad \text{and} \quad t = \min\left\{i' \mid \sum_{i=0}^{i'} \#L(d,r,i) > n\right\}. \quad (2)$$

Notation: Given $\mathbf{u}, \mathbf{v} \in \mathbb{F}_q^n$, we define the Schur product as the component wise product on $\mathbb{F}_q^n$, i.e. $(\mathbf{u} * \mathbf{v})_i = u_i v_i$ and $(\mathbf{u}^{*j})_i = u_i^j$, for $j \geq 1$. We will write $\mathbf{u}^{*0}$ for the word with all the components equal to 1.

Initialization: Let $\mathbf{y} \in \mathbb{F}_q^n$ be the received word, and define r and t according to the conditions (2).

Step 1 For $0 \leq i \leq t$, find $Q_i \in \mathbb{F}_q[L(d,r,i)]$, not all zero, such that $\sum_{i=0}^{t} \text{ev}(Q_i) * \mathbf{y}^{*i} = 0$.

Step 2 Factorize $Q(Y) = \sum_{i=0}^{t} Q_i Y^i \in \mathbb{F}_q[\mathbf{X}, Y]$ and detect all possible $f \in \mathbb{F}_q[\mathbf{X}]$ such that $(Y - f) | Q(Y)$. This can be done by the method of Wu [5].

Step 3 Return $\{\text{ev}(f) \in \mathcal{C}_H \mid f$ is a solution of Step 2$\}$.

The list $\{\text{ev}(f) \in \mathbb{F}_q[\mathbf{X}] \mid f$ is a solution of Step 2$\}$ is a list of at most t elements that contains all the codewords $\mathbf{x}$ in $\mathcal{C}_H$ such that $d_H(\mathbf{x}, \mathbf{y}) \leq r$.

For completion, we write the proof of the last algorithm adapting it to the notation proposed in the previous lines.

Theorem 2 ([3, Theorem 4]). *The last algorithm gives the claimed output.*

Proof. Since $\sum_{i=0}^{t} \#L(d,r,i) > n$, then the equation $\sum_{i=0}^{t} \text{ev}(Q_i) * \mathbf{y}^{*i} = 0$ has more indeterminates than equations and then a non-zero solution exists. Now, suppose that there exists $\mathbf{x} = \text{ev}(f) \in \mathcal{C}_H$ such that $d_H(\mathbf{y}, \mathbf{x}) \leq r$. Since $f \in H$, then all the monomials in the support of f^i are in iH. Then, all the monomials $X^{\mathbf{a}}$ appearing in $Q_i f^i$ satisfies $\mathbf{a} \in H_r$ by definition of $L(d,r,i)$. This implies that $\text{ev}(Q(f)) \in \mathcal{C}_{H_r}$ and

$$\text{w}_H(\text{ev}(Q(f))) \geq r + 1 \quad \text{or} \quad Q(f) = 0. \quad (3)$$

On the other hand, since $d_H(\mathbf{y}, \mathbf{x}) \leq r$, we have that $\sum_{i=0}^{t} \text{ev}(Q_i) * \mathbf{y}^{*i} = 0$ and $\text{ev}(Q(f))$ can differ in at most r distinct positions. Thus $\text{w}_H(\text{ev}(Q(f))) \leq r$. By Eq. (3) we conclude that $Q(f) = 0$, which means $(Y - f) | Q(Y)$. □

To use the previous algorithm and to know the number of errors that we can correct, we just need to compute $L(d,r,i)$ along with their sizes. In general, it is not clear what is the form of $L(d,r,i)$, but with the following results we can estimate their sizes, which is one of the main contributions of this section.

Proposition 5. *Let $\mathcal{C}_A = \text{Hyp}_q(d_A, m)$ and $\mathcal{C}_B = \text{Hyp}_q(d_B, m)$. Then $\mathcal{C}_{A+B} \subset \text{Hyp}_q(d_A + d_B - q^m, m)$.*

Proof. Take $f \in \mathbb{F}_q[A]$ and $g \in \mathbb{F}_q[B]$. Denote the set of zeros of f (resp. g) in $\mathbb{F}_q^m$ by $Z(f)$ (resp. $Z(g)$). We know that $|Z(f)| \leq q^m - d_A$ and $|Z(g)| \leq q^m - d_B$. This implies that $|Z(fg)| \leq |Z(f)| + |Z(g)| \leq 2q^m - d_A - d_B$. In other words, for any $fg \in \mathcal{C}_{A+B}$, we have that $w_H(\text{ev}(f) * \text{ev}(g)) \geq d_A + d_B - q^m$. Then $\delta(\mathcal{C}_{A+B}) \geq d_A + d_B - q^m$. As $\mathcal{C}_{A+B}$ is a monomial code, its minimum distance is the minimum of the footprints of its defining monomials. Thus we obtain the conclusion. □

With the above result we can bound the size of the set $L(d, r, 1)$ and so, we can bound the number of errors that we can correct with the proposed algorithm when we adapt it to unique decoding.

Corollary 1. *Let $\mathcal{C} = \text{Hyp}_q(d, m)$ and $d \geq r \geq 1$. If $\mathcal{C}_{H_1} = \text{Hyp}_q(q^m + r - d + 1, m)$, then $H_1 \subseteq L(d, r, 1)$.*

The number of errors that we can uniquely decode with the proposed algorithm is given by an easy-to-check formula.

Corollary 2. *If $\#H_r + \#H_1 > n$, then the algorithm can correct up to r errors solving a linear equation in $(\mathbb{F}_q[\mathbf{X}])[Y]$.*

If $m = 2$ we can do even better. In the case of two variables, we can not only bound the size of the set $L(d, r, 1)$ but we can also know exactly what monomial code is associated with such subset.

Proposition 6. *Let $\mathcal{C}_H = \text{Hyp}_q(d, 2)$, with $d > q$. Take $a = \left\lfloor q - \frac{d}{q} \right\rfloor$ and $b = q - \frac{d}{q-a}$. For $r < a - b + 1$, we have $\mathcal{C}_{L(d,r,1)} = \text{Cube}_q(q - 1 - a, 2)$.*

Proof. Take $c = q - 1 - a$. Observe that $(0, a) \in H$ but for any $a < i_2 \in \mathbb{Z}$, $(0, i_2) \notin H$. Similarly, $(q - a)(q - b) \geq d$ but for any $i_1 > b$, $(q - a)(q - i_1) < d$. As

$$(q - c - a)(q - c - b) = a - b + 1 > r,$$

and since $\{(i_1, i_2) \mid i_1 + i_2 \leq a + b\}$ and H_r are both convex sets, then for any $(i_1, i_2) \in H$ such that $i_1 + i_2 \leq a + b$, we have $(i_1 + c, i_2 + c) \in H_r$.

Now, suppose that $(i_1, i_2) \in H$ but $i_1 + i_2 > a + b$. Then we have

$$\begin{aligned}
(q - i_1 - c)(q - i_2 - c) &= (q - i_1)(q - i_2) + c(i_1 + i_2 - 2q) + c^2 \\
&\geq (q - a)(q - b) + c(i_1 + i_2 - 2q) + c^2 \\
&> (q - a)(q - b) + c(a + b - 2q) + c^2 \\
&= (q - a - c)(q - b - c) \\
&= a - b + 1 \\
&> r.
\end{aligned}$$

This means that $(i_1 + c, i_2 + c) \in H_r$. Then we have $c + H \subset H_r$. By Definition 3, we can easily see that if (i_1, i_2) satisfies the property that $i_1, i_2 \leq c$, then $(i_1, i_2) \in L(d, r, 1)$.

Finally, we can assure $L(d, r, 1) = \{(i_1, i_2) \mid i_1, i_2 \leq c\}$, otherwise it would exist some $(i_1, i_2) \in L(d, r, 1)$ with $i_1 > c$ such that $i_1 + a > q - 1$ and $(i_1 + c, i_2 + c) \in H_r$, which is a contradiction. □

The last result can be generalized for all the sets $L(d,r,i)$.

Corollary 3. *Let* $\mathcal{C}_H = \mathrm{Hyp}_q(d,2)$, $d > q$, *and* a, b *and* r *as before. Then* $\mathcal{C}_{L(d,r,i)} = \mathrm{Cube}_q(q - 1 - ia, 2)$.

Proof. We know the case $i = 1$. Assume the result is true for $i \in \mathbb{N}$. As $L(d, r, i+1) + H \subseteq L(d, r, i) = \mathrm{Cube}_q(q - 1 - ia, 2)$ and $\mathrm{Hyp}_q(d, 2) \subseteq \mathrm{Cube}_q(a, 2)$ [2, Theorem 4.3], then we have that for any $(i_1, i_2) \in H$, $i_1, i_2 \leq a$. This implies that

$$(q - 1 - (i+1)a + i_1, q - 1 - (i+1)a + i_2) \leq (q - 1 - ia, q - 1 - ia).$$

We conclude that $\{(i_1, i_2) \mid i_1, i_2 \leq q - 1 - (i+1)a\} \subseteq L(d, r, i+1)$. The equality follows from the fact that $(a, 0)$ is a point in H. □

Remark 3. Using the previous results, for $m = 2$ we have the following bound for the number of errors that our algorithm uniquely corrects. Take $t = \lfloor \frac{q-1}{a} \rfloor$ and $r < a - b + 1$, with a and b as before for $\mathcal{C}_H = \mathrm{Hyp}_q(d, 2)$. If $\#H_r + \sum_{i=1}^{t}(q - 1 - ia)^2 > n$, then the algorithm can correct up to r errors.

Example 4. Consider $\mathcal{C} = \mathrm{Hyp}_q(d, 2)$ with $q = 16$ and $d = 81$. See Fig. 3 for a representation of this example. Take $r = 8$. Brute force computation on Geil and Matsumoto's algorithm gives that:

- $\mathcal{C}_{L(d,r,0)} = \mathrm{Hyp}_q(r+1, 2)$, which coincide with Definition 3.
- Moreover, $\mathcal{C}_{L(d,r,1)} = \mathrm{Cube}_q(5, 2)$, which matches with Proposition 6 since

$$\mathrm{Cube}_q(5,2) = \mathrm{Cube}_q(q - 1 - a, 2) \text{ with } a = \left\lfloor q - \frac{d}{q} \right\rfloor = 10.$$

- Finally, $\mathcal{C}_{L(d,r,2)} = \{0\}$.

Moreover, following Remark 3, $r = 8$ is the maximum number of errors that we can correct with this algorithm since

$$r < a - b + 1, \text{ where } a = \left\lfloor q - \frac{d}{q} \right\rfloor = 10 \quad \text{and} \quad b = q - \frac{d}{q - a} = 2.5.$$

Observe that $\#H_r + \sum_{i=1}^{t}(q - 1 - ia)^2 = \#L(d,r,0) + \#L(d,r,1) > n = 256$, where $t = \lfloor \frac{q-1}{a} \rfloor = 1$ and $L(d,r,0)$ and $L(d,r,1)$ are the set of lattice points below the green and the black curve of Fig. 3, respectively. The sets $L(d,r,0)$ and $L(d,r,1)$ are, in general, difficult to describe. Using the results of this section we explicitly describe these sets when $m = 2$, and we provide a lower bound on their sizes for the cases when $m > 2$.

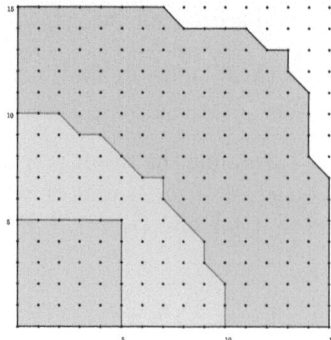

Fig. 3. Let $q = 16$ and $m = 2$. In this Figure the code $\mathcal{C} = \mathrm{Hyp}_q(81, 2)$ is equal to $\mathcal{C}_H$, $\mathcal{C}_{L(81,8,0)} = \mathrm{Hyp}_q(9, 2)$ is equal to $\mathcal{C}_{L_0}$ and $\mathcal{C}_{L(81,8,1)} = \mathrm{Cube}_q(5, 2)$ is equal to $\mathcal{C}_{L_1}$ where H, L_0 and L_1 are the sets of lattice points below the red, the black and the green curve, respectively. (Color figure online)

5 Comparisons and Conclusions

Table 1 compares the performance of the five decoding algorithms proposed in this paper for the hyperbolic code $\mathcal{C} = \mathrm{Hyp}_q(d, m)$, where $q = 32, m = 2$ and d takes different values. Table 1 is composed by 6 blocks, one for each value of d. Each block contains 5 lines, which represent the following:

- First line refers to the algorithm of Sect. 2.1. Here, we compute the smallest integer s such that $\mathcal{C} \subseteq \mathrm{RM}_q(s, m)$. Then, we use Pellikaan-Wu list-decoding algorithm for Reed-Muller codes to decode $\mathcal{C}$.
- Second line refers to the algorithm of Sect. 2.2. Here, we compute the largest integer s such that $\mathrm{RM}_q(s, m) \subseteq \mathcal{C}$. Then, we use the Pellikaan-Wu list-decoding algorithm for Reed-Muller codes to decode $\mathcal{C}$ plus some brute force.
- Third line refers to the algorithm of Sect. 2.3, an intermediate case between the above two options. In this case, we use again the Pellikaan-Wu list-decoding algorithm for Reed-Muller codes to decode $\mathcal{C}$.
- Fourth line refers to the algorithm of Sect. 3. More precisely, we compute the smaller integer s such that $\mathcal{C} \subseteq \mathrm{Cube}_q(s, m)$. Then, we use the algorithm described in Sect. 3 for cube codes to decode $\mathcal{C}$.
- Fifth line refers to the specific algorithm known for $\mathcal{C}$ described in Sect. 4.

The third column describes the number of calls to the corresponding decoder. The last column represents the minimum distance of the auxiliary code that we are using in each case.

In Table 1 we observe that the algorithm with the greatest error correcting capability is always achieved by the method given in the second line. However, the huge amount of calls to the decoder makes it highly impractical. The third method always corrects more errors than the first one, but requires more calls to the decoder. Concerning the third, fourth and fifth method, we find instances

Table 1. Comparison between the five different algorithms described above to decode $\text{Hyp}_{32}(d, 2)$, for different values of d.

		Error correcting capability	Number of calls	Called algorithm	Minimum distance
$d = 257$		$E_1 = 16$	1	$\text{Dec}(\text{RM}_q(31, 2))$	$\delta(\text{RM}_q(31, 2)) = 32$
		$E_2 = 155$	q^{134}	$\text{Dec}(\text{RM}_q(23, 2))$	$\delta(\text{RM}_q(23, 2)) = 288$
		$E_3 = 32$	q^8	$\text{Dec}(\text{RM}_q(30, 2))$	$\delta(\text{RM}_q(30, 2)) = 64$
		$E_4 = 24$	56	$\text{Dec}(\text{RS}_q(23))$	$\delta(\text{Cube}_q(24, 2)) = 81$
		$E_5 = 23$	1	$\text{Dec}(\text{Hyp}_q(d, 2))$	$\delta(\text{Hyp}_q(d, 2)) = 257$
$d = 225$		$E_1 = 14$	1	$\text{Dec}(\text{RM}_q(34, 2))$	$\delta(\text{RM}_q(34, 2)) = 29$
		$E_2 = 137$	q^{156}	$\text{Dec}(\text{RM}_q(24, 2))$	$\delta(\text{RM}_q(24, 2)) = 256$
		$E_3 = 15$	q	$\text{Dec}(\text{RM}_q(33, 2))$	$\delta(\text{RM}_q(33, 2)) = 30$
		$E_4 = 15$	57	$\text{Dec}(\text{RS}_q(24))$	$\delta(\text{Cube}_q(24, 2)) = 64$
		$E_5 = 19$	1	$\text{Dec}(\text{Hyp}_q(d, 2))$	$\delta(\text{Hyp}_q(d, 2)) = 225$
$d = 193$		$E_1 = 13$	1	$\text{Dec}(\text{RM}_q(36, 2))$	$\delta(\text{RM}_q(36, 2)) = 27$
		$E_2 = 118$	q^{182}	$\text{Dec}(\text{RM}_q(25, 2))$	$\delta(\text{RM}_q(25, 2)) = 224$
		$E_3 = 14$	q^3	$\text{Dec}(\text{RM}_q(35, 2))$	$\delta(\text{RM}_q(35, 2)) = 28$
		$E_4 = 15$	58	$\text{Dec}(\text{Cube}_q(25, 2))$	$\delta(\text{Cube}_q(25, 2)) = 49$
		$E_5 = 15$	1	$\text{Dec}(\text{Hyp}_q(d, 2))$	$\delta(\text{Hyp}_q(d, 2)) = 193$
$d = 150$		$E_1 = 12$	1	$\text{Dec}(\text{RM}_q(39, 2))$	$\delta(\text{RM}_q(39, 2)) = 24$
		$E_2 = 83$	q^{212}	$\text{Dec}(\text{RM}_q(27, 2))$	$\delta(\text{RM}_q(27, 2)) = 160$
		$E_3 = 12$	q^6	$\text{Dec}(\text{RM}_q(38, 2))$	$\delta(\text{RM}_q(38, 2)) = 25$
		$E_4 = 8$	60	$\text{Dec}(\text{RS}_q(27))$	$\delta(\text{Cube}_q(27, 2)) = 25$
		$E_5 = 9$	1	$\text{Dec}(\text{Hyp}_q(d, 2))$	$\delta(\text{Hyp}_q(d, 2)) = 150$
$d = 65$		$E_1 = 8$	1	$\text{Dec}(\text{RM}_q(47, 2))$	$\delta(\text{RM}_q(47, 2)) = 16$
		$E_2 = 49$	q^{343}	$\text{Dec}(\text{RM}_q(29, 2))$	$\delta(\text{RM}_q(29, 2)) = 96$
		$E_3 = 8$	q^6	$\text{Dec}(\text{RM}_q(46, 2))$	$\delta(\text{RM}_q(46, 2)) = 17$
		$E_4 = 3$	62	$\text{Dec}(\text{RS}_q(29))$	$\delta(\text{Cube}_q(29, 2)) = 9$
		$E_5 = 5$	1	$\text{Dec}(\text{Hyp}_q(d, 2))$	$\delta(\text{Hyp}_q(d, 2)) = 65$
$d = 15$		$E_1 = 3$	1	$\text{Dec}(\text{RM}_q(56, 2))$	$\delta(\text{RM}_q(56, 2)) = 7$
		$E_2 = 7$	q^{64}	$\text{Dec}(\text{RM}_q(48, 2))$	$\delta(\text{RM}_q(48, 2)) = 15$
		$E_3 = 4$	q^3	$\text{Dec}(\text{RM}_q(55, 2))$	$\delta(\text{RM}_q(55, 2)) = 8$
		$E_4 = 0$	64	$\text{Dec}(\text{RS}_q(31))$	$\delta(\text{Cube}_q(31, 2)) = 1$
		$E_5 = 0$	1	$\text{Dec}(\text{Hyp}_q(d, 2))$	$\delta(\text{Hyp}_q(d, 2)) = 15$

where each of the methods outperforms the others. All the algorithms we propose except the fifth one, rely on a known decoder for either a Reed-Muller or a cube code. As a consequence, a better decoder for any of these codes would imply better error correction capability. Interestingly, when decoding in terms of cube codes, we reduce many times the problem to a code with a promisingly high minimum distance, but then we use a decoding algorithm with poor error correcting capability. We consider that it is an interesting problem to find better decoding algorithms for cube codes which, in particular, will lead to better decoding algorithms for hyperbolic codes.

References

1. Berlekamp, E.: Nonbinary BCH decoding (abstr.). IEEE Trans. Inf. Theory **14**(2), 242 (1968)
2. Camps-Moreno, E., García-Marco, I., López, H.H., Márquez-Corbella, I., ínez-Moro, E.M., Sarmiento, E.: On the generalized Hamming weights of hyperbolic codes. J. Algebra Appl. **23**(7), Paper No. 2550062 (2024)
3. Geil, O., Matsumoto, R.: Generalized Sudan's list decoding for order domain codes. In: Applied Algebra, Algebraic Algorithms and Error-Correcting Codes. Lecture Notes in Computer Science, vol. 4851, pp. 50–59. Springer, Berlin (2007)
4. Sugiyama, Y., Kasahara, M., Hirasawa, S., Namekawa, T.: A method for solving key equation for decoding Goppa codes. Inf. Control **27**, 87–99 (1975)
5. Wu, X.-W.: An algorithm for finding the roots of the polynomials over order domains. In: Proceedings IEEE International Symposium on Information Theory, p. 202 (2002)

Fast Decoding of Group Testing Results from Reed-Solomon d-Disjunct Matrices

Dongxia Luo and Lucia Moura[✉]

University of Ottawa, Ottawa, Canada
{dluo093,lmoura}@uottawa.ca

Abstract. Non-adaptive combinatorial group testing has applications in disease screening as well as in many problems in digital security and communications. Matrices that are d-disjunct (also called d-cover-free) can be built using codes and allow for the detection of d defective items using group testing. In this paper, we study d-disjunct matrices built from Reed-Solomon codes, and design a specialized algorithm for decoding the results of group testing using these matrices. We do an experimental comparison between our method and the naive one that only uses the d-disjunct property of the matrix, and show that the former outperforms the latter as the size of the problem grows.

Keywords: combinatorial group testing · cover-free family · Reed-Solomon code · polynomials over finite fields

1 Introduction

Combinatorial group testing (CGT) is concerned with finding d defective items from a set of n items by combining the items into groups or pools that are tested together. The result of a test (testing items in a pool) is positive if at least one of the items in the pool is defective, and negative, otherwise. The objective is to identify the positive items with the minimum number of tests. Group testing is useful for disease screening, clone libraries for a DNA sequence, as well as many problems in cryptography and communications [2,6,7].

Here we focus on non-adaptive group testing. In adaptive group testing the results of previous tests can be used to determine the next test, while in non-adaptive group testing all tests are decided a priori. The case of adaptive group testings is solved optimally by the binary splitting algorithm [6,7]. However, many applications require non-adaptive group testing, which has the additional advantage of allowing for tests to be run in parallel, so there is active research in non-adaptive group testing [2,6].

A non-adaptive group testing scheme is represented by a $t \times n$ binary matrix where each column is associated with an item, each row is associated with a test (pool) and corresponds to the characteristic vector of a test. To be able to detect up to d defectives, the matrix must be $\bar{d}$-separable: the boolean sum of any set of up to d columns must be distinct. A stronger requirement is that the matrix be d-disjunct: the boolean sum of any d columns must not "contain" any other column (i.e. there must be a row where the other column has a 1 while each of the d columns has a 0). The advantage of d-disjunct matrices is that determining the set of defectives from the vector of test results can be done in time $O(tn)$. Indeed, the d-disjunct property guarantees that it is sufficient to compute the union of the items appearing in negative tests to obtain the set of negative items U, so the set of defective items is simply $\{1,\ldots,n\} \setminus U$. The d-disjunct matrices have been studied under several names such as d-cover-free families and superimposed codes, and their study was pioneered by Kautz and Singleton [9]; see [7] for a recent survey.

One of the main objectives in group testing is to minimize the number of tests. Given d and n, let $t(d,n)$ be the minimum number of rows (tests) in a d-disjunct matrix with n columns. It has been shown that $\Omega(\frac{d^2 \log n}{\log d}) \leq t(d,n) \leq O(d^2 \log n)$. The lower bound was determined in [3,5,12] and the upper is achieved by the deterministic algorithm by Porat and Rotschild [11]. Closing the gap between lower and upper bound is an important open question.

Another desired objective in group testing is that the d-disjunct matrix M be constructed efficiently. A construction is *explicit* if the matrix can be computed in time polynomial in t, n and a construction is *strongly explicit*, if each position of the matrix $M_{i,j}$ can be computed in time polynomial with $t, \log n$. The construction in [11] is explicit and produces the best known number of rows t, but it is not strongly explicit.

Yet another important objective of group testing is that the *decoding*, that is the determination of the defective items from the vector of test results, be done efficiently. While any d-disjunct matrix enables a decoding in time $O(tn)$, when n gets very large, it is desirable to have a more efficient decoding, i.e. a decoding in time $poly(t)$ [8].

In this paper, we study decoding algorithms for d-disjunct matrices built from Reed Solomon codes. The construction of d-disjunct matrices from Reed-Solomon codes is very simple, and given next.

Construction 1. *Let n and $d \geq 2$ be integers. Let q be a prime power, $N \leq q$ and $k = \lceil \log_q n \rceil$ such that $d(k-1) \leq N - 1 \leq q - 1$. Define a $q^2 \times q^k$ binary matrix R with rows indexed by $(x,y) \in F_q \times F_q$ and columns indexed by all polynomials f of degree at most $k - 1$, such that $R_{(x,y),f} = 1$ if and only if $f(x) = y$. Take any $X \subseteq F_q$, $|X| = N$, and only keep the rows (x,y) with $x \in X$. The constructed matrix R has $t = Nq \leq q^2$ rows and q^k columns.*

It is well known that Construction 1 yields a d-disjunct matrix [4,9]; a proof why Nq rows are sufficient can be found in [7,13]. It is easy to see by Construction 1 that this construction is strongly explicit. The growth of the number of rows t with respect to n and d depends on parameter choices, but specific choices of q and k can be used to obtain matrices from Construction 1 with $t = O((\frac{d \log n}{\log d})^2)$ (see Theorem 1), not too far from current upper and lower bounds on $t(d,n)$.

The easy explicit construction combined with the reasonable number of tests suggests that disjunct matrices from Reed-Solomon codes are of practical relevance for group testing applications. Thus, it is worth to look for algorithms for group test decoding from these matrices that are more efficient than the naive ones, which works for any d-disjunct families in time $O(tn)$.

This paper explores algorithms for the decoding problem for the d-disjunct matrices R given in Construction 1. Our focus is to experimentally compare the naive algorithm that works for any d-disjunct matrix with an algorithm specialized in the structure of the d-disjunct families built from Reed-Solomon. Indeed, designing the latter algorithm involves solving the following problem, which may be of independent interest.

Problem 1. Let n and d and parameters N, q, k satisfying the conditions of Construction 1. Determine the unique set of $s \leq d$ polynomials $f_1, f_2, \ldots, f_s$ of degree $k-1$ over F_q, given the sets of their evaluations on each point in $X \subseteq F_q$, $|X| = N$, more precisely, given $S_x = \{f_1(x), f_2(x), \ldots, f_s(x)\}$, for each $x \in X$.

In Sect. 3, we give an algorithm for solving Problem 1 which is specialized for the case of q prime. Restricting the analysis to q prime does not affect much the desired proprieties of Construction 1 for d-disjunct matrices, since there is some freedom on the choice of q and k and plenty of prime numbers q available to obtain $n \leq q^k$, $n \approx q^k$ and good number of tests (see Theorem 1).

In Sect. 4, we describe two implementations of the naive algorithm for group testing decoding given any d-disjunct matrix M and a vector $y \in \{0,1\}^t$ of test results. Each implementation uses a different data structure to store the matrix.

In Sect. 5, we experimentally compare the algorithm for group testing decoding using our specialized version that solves Problem 1 with the variations of the naive algorithm for a d-(t, n) matrix built with Construction 1. Our experiments explore a wide range of parameter combinations for satisfying the hypothesis of Construction 1. The naive algorithms are implemented using two different data structures to store the rows of the matrix: a compact bit-array and a list of indices of non-zero entries. We show that the compact bit-array implementation yields a faster naive algorithm than the list of indices. We also experimentally show that the former naive algorithm is the fastest of the tree algorithms when the number n of items is not so large (up to tens of thousands). Finally, we show that our specialized method outperforms the naive ones for larger values of n, starting on hundreds of thousands, and as n moves close and beyond 1 million, our algorithm is 20–100 times faster. The larger experiments are performed only with our fast algorithm to study the effect of various parameters on its running time.

The present paper highlights the practical aspects of applying Reed-Solomon d-disjunct matrices in group testing and should be of interest for practitioners interested in their application.

2   Preliminaries and Background

Solving group testing using d-disjunct matrices (also known as d-cover free families) and codes is well known since the seminal work of Kautz and Singleton [9]. Surveys [7,13] contain proofs of the results in this section.

Definition 1 (d-disjunct matrix). *Given $d < t \leq n$ positive integers, a d-(t,n) disjunct matrix is a $t \times n$ binary matrix M such that any set of $d+1$ columns contains a permutation sub-matrix of dimension $d+1$.*

Á 1-disjunct matrix is equivalent to the incidence matrix of a Sperner family, so from Sperner's theorem we can get the optimal number of rows for given n:

$$t(1,n) = \min\left\{t : n \leq \binom{t}{\lfloor t/2 \rfloor}\right\}.$$

Thus we are only concerned with $d \geq 2$.

Let M be a d-(t,n) disjunct matrix and $y \in \{0,1\}^t$ be the result of tests for matrix M i.e. $y_i = 1$ if and only if the test corresponding to row M_i is positive (contains defectives). An algorithm to detect defectives only needs to compute a vector in $\{0,1\}^n$ that is the logical-OR of all rows M_i such that $y_i = 0$ to obtain the negative items; complementing this vector gives the positive/defective items. The correctness is immediate from Definition 1, since a negative item always appears in a row where the d positive items do not appear.

Definition 2 (code). *An (N,n,q) code $\mathcal{C}$ of length N over a q-ary alphabet Q consists of n codewords $c = (c_1,...,c_N) \subseteq Q^N$. The code $\mathcal{C}$ has minimum distance D if any two codewords in $\mathcal{C}$ differ in at least D positions.*

Proposition 1 (d-disjunct matrices from codes). *Let $d < n$ be positive integers. If there exists an (N,n,q)-code $\mathcal{C}$ with minimum distance D, then there exists a d-(Nq,n) disjunct matrix, for any $d \leq \lfloor \frac{N-1}{N-D} \rfloor$.*

The following proposition is a consequence of Proposition 1 using the fact that Construction 1 uses a code with minimum distance $N - k + 1$.

Proposition 2. *Let $2 \leq d < n$. Let q be a prime power and k be an integer satisfying $k = \lceil \log_q n \rceil$ and $d(k-1) \leq q-1$. Take $N = d(k-1) + 1$. Then, Construction 1 gives a d-(Nq,q^k) disjunct matrix, and (by possibly removing some columns columns) a d-(Nq,n) disjunct matrix.*

The next known theorem shows that appropriate choices of parameters in Construction 1 yields d-(t,n) disjunct matrices with $t \in O(d^2(\frac{\log n}{\log d})^2)$. This is not too far from the best known bounds $\Omega(\frac{d^2 \log n}{\log d}) \leq t(d,n) \leq O(d^2 \log n)$. A proof of Theorem 1 can be found in [1, Section 7.2.2].

Theorem 1. *Let $d < n$ be positive numbers. There is a family of d-(t,n) disjunct matrices using Construction 1 satisfying t in $O(d^2(\frac{\log n}{\log d})^2)$.*

Proof (sketch). Take $q \approx 2d\frac{\log n}{\log d}$ a prime number (By Bertrand's postulate there is always a prime between this number and its double); take $k = \lceil \log_q n \rceil$ and $N = d(k-1) + 1$. We use Construction 1 and for large enough parameters this family is d-disjunct as

$$\left\lfloor \frac{q-1}{k-1} \right\rfloor \geq \left\lfloor \frac{2d\frac{\log n}{\log d} - 1}{\frac{\log n}{\log q} + 1 - 1} \right\rfloor = \left\lfloor 2d\frac{\log q}{\log d} - \frac{\log q}{\log n} \right\rfloor \geq d,$$

and $t = Nq \leq q^2 \approx (2d\frac{\log n}{\log q})^2$. $\square$

2.1 A Recurrence Relation to Evaluate Polynomials over Prime Fields

The following lemma is stated without proof, which is available in [10].

Lemma 1. *Let k be an integer and use the convention that $0^0 = 1$. Then, $\sum_{a=0}^{k}(-1)^a \binom{k}{a} a^i = 0$ for $0 \leq i < k$.*

Let p be a prime, $k \geq 2$, and f be a polynomial of degree at most $k-1$ over $\mathbb{F}_p$. The next theorem shows how to recursively evaluate a polynomial f on $N \leq p$ consecutive points $\{0, \ldots, N-1\} \subseteq \mathbb{F}_p$ given their evaluations $f(0), \ldots, f(k-1)$. Indeed, the equations in Lemma 1 are valid in $\mathbb{Z}$, so they are also valid in $\mathbb{Z}_p$.

Theorem 2. *Let p be a prime number and let f be a polynomial over $\mathbb{F}_p$ of degree at most $k - 1$. If $2 \leq k < p$, then*

$$\sum_{a=0}^{k}(-1)^a \binom{k}{a} f(x-a) = 0, x \in \mathbb{F}_p. \tag{1}$$

Alternatively, we can express the recursive functions as:

$$f(x) = \sum_{a=1}^{k}(-1)^{a-1} \binom{k}{a} f(x-a). \tag{2}$$

Proof. Write $f(x) = \alpha_{k-1} x^{k-1} + \ldots + \alpha_1 x + \alpha_0$. As we expand the recursive formula, we have $\sum_{a=0}^{k}(-1)^a \binom{k}{a} f(x-a) = \binom{k}{0} f(x) - \binom{k}{1} f(x-1) + \binom{k}{2} f(x-2) - \ldots + (-1)^k \binom{k}{k} f(x-k)$. Moreover, we have the following set of equations:

$$f(x) = \alpha_{k-1} x^{k-1} + \alpha_{k-2} x^{k-2} + \alpha_{k-3} x^{k-3} + \ldots + \alpha_1 x + \alpha_0,$$
$$f(x-1) = \alpha_{k-1}(x-1)^{k-1} + \alpha_{k-2}(x-1)^{k-2} + \ldots + \alpha_1(x-1) + \alpha_0,$$
$$\vdots$$
$$f(x-k) = \alpha_{k-1}(x-k)^{k-1} + \alpha_{k-2}(x-k)^{k-2} + \ldots + \alpha_1(x-k) + \alpha_0.$$

Substituting these equations into the left-hand side of Eq. 1, we obtain

$$\sum_{a=0}^{k}(-1)^a \binom{k}{a} f(x-a) = \sum_{j=0}^{k-1} \alpha_j \sum_{a=0}^{k}(-1)^a \binom{k}{a}(x-a)^j.$$

We complete the proof by showing $\sum_{a=0}^{k}(-1)^a \binom{k}{a}(x-a)^j = 0, 0 \leq j < k$:

$$\sum_{a=0}^{k}(-1)^a \binom{k}{a}(x-a)^j$$
$$= \sum_{a=0}^{k}(-1)^a \binom{k}{a} \sum_{i=0}^{j} \binom{j}{i}(-a)^i(x^{j-i}), \text{ using binomial expansion,}$$
$$= \sum_{i=0}^{j} \binom{j}{i}(x^{j-i})(-1)^i \sum_{a=0}^{k}(-1)^a \binom{k}{a} a^i, \text{ rearranging,}$$
$$= 0, \text{ by Lemma 1.}$$

□

3 Decoding Group Testing from d-Disjunct Matrices from Reed-Solomon Codes

In this section, we provide a decoding algorithm of group testing results given by $y \in \{0,1\}^t$ for a d-(t,n) disjunct matrix M built from Construction 1. This algorithm requires that q is a prime number and returns a set of defective items. This is equivalent to solving Problem 1, which we rephrase next:

> Let n and d and parameters N, q, k satisfying the conditions of Construction 1, q a prime number. Given a list of N sets $S = (S_0, S_1, S_2, \ldots, S_{N-1})$, where $S_i \subseteq F_q, |S_i| \le d$, determine the unique set of $s \le d$ polynomials $f_1, f_2, \ldots, f_s$ of degree at most $k-1$ over $F_q = Z_q$, such that $S_i = \{f_1(i), f_2(i), \ldots, f_s(i)\}$ for all $0 \le i \le N-1$.

Because the parameters satisfy the hypotheses of Proposition 2, we know the corresponding group testing matrix is d-disjunct, which guarantees the uniqueness of the set of up to d polynomials to be determined.

Algorithm FindPolynomials(k, N, q, d, S) given in Algorithm 1 is the algorithm to find all the desired polynomials corresponding to defective elements. After setting up initialization of auxiliary variables in lines 1 to 17, it proceeds to the main loop in lines 18 to 68. The main loop will discover one polynomial f at a time as follows. It first selects $p \in \{0, \ldots, N-1\}$ such that S_p has at least one unused value, i.e. a value $y \in S_i$ that has not been the image of p for any polynomial previously found, and it will fix $f(p) = y$. Then, a set of k consecutive integers in $\{0, \ldots, N-1\}$ starting at value startPos and including p is determined; to simplify the explanation suppose $p = $ startPos, and the values are $p, p+1, \ldots, p+k-1$. We need to iterate over all combinations of $(f(p+1), f(p+2), \ldots, f(p+k-1)) \in S_{p+1} \times S_{p+2} \times \cdots \times S_{p+k-1}$. Each of these combinations together with $f(p) = y$ determines a unique polynomial of degree at most $k-1$. This polynomial is tested to see if it is valid by algorithm isValidPolynomial($k, q, N, S, codeword, startPos$) given in Algorithm 2, which uses Theorem 2 and the chosen k values of $f(p+i)$ to efficiently evaluate the polynomial in the rest of the values $x \in \{0, \ldots, N-1\} \setminus \{p, \ldots, p+k-1\}$. Whenever it is found that $f(x) \notin S_x$ for x in this range, the polynomial is deemed not valid, otherwise it is valid and the values of $f(i)$ are stored in codeword[i] and saved to the list of polynomials listPoly. We repeat this process until there are no unused elements remaining in any of the S_i, which means we have successfully found all the polynomials (corresponding to the defective items).

All the details of this algorithm is given in Algorithms 1 and 2, which refer to other auxiliary algorithms. Algorithms successor and sortList are not provided, but are simple to describe. Algorithm successor(T, m, k) finds the successor of a mixed-radix k-tuple T in lexicographical order in $O(k)$. Algorithm sortList(A, Unused, $k, p, startPos$) rearranges the elements of a d-tuple $A_i, 1 \le i \le k$ so that the indices of list Unused$_i$ that are true are placed in

A_i before the indices that are false; this can be done in $O(kd)$. Note that a lot of bookkeeping is in place to keep track of the unused elements in each S_i, $0 \leq i \leq N-1$, and to accomplish all the tasks efficiently. Keeping track of unused elements in each S_i is useful not only to determine the value p at each iteration but also to employ a heuristic when trying different k-tuples of values for $(f(p+1), f(p+2), \ldots, f(p+k-1))$. Values of S_i can be used by more than one polynomial, i.e. two valid polynomials f_1, f_2 may have $f_1(i) = f_2(i)$; however, even though "repeats" are allowed, our heuristic that decides the order in which values are tried uses unused values of S_i first. This heuristic is accomplished by a rearrangement of unused indices moved to the front in line 43 by calling Algorithm sortList followed by the processing of tuples T in lexicographical order by invoking the algorithm successor(T, m, k) in line 52.

It is easy to verify the correctness of Algorithm 1.

Proposition 3. *Let $n > d \geq 2$ be integers, and let N, q, k be parameters satisfying the conditions of Construction 1, where q is a prime number. Let $S = (S_0, S_1, S_2, \ldots, S_{N-1})$, be a list of sets, where $S_i \subseteq F_q, |S_i| \leq d$, for all $0 \leq i \leq N-1$. Then Algorithm FindPoynomials(k, N, q, d, S) given in Algorithm 1 correctly solves Problem 1.*

Algorithm 1. FindPolynomials (k, N, q, d, S)

Input: k (the degree of the polynomials is $\leq k-1$), N(codeword length), q (a prime number), d (upper bound on number of polynomials), $S = (S_0, S_1, S_2, \ldots, S_{N-1})$
Output: listPoly (a list of polynomials satisfying the requirements of Problem 1)
1: ▷ **Initialization of auxiliary variables:**
2: Unused = (Unused$_0$,...,Unused$_{N-1}$) a list of boolean lists with the same sizes as in S with all values initialized to true;
3: Count = an array of $N+1$ integers; ▷ stores the number of unused elements in S_i
4: total = 0;
5: **for** $i = 0$ to $N-1$ **do**
6:     Count[i] = $|S_i|$;   ▷ Initialize as the current number of unused
7:     total $+= |S_i|$;
8: **end for**
9: Count[N] = total;   ▷ Count[N] stores the total number of unused elements in S
10: binom1,binom2 = 2 arrays with k integers;   ▷ Precomputed binomials
11: **for** $a = 1$ to k **do**
12:     binom1[$a-1$] = $(-1)^{a-1} \binom{k}{a}$;
13:     binom2[$a-1$] = $(-1)^{a-1} \binom{k}{a-1}$;
14: **end for**
15: listPoly = an empty list of codewords ▷ Store polynomials when they are found;
16: numPolyFound = 0;   ▷ Keep track of number of polynomials found
17: ▷ (Algorithm continues in the next page with the main loop to find polynomials)

18: **while** Count$[N] \neq 0$ **do** ▷ There are unused elements, so not all polys found
19:     $p = -1$;   ▷ Index representing the first S_i that still has unused elements
20:     **for** $i = 0$ to $N-1$ **do**   ▷ Find p
21:         **if** Count$[i] \neq 0$ **then**
22:             $p = i$; **break**;
23:         **end if**
24:     **end for**
25:     ▷ Find the starting position for k consecutive integers containing p in $[0, N-1]$
26:     **if** $p+k \leq N$ **then**
27:         startPos $= p$;
28:     **else**
29:         startPos $= N - k$;
30:     **end if**
31:     ▷ $m[i]$ stores the number of elements to consider from $S_{\text{startPos}+i}$, $1 \leq i \leq k$.
32:     $m =$ an array with size k;
33:     **for** $i = 0$ to $k-1$ **do**
34:         **if** i+startPos $==$ p **then**
35:             $m[i] = 1$;   ▷ An unused element of S_p is fixed
36:         **else if** Count$[i$+startPos$] == d-$numPolyFound **then**
37:             $m[i] = $ count$[i$+startPos$]$; ▷ Only unused in $S_{\text{startPos}+i}$ are considered
38:         **else**
39:             $m[i] = |S_{i+\text{startPos}}|$;   ▷ All elements in $S_{\text{startPos}+i}$ are considered
40:         **end if**
41:     **end for**
42:     $A =$ a list of k arrays A_i of size $m[i]$ containing indices of S;
43:     sortList(A,Unused,k,p,startPos); ▷ Rearrange indexes with unused in S_i first
44:     $T = [0, 0, \ldots, 0]$, an array with size k, iterating over tuples $T[i] \in [0, m[i]]$
45:     codeword $=$ an array with size N; ▷ codeword corresponding to a polynomial
46:     found $=$ false; hasSuccessorEnd $=$ true;
47:     **while** (!found) && (hasSuccessor) **do**
48:         **for** $i = 0$ to $k-1$ **do**   ▷ Get the value from T and A
49:             codeword$[i] = S_{i+\text{startPos}}$.get($A_i[T[i]]$);
50:         **end for**
51:         found $=$ isValidPolynomial(k, q, N, S, binom1,binom2,codeword,startPos);
52:         hasSuccessor $=$successor(T, m, k);   ▷ T gets next tuple, or false if last
53:     **end while**
54:     **if** !found **then**
55:         exit with an error; ▷ It never happens for correct inputs; must find poly
56:     **end if**
57:     **for** $i = 0$ to $N-1$ **do**   ▷ Poly used $f(i)=$codeword$[i]$ in S_i, update Unused
58:         index $=$ index of (codeword$[i]$) in S_i;
59:         **if** Unused$_i$[index] **then**   ▷ Value used for the first time
60:             Unused$_i$[index] $=$ false;   ▷ Updated Unused
61:             Count$[i] =$ Count$[i] - 1$;   ▷ Decrement unused count for S_i
62:             Count$[N] =$ Count$[N] - 1$; ▷ Decrement total count of unused elements
63:         **end if**
64:     **end for**
65:     numPolyFound++;
66:     listPoly.add(codeword); ▷ Store codeword corresponding to polynomial found
67: **end while**
68: **return** listPoly;

Algorithm 2. isValidPolynomial(k, q, N, S, codeword, startPos)

Input: as above; global arrays binom1, binom2 computed once in Algorithm 1
Output: a boolean result showing if a codeword of length N is a valid polynomial.

current = $k+$ startPos;
for $i =$ current to $N - 1$ **do**
    $f = 0$;
    for $j = 1$ to k **do**
        previous $= i - j$;
        $f+ =$ binom1$[j - 1] \times$ codeword[previous];
    end for
    if S_i.contains(f) **then**
        codeword[i] $=f$;   ▷ Update codeword
    else
        return false;
    end if
end for
current = startPos-1;
for $i =$ current to 0 **do**
    $f = 0$;
    flag $= i + k$;
    for $j = 0$ to $k - 1$ **do**
        previous $= i - j$;
        $f+ =$ binom2$[j] \times$ codeword[flag];
        flag$--$;
    end for
    if S_i.contains(f) **then**
        codeword[i] $=f$;   ▷ Update codeword
    else
        return false;
    end if
end for
return true;

Proposition 4. *Algorithm 1 runs in time $O(Nkd^k)$.*

Proof. The main loop in line 18 runs at most d times, and we claim that each iteration takes time $O(Nkd^{k-1})$. The crucial part to analyse is the inner loop in lines 47 to 53 that dominates the running time. This inner loop iterates over at most d^{k-1} tuples T in the first iteration of the main loop, at most $(d-1)^{k-1}$ tuples T in the second iteration of the main loop, until iterating over 1 tuple T in the dth iteration; at each iteration of the inner loop one call to isValidPolynomial is made. In total we have less than d^k calls to isValidPolynomial, which in turn runs in $O(Nk)$ (see Algorithm 2), completing the proof. □

Finally, we describe Algorithm RSDECODE which computes the defective items for group testing from the d-disjunct matrix constructed in Construction 1. From the list of failing tests it is easy to obtain the sets S_i, for all $1 \leq i \leq N$ in time $O(dN)$. Then, Algorithm 1 is run which returns the list of codewords (storing the evaluation vectors for the polynomials corresponding to defective items), in time $O(Nkd^k)$ (see Proposition 4). It remains to explain how to compute the defective items from the list of codewords, which stores the evaluation vectors for the polynomials corresponding to defective items. This is accomplished by solving a $k \times k$ system of linear equations to determine the coefficients of each polynomial from its k evaluations. Because the same system of equations is used for all polynomials, this task can be done by doing one matrix inversion and d multiplications of this matrix by a vector of length k, taking time $O(k^3 + dk^2)$. Then, each tuple of length k containing the coefficients of each polynomial is computed, and "unranked" as to obtain its index in $\{1, \ldots, n\}$; each unranking can be done in time $O(k)$, using time $O(dk)$ for all polynomials. In total, the defective items are computed in time $O(k^3 + dk^2)$. Therefore, RSDECODE time is dominated by the time of Algorithm 1, and runs in $O(Nkd^k)$. Note that the time complexity of computing the coefficients of the polynomials corresponding to defective items could be reduced to $O(dk \log k)$ by using Lagrange interpolation with FFT. However, this would not affect the total running time which is dominated by $O(Nkd^k)$.

4 Two Versions of the Naive Method to Decode d-Disjunct Matrices

In this section, we discuss two implementations of the naive algorithm to decode the test results based on general d-disjunct matrix, described next.

Algorithm 3. DMNAIVEDECODE(M, y)

Input: a d-(t, n) disjunct matrix M, a vector of test results $y \in \{0, 1\}^t$
Output: the set of defective items
  Set $U = \emptyset$ ▷ a set to store the union of the non-defective items
  for $i = 1$ to t do
    if ($y_i == 0$) then   ▷ if test i is negative
      $U = U \cup \{j : M[i][j] = 1\}$ ▷ all items in test i are non-defective
    end if
  end for
  return $\{1, \ldots, n\} \setminus U$ ▷ complementing set U gives the set of defective items

---

The following two implementations of DMNAIVEDECODE are considered, each one employing a different data structure for storing the rows of M:

1. DMCOMPACT: In this data structure, each row of M corresponds to a set implemented as a bit array stored as an array of unsigned integers of B bits each. Each array has size n/B; we used $B = 64$. The union of two sets is done via n/B bitwise-OR operations for each pair of integers. Complementation

is done with bitwise complement. The running time for the naive algorithm using this data structure is $O(\frac{nt}{B})$, which is still $O(nt)$ for constant B. However, in practice, this linear reduction on running time makes a difference.

2. DMLIST: In this data structure, each row of M corresponds to a set implemented as a sorted list of integers in $\{1, \ldots, n\}$. The union of two sets A and B can be done by merging the two sorted lists in time proportional to $O(|A| + |B|)$. The complement of a set can be computed in time $O(n)$. Suppose each row of M has a constant number L of 1 s; in this case we have at most t lists of size L to merge. If we implement Algorithm 3 as given with this data structure we obtain a running time of at most $2L + 3L + \ldots + tL + n = O(t^2 L + n)$. However, if we combine the lists in a *bottom-up merge*, which first combines the pairs of lists of size L, then pairs of lists of size 2L, etc. until all lists are merged, then the time complexity is $O(t \log tL + n)$. Note that this analysis is a bit rough, since we are upper bounding the size of the union of two sets with the sum of the cardinalities of the sets. For both variations, in the worst case, we still have time $O(tn)$ steps, for example when each of the t tests has L linear with n, but in practice this implementation can be beneficial in some cases. We experimented with both variations and opted for the one that uses the bottom-up merge.

In our experiments, we compare both implementations of the naive algorithm, DMCOMPACT and DMLIST, for matrix M coming from Reed-Solomon code. We make some observations based on the parameters q, k, d, N used in Construction 1. First, we observe that by construction, the number of rows is $t = Nq$, the number of columns is $n = q^k$, the number of 1's per row/test is $L = q^{k-1} = n/q$, and the number of 1s per column is $N \leq q$. The number of passing test z if we have exactly $d' \leq d$ defectives is in the range $N(q-d') \leq z \leq N(q-1) - d' + 1$.

The bottom line is that the worst-case running time of both DMCOMPACT and DMLIST is $O(tn)$, which for Construction 1 translates as $O(Nq^{k+1})$. By Proposition 4, RSDECODE has running time in $O(Nkd^k)$. As $d \leq \frac{N-1}{k-1} \leq \frac{q-1}{k-1}$, we expect RSDECODE to be faster as the parameters grow. However, since these values are rough upper bounds and asymptotic, the experiments can give us more information about their practical relative performance.

5 Experimental Results

For the experiments, we run all the tests in Java with a laptop running MacOS with 8 GB of random-access memory, and the processor speed reaches a maximum of 3.49 GHz with a M2 chip. Unless specified otherwise, each test is iterated 50 times, the list of defective items is randomly generated for each test and the average data is recorded. All execution times are reported in nanoseconds.

5.1 Comparison Between Three Decoding Methods

As discussed in the Sect. 4, we have two naive algorithms for identifying defective items from d-disjunct matrices: DMLIST and DMCOMPACT. Additionally, we have a direct decoding algorithm from Reed-Solomon codes, called RSDECODE, as described in Sect. 3.

With this in mind, our objective in this section is to determine the fastest method among DMCOMPACT, DMLIST, and RSDECODE for locating defective items. All the methods start by receiving a vector $y \in \{0,1\}^t$ with the outcome of the tests, where $y_i = 1$ if and only if test i is positive.

Table 1 presents four ranges for n: hundreds, thousands, tens of thousands, and hundreds of thousands, respectively. Within each range, we start with $d = 2$ and gradually increase it. We select combinations of d and n, and optimize the other parameters k, N, q to minimize t. More precisely, minimize $t = Nq$ subject to the constraints that q is prime, $k \geq 2$, $q^k \geq n$, $N = d(k-1) + 1 \leq q$, by inspection. For each combination, we measure the time taken by each method in nanoseconds and the winning time is highlighted in bold. In Table 1, each row is the average of 50 different random tests, except for the last row, where we only run 10 times, due to the slow running times of the naive algorithms.

Table 1. Comparison of execution time between three decoding methods

d	n	k	N	q	t	DMCOMPACT	DMLIST (ns)	RSDECODE (ns)
2	125	3	5	5	25	**112,571**	421,594	209,695
3	121	2	4	11	44	**53,645**	347,043	249,892
5	121	2	6	11	66	**41,646**	421,685	473,245
2	2,401	4	7	7	49	**150,921**	1,455,791	252,119
3	1,331	3	7	11	77	**182,709**	1,472,584	357,295
5	1,331	3	11	11	121	**138,018**	1,722,737	813,221
15	1,369	2	16	37	592	**238,094**	2,018,606	1,013,259
20	1,369	2	21	37	777	**253,320**	2,183,055	1,587,777
2	14,641	4	7	11	77	**215,167**	4,371,045	280,464
3	14,641	4	10	11	110	**207,595**	6,273,374	459,406
15	29,791	3	31	31	961	**431,883**	28,301,262	2,240,591
20	68,921	3	41	41	1,681	**1,044,236**	81,470,088	2,500,099
35	10,201	2	36	101	3,636	**526,407**	20,512,707	3,532,105
2	161,051	5	9	11	99	472,149	41,150,519	**309,830**
3	371,293	5	13	13	169	2,584,545	139,763,133	**650,857**
15	103,823	3	31	47	1,457	**1,310,525**	105,001,854	1,863,915
25	148,877	3	51	53	2,703	88,365,932	319,966,445	**4,022,617**
45	912,673	3	91	97	8,827	212,022,368	8,334,310,257	**10,014,493**

From Table 1, several key observations can be made. First, DMCOMPACT always performs better when compared to DMLIST. Additionally, RSDECODE is faster than DMLIST in all cases except when $d = 5$ and $n = 121$. Moreover, when n is between 100 and 100,000, DMCOMPACT is always the fastest.

However, beyond this range, RSDECODE achieves the shortest execution time. Meanwhile, while the execution time of DMCOMPACT and DMLIST increases as n increases, this trend is not observed in RSDECODE. Even with larger values of n, RSDECODE can perform better with a smaller value of d. This behavior can be explained due to the fact that the execution time for RSDECODE depends on d^k, rather than $n = q^k$. With that being said, during group testing, one should consider using DMCOMPACT to locate defectives when n is relatively small. Conversely, as n grows larger, RSDECODE becomes the best choice among the three methods.

Theorem 1 shows how we can obtain an asymptotically small t that is polynomial on d and $\log n$. Table 2 uses parameters derived from Theorem 1. We know that:

$$q \approx \frac{2 \times d \times \log n}{\log d} \approx \frac{2 \times d \times \log q^k}{\log d} \approx \frac{2 \times d \times k \times \log q}{\log d}. \tag{3}$$

Moving $\log q$ to the other side in Eq. 3 results in the following:

$$\frac{q}{\log q} \approx \frac{2 \times d \times k}{\log d}. \tag{4}$$

Hence, in Table 2, our test approach is to compare the three algorithms with parameters chosen as follows: first, we set $k = 3$, and select values of d to be 2, 3, 5. We then calculate N as $N = d(k-1) + 1$, and determine q using Eq. 4. After that, we compute t and n from these three parameters and conduct tests. Again, the winning method for each set of parameters is highlighted in bold.

Table 2. Performance of three methods using parameters from Theorem 1

d	k	N	q	n	t	DMCOMPACT (ns)	DMLIST (ns)	RSDECODE (ns)
2	3	5	79	493,039	395	1,746,554	116,839,654	**255,444**
5	3	11	83	571,787	913	89,719,040	412,072,578	**547,321**
8	3	17	107	1,225,043	1,819	182,862,278	6,996,290,225	**951,989**

The results from Table 2 align with the observations from Table 1. As n is relatively large, RSDECODE always takes the least amount of time to locate defective items, while DMLIST is always the slowest among the three. Moreover, as n grows, the execution time difference between RSDECODE and the other two methods also increases. Furthermore, the value of t in this table remains small with large values of n, which successfully achieves our goal of maintaining t small.

5.2 Performance of RSDECODE

This section aims to analyze the influence of parameter choices on the efficiency of RSDECODE. It consists of three parts describing different approaches to setting up parameters, followed by performance testing based on these parameters.

Parameters from Theorem 1. In this section, we adopt the same test approach as Table 2 by using Theorem 1. Nevertheless, unlike before, we are no longer restricting ourselves to n around 1,000,000, since we do not have to test the time-consuming naive algorithms. Moreover, on top of setting $k = 3$, we also introduce two more parts, by setting k to be 5 and 7, respectively. In order to evaluate the performance, we not only test the execution time, but also count the numbers of times the algorithm calls "isValidPolynomial" method (column labeled "#calls ivp"), as detailed in Sect. 3.

Table 3. Performance of RSDECODE using parameters from Theorem 1

d	k	N	q	n	t	time (ns)	#calls ivp
2	3	5	79	493,039	395	255,444	2.0
5	3	11	83	571,787	913	547,321	6.18
8	3	17	107	1,225,043	1,819	951,989	16.28
15	3	31	173	5,177,717	5,363	2,164,814	75.98
25	3	51	263	18,191,447	13,413	3,839,481	378.76
40	3	81	389	58,863,869	31,509	8,347,262	1999.92
2	5	9	149	73,439,775,749	1,341	493,653	2.0
5	5	21	157	95,388,992,557	3,297	952,205	13.62
8	5	33	211	418,227,202,051	6,963	2,302,795	91.64
15	5	61	331	3,973,195,810,651	20,191	4,804,404	2677.46
25	5	101	479	25,216,079,618,399	48,379	11,410,024	34411.2
40	5	161	719	192,151,797,699,599	115,759	59,413,144	604302.26
2	7	13	223	27,424,204,663,190,048	2,899	465,962	2.0
5	7	31	239	44,543,599,279,432,080	7,409	1,540,809	68.82
8	7	49	311	281,399,112,371,155,264	15,239	3,823,815	3851.14
15	7	91	479	5,785,602,523,725,084,672	43,589	14,584,799	87974.48
25	7	151	719	9.9334985e19	108,569	518,309,039	9676274.02
40	7	241	1,061	1.513588e21	255,701	8,499,031,204	1.017418919e8

The Table 3 illustrates the advantages of using RSDECODE. Unlike the naive algorithms, which struggle with large datasets, RSDECODE can detect defectives with extremely large values of n while keeping a relatively small t. In the table, the number of calls to ivp tends to increase as d increases. In the worst-case scenario, we would expect to have a maximum of d^{k-1} calls to find the first polynomial, followed by $(d-1)^{k-1}$ calls for the second polynomial, and so on. However, the average number of calls to ivp never even reaches d^{k-1}, indicating that RSDECODE can likely perform much better than the worst-case scenario. Additionally, the largest execution time recorded in these tables is about 8.5 s when $n = 1.513588 \times 10^{21}$, which is quite reasonable. Therefore, RSDECODE

Table 4. Performance of RSDECODE with different parameters

d	k	N	q	n	t	time (ns)	#calls ivp
3	5	13	13	371,293	169	538,069	4.76
3	5	13	59	714,924,299	767	584,028	3.22
3	5	13	97	8,587,340,257	1,261	650,068	3.94
3	6	11	11	1,771,561	121	566,304	6.26
3	6	11	59	13,841,287,201	539	635,336	5.46
3	6	11	97	832,972,004,929	1,067	623,955	4.0
3	7	19	19	893,871,739	361	736,800	10.86
3	7	19	59	2,488,651,484,819	1,121	775,912	8.48
3	7	19	97	80,798,284,478,113	1,843	856,450	5.3
3	8	22	23	78,310,985,281	506	893,042	10.98
3	8	22	59	146,830,437,604,321	1,298	945,186	9.74
3	8	22	97	7,837,433,594,376,961	2,134	1,046,835	7.96
8	3	17	37	50,653	629	890,610	27.68
8	3	17	107	1,225,043	1,819	951,989	16.28
8	3	17	211	9,393,931	3,587	893,622	11.36
8	5	33	37	69,343,957	1,221	2,089,694	549.04
8	5	33	211	418,227,202,051	6,963	2,302,795	91.64
8	5	33	311	2,909,390,022,551	10,263	1,739,017	54.72
8	7	49	53	1,174,711,139,837	2,597	4,128,484	9760.5
8	7	49	311	281,399,112,371,155,264	15,239	3,823,815	3851.14
8	7	49	503	8,146,590,668,153,304,064	24,647	3,712,977	1623.4
40	3	81	89	704,969	7,209	8,151,960	6017.4
40	3	81	389	58,863,869	31,509	8,347,262	1999.92
40	3	81	503	127,263,527	40,743	8,300,618	1227.98
40	5	161	163	115,063,617,043	26,243	322,625,157	4610179.46
40	5	161	311	2,909,390,022,551	50,071	158,320,856	2123141.94
40	5	161	719	192,151,797,699,599	115,759	59,413,144	604302.26
40	7	241	1061	1.513588e21	255,701	8,499,031,204	1.017418919e8
40	7	241	1297	6.1741878e21	312,577	6,214,728,079	9.78794511e7
40	7	241	2857	1.5537162e24	688,537	2,044,789,696	3.39832409e7

proves to be capable of handling large datasets. On top of that, the difference in k across the table affects the number of calls. As k increases, so does the number of calls to ivp and the running time.

Parameters Giving Optimal t Given n and d. As previously discussed in Sect. 5.1, the parameters k, q, and N presented in Table 1 are derived from fixing d and n and then determining which combination yields the smallest value of t.

From Table 1, we notice that as n increases, the execution time for RSDECODE also increases. Nevertheless, this increase is significantly slower compared to the growth of n. In the table, n ranges from 100 to approximately 900,000, indicating a growth of 9000 times. In contrast, the execution time of RSDECODE only grows by a factor of 50.

Analysing the Effect of Different Parameters on the Running Time. This section aims to analyze the performance of RSDECODE by choosing combinations of d and k. In Table 4, we choose d to be 3, 8 and 40, and we select k between 3 and 8, and determine N as $N = d(k-1) + 1$. After that, we choose increasing values of q bigger than N to run our tests, making $n = q^k$ grow. Again, all the tests are measured in both execution time and number of calls.

Based on Table 4, it is evident that as d increases, the number of calls to ivp also grows. On the other hand, as q grows for the same set of d, k and N, we observe that the numbers of calls to ivp decreases, resulting in shorter execution time. We are not quite sure what is the reason, but it appears that larger fields are making us get faster to the combinations of k-tuples in $S_p \times S_{p+1} \times \ldots \times S_{p+k-1}$ that leads to the right polynomials, causing less calls to ivp. This is an interesting observation with respect to Problem 1 that could be further investigated.

References

1. Bandeira, A.: Ten lectures and forty-two open problems in the mathematics of data science (2015). https://ocw.mit.edu/courses/18-s096-topics-in-mathematics-of-data-science-fall-2015/resources/mit18_s096f15_tenlec/. openCourseWare
2. Du, D., Hwang, F.K., Hwang, F.: Combinatorial Group Testing and its Applications, vol. 12. World Scientific (2000)
3. D'yachkov, A.G., Rykov, V.V.: Bounds on the length of disjunctive codes. Probl. Peredachi Inform. **18**(3), 7–13 (1982)
4. Erdős, P., Frankl, P., Füredi, Z.: Families of finite sets in which no set is covered by the union of R others. Israel J. Math. **51**(1–2), 79–89 (1985)
5. Füredi, Z.: ONR-cover-free families. J. Combin. Theory Ser. A **73**(1), 172–173 (1996)
6. Hwang, F.K.M., Du, D.Z.: Pooling Designs and Nonadaptive Group Testing: Important Tools for DNA Sequencing, vol. 18. World Scientific (2006)
7. Idalino, T., Moura, L.: A survey of cover-free families: constructions, applications, and generalizations. In: Colbourn, C., Dinitz, J. (eds.) New Advances in Designs, Codes and Cryptography, NADCC 2022, Fields Institute Communications, vol. 86, pp. 195–239. Springer (2024)
8. Indyk, P., Ngo, H., Rudra, A.: Efficiently decodable non-adaptive group testing. In: Proceedings of the Twenty-First Annual ACM-SIAM Symposium on Discrete Algorithms, pp. 1126–1142 (2010)
9. Kautz, W., Singleton, R.: Nonrandom binary superimposed codes. IEEE Trans. Inf. Theory **10**(4), 363–377 (1964)
10. Luo, D.: Modification-tolerant digital signatures using combinatorial group testing: theory, algorithms and implementation (2024). MSc thesis, in preparation

11. Porat, E., Rothschild, A.: Explicit nonadaptive combinatorial group testing schemes. IEEE Trans. Inf. Theory **57**(12), 7982–7989 (2011)
12. Ruszinkó, M.: On the upper bound of the size of the r-cover-free families. J. Combin. Theory Ser. A **66**(2), 302–310 (1994)
13. Wei, R.: On cover-free families. Technical report, Lakehead University (2006). https://arxiv.org/abs/2303.17524

Quantum CSS Duadic and Triadic Codes: New Insights and Properties

Reza Dastbasteh[(✉)], Olatz Sanz Larrarte, Josu Etxezarreta Martinez, Antonio deMarti iOlius, Javier Oliva del Moral, and Pedro Crespo Bofill

Tecnun - University of Navarra, Donostia-San Sebastian, Spain
{rdastbasteh,osanzl,jetxezarreta,ademartio,jolivam,pcrespo}@unav.es

Abstract. In this study, we investigate the construction of quantum CSS duadic codes with dimensions greater than one. We introduce a method for extending smaller splittings of quantum duadic codes to create larger, potentially degenerate quantum duadic codes. Furthermore, we present a technique for computing or bounding the minimum distances of quantum codes constructed through this approach. Additionally, we introduce quantum CSS triadic codes, a family of quantum codes with a rate of at least $\frac{1}{3}$.

Keywords: quantum code · duadic code · triadic code · degenerate code · minimum distance

1 Introduction

Quantum error-correcting codes, also known as quantum codes, serve to protect quantum information from the detrimental effects of noise, such as decoherence, during its transmission, storage or processing. Analogous to classical error-correcting codes, quantum codes play an important role in ensuring the fidelity of quantum information. Among the various techniques employed for constructing quantum codes, the stabilizer construction stands out as one of the most widely used [5,11]. Indeed, this approach has been instrumental in constructing numerous families of quantum codes. A notable instance of the stabilizer construction is the class of Calderbank-Steane-Shor (CSS) codes. The binary CSS construction offers a straightforward method for generating binary quantum codes by utilizing a binary code and one of its subcodes.

In [1], it was demonstrated that the CSS construction is well-suited for binary duadic codes, giving rise to an infinite class of quantum duadic codes. A notable property of quantum duadic codes is the presence of an infinite subclass of degenerate codes. In general, a quantum stabilizer code is called degenerate if its defining stabilizer group contains a non-trivial error with weight smaller than the minimum distance that, thus, acts trivially on the code [5]. It is important to note that error patterns resulting in non-trivial syndromes necessitate proactive correction of the physical qubits in the quantum code. This correction process may introduce additional errors into the system, stemming from the noise introduced by quantum gates. Consequently, the construction of highly degenerate codes can potentially reduce the requirement for active error correction. Moreover, codes with degenerate low-weight errors have the potential to perform better since many of the high probability errors do not require to be corrected [8,19].

One of the primary challenges in constructing degenerate codes lies in determining the true minimum distance or establishing bounds on the minimum distances of these codes. In the case of the CSS duadic quantum codes presented in [1], all the degenerate codes have a composite length, dimension one, and satisfy a square root minimum distance lower bound.

In this work, inspired by the methodology outlined in [1,8], we explore the construction of binary quantum duadic codes with larger dimensions using the CSS construction. This is enabled by allowing the multipliers of duadic codes to fix more than one element. Furthermore, we investigate the existence of such codes and propose a method for extending two smaller length splittings to a larger length splitting. This approach results in a class of degenerate binary quantum codes. Another advantage of the extended splitting method is that it allows us to compute or bound the true minimum distance of quantum codes constructed in this way. Moreover, we introduce quantum triadic codes, which can be used to build binary quantum codes with rates greater than or equal to $\frac{1}{3}$. Constructing codes with high rates is specially relevant in order to reduce resource consumption. While these constructions might not be the most suitable for integration in state-of-the-art quantum processors as a result of connectivity constraints, they can be useful for future processors or quantum communications [28]. Finally, we present the parameters for several quantum duadic and triadic codes constructed using our approach.

This paper is structured as follows. In Sect. 2, we give an overview of the structure and properties of cyclic, duadic, and quantum stabilizer codes. Section 3 discusses the existence of quantum CSS duadic codes with dimensions larger than one, along with their construction. In Sect. 4, we explore the properties of quantum CSS duadic codes constructed from extended splittings, their degenerate subclass, and methods for computing or bounding their minimum distances. Section 5 introduces quantum triadic codes and their properties. Finally, Sect. 6 summarizes our results.

2 Preliminaries

Let $\mathbb{F}_2$ be the binary field. Throughout this work, whenever we discuss cyclic or duadic codes, n is a positive odd integer.

2.1 Cyclic and Duadic Codes

Duadic codes encompass a variety of well-known codes such as quadratic residue codes, Golay codes, and numerous Reed-Muller and Reed-Solomon codes, all of which have rich algebraic structure and can be used to construct good classical and quantum codes [8,9,14,24,26], and [16, Section 6.5]. Given that they are cyclic, we will commence with a brief overview of cyclic codes.

A linear code $C \subseteq \mathbb{F}_2^n$ is called *cyclic* if for every $c = (c_0, c_1, \ldots, c_{n-1}) \in C$, the vector $(c_{n-1}, c_0, \ldots, c_{n-2})$ obtained by a cyclic shift of the coordinates of c is also in C. It is well known that there exists a one-to-one correspondence between binary cyclic codes of length n and ideals of the ring $\mathbb{F}_2[x]/\langle x^n - 1 \rangle$, for example see [16, Section 4.2]. Thus each cyclic code can be uniquely represented by a monic polynomial $g(x)$, where $g(x)$ is the minimal degree generator of the corresponding ideal. The polynomial $g(x)$ is called the *generator polynomial* of such cyclic code. Let α be a primitive n-th root of unity in a finite field extension of $\mathbb{F}_2$. Alternatively, such cyclic code can be represented by its unique *defining set*

$$\{t : 0 \leq t \leq n - 1 \text{ and } g(\alpha^t) = 0\}. \tag{1}$$

As we will see, many of the characteristics of cyclic codes are easily determined by the notion of a defining set. For each $s \in \mathbb{Z}/n\mathbb{Z}$, the *2-cyclotomic coset* modulo n containing s is defined as

$$Z(s) = \{(s2^j) \mod n : 0 \leq j \leq m - 1\}, \tag{2}$$

where m is the smallest positive integer such that $s2^m \equiv s \pmod{n}$. All different 2-cyclotomic cosets partition $\mathbb{Z}/n\mathbb{Z}$, and the defining set of a linear cyclic code is a union of cyclotomic cosets. A *multiplier* μ_b on $\mathbb{Z}/n\mathbb{Z}$ is defined by $\mu_b(x) = (bx) \mod n$, for some integer b such that $\gcd(n, b) = 1$. Note also that if $A \subseteq \mathbb{Z}/n\mathbb{Z}$, then $\mu_b(A) = bA = \{(ba) \mod n : a \in A\}$. Multipliers act as special permutations on $\mathbb{Z}/n\mathbb{Z}$ and they are essential to define duadic and related codes. If C is a cyclic code with the defining set A, then $\mu_b(C)$, i.e., the code obtained by permuting the coordinates of C by applying μ_b, is a cyclic code with the defining set $\mu_{b^{-1}}(A) = b^{-1}A$.

Recall that the Euclidean inner product of $u = (u_0, u_1, \ldots, u_{n-1})$ and $v = (v_0, v_1, \ldots, v_{n-1})$ is defined by

$$u \cdot v = \sum_{i=0}^{n-1} u_i v_i \in \mathbb{F}_2.$$

The (Euclidean) dual of a binary cyclic code C of length n is defined by

$$C^\perp = \{u \in \mathbb{F}_2^n : u \cdot c = 0 \text{ for each } c \in C\}.$$

If a cyclic code C has defining set A, then the set

$$(\mathbb{Z}/n\mathbb{Z}) \setminus \mu_{-1}(A) = (\mathbb{Z}/n\mathbb{Z}) \setminus (-A) = (\mathbb{Z}/n\mathbb{Z}) \setminus \{(-a) \mod n : a \in A\}$$

is the defining set of $C^\perp$ [16, Theorem 4.4.9 (iv)] (in [16] $\mathbb{Z}/n\mathbb{Z}$ and A are denoted by $\mathcal{N}$ and T, respectively). One of our main topics of interest in this paper are dual-containing cyclic codes, i.e., when $C^\perp \subseteq C$. The following theorem determines when a binary cyclic code is dual-containing.

Theorem 1. *Let $C \subseteq \mathbb{F}_2^n$ be a binary cyclic code with the defining set A. Then $C^\perp \subseteq C$ if and only if $A \cap -A = \emptyset$.*

Proof. The proof follows from [16, Theorem 4.4.11], and here we only give a short proof for clarification. The code $C^\perp$ has the defining set $(\mathbb{Z}/n\mathbb{Z}) \setminus (-A)$. Then we have $C^\perp \subseteq C$ if and only if $A \subseteq ((\mathbb{Z}/n\mathbb{Z}) \setminus (-A))$, which happens if and only if $A \cap -A = \emptyset$. □

Recall that the *(Hamming) weight* of a binary vector v is defined as the number of its non-zero coordinates, denoted by $\mathrm{wt}(v)$. The *minimum (Hamming) distance* of a binary code C is defined as

$$d(C) = \min\{\mathrm{wt}(c) : 0 \neq c \in C\}.$$

Another feature of cyclic codes is that their minimum distance is lower bounded by the Bose-Chaudhuri-Hocquenghem (BCH) bound. For integers $a \leq b$, we denote the consecutive set $\{a, a+1, \ldots, b\}$ by $[a, b]$.

Theorem 2. *[2, 15] Let C be a binary cyclic code of length n with the defining set A. If $[a, b] \subseteq A$, then $d(C) \geq (b - a) + 2$.*

Duadic codes are an important subclass of cyclic codes with rich algebraic and combinatorial properties [20,22,23]. Quadratic residue (QR) codes are also a special case of duadic codes, and many examples of linear codes with optimal parameters are constructed from duadic codes. For a more extensive discussion of duadic code see [16, Chapter 6] and [10, Section 2.7]. In what follows, we provide a more general definition of binary duadic codes based on the concept introduced in Definition 2.6 of [8], which allows multipliers to fix more than one element.

Definition 1. *Let $X, S_1, S_2 \subseteq \mathbb{Z}/n\mathbb{Z}$ be unions of 2-cyclotomic cosets such that*

1. *$X \cup S_1 \cup S_2 = \mathbb{Z}/n\mathbb{Z}$,*
2. *S_1, S_2, and X are non-empty and disjoint,*
3. *there exists a multiplier μ_b such that $\mu_b(S_1) = S_2$, $\mu_b(S_2) = S_1$, and $\mu_b(Z(s)) = Z(s)$ for each $s \in X$.*

Then the triple (X, S_1, S_2) is called a splitting *(2-splitting) of $\mathbb{Z}/n\mathbb{Z}$ that is given by μ_b.*

Let $X \subseteq \mathbb{Z}/n\mathbb{Z}$. Recall that α is a primitive n-th root of unity in a finite field extension of $\mathbb{F}_2$. A vector $(c_0, c_1, \ldots, c_{n-1}) \in \mathbb{F}_2^n$ is called *even-like with respect to X* provided that $c(\alpha^s) = 0$ for each $s \in X$, where $c(x) = \sum_{i=0}^{n-1} c_i x^i \in \mathbb{F}_2[x]$. We call a cyclic code *even-like with respect to X* if it has only even-like codewords with respect to X. Otherwise we call it an *odd-like code with respect to X*. In particular, a linear cyclic code with the defining set A is even-like with respect to any $X \subseteq A$. Whenever the set X is clear from the context, we simply call the mentioned codes even-like and odd-like.

Definition 2. Let n be a positive odd integer and (X, S_1, S_2) be a splitting of $\mathbb{Z}/n\mathbb{Z}$. Then binary cyclic codes of length n with the defining sets S_1 and S_2 (respectively $S_1 \cup X$ and $S_2 \cup X$) are called odd-like (respectively even-like) duadic codes.

Previously, the focus was primarily on duadic codes with a splitting that satisfies $X = \{0\}$, and less attention was given to duadic codes with $|X| > 1$. However, as we will explore later, the latter codes can also be used to construct families of classical and quantum codes with similar properties.

Let C and D be odd-like and even-like binary duadic codes of length n with the defining sets S_1 and $S_1 \cup X$, respectively. The *minimum odd-like weight* of C (with respect to X) is defined by

$$d_o(C) = \min\{\text{wt}(c) : c \in C \setminus D\}.$$

The next theorem highlights some properties of binary duadic codes in this general setting.

Theorem 3. Let n be a positive odd integer and (X, S_1, S_2) be a splitting of $\mathbb{Z}/n\mathbb{Z}$ given by μ_b. Let C_1 and C_2 (respectively D_1 and D_2) be odd-like (respectively even-like) duadic codes of length n with the defining sets S_1 and S_2 (respectively $S_1 \cup X$ and $S_2 \cup X$). Then the following hold for each $1 \leq i \leq 2$:

1. $\mu_b(C_1) = C_2$ and $\mu_b(D_1) = D_2$ (the pairs of duadic codes are permutation equivalent).
2. $\dim(C_i) = \frac{n+|X|}{2}$ and $\dim(D_i) = \frac{n-|X|}{2}$.
3. $C_1 + C_2 = \mathbb{F}_2^n$ and $D_1 \cap D_2 = \{(0,0,\ldots,0)\}$.
4. $\mu_{-1}(S_1) = S_2$ (or equivalently $\mu_{-1}(S_2) = S_1$) if and only if $D_1^\perp = C_1$ and $D_2^\perp = C_2$. Also, $\mu_{-1}(S_i) = S_i$ if and only if $D_1^\perp = C_2$ and $D_2^\perp = C_1$. Moreover, the codes $(D_1 + D_2)^\perp$ and $C_1 \cap C_2$ are permutation equivalent.
5. $d_o(C_i)^2 \geq d(C_1 \cap C_2)$ (the code $C_1 \cap C_2$ has the defining set $(\mathbb{Z}/n\mathbb{Z}) \setminus (X)$).

Proof. The proof is similar to that of [20, Theorem 3],[8, Theorem 4.1], and [16, Theorems 6.4.2 and 6.4.3], and we only give a remark about part (4). The code $(D_1 + D_2)^\perp$ and $C_1 \cap C_2$ have the defining sets $A = (\mathbb{Z}/n\mathbb{Z}) \setminus (-X)$ and $-A = (\mathbb{Z}/n\mathbb{Z}) \setminus (X)$. Thus, they are permutation equivalent (see, for example, [7, Theorem 1.6.4]). □

In Sect. 3, we investigate the existence of binary duadic codes when $|X| > 1$.

2.2 Binary Quantum Codes

In this section, $\mathcal{H} = \mathbb{C}$ is the complex Hilbert space with the inner product $\langle a, b \rangle = \sum_{i=1}^{n} \overline{a_i} b_i$, where $a, b \in \mathcal{H}^n$ and $\overline{a_i}$ is the complex conjugate of a. The basic unit of quantum information, i.e., the qubit, lies in the two-dimensional complex Hilbert space, $\mathcal{H}^2$. The set of one qubit Pauli matrices are defined as

$$I = \begin{bmatrix} 1 & 0 \\ 0 & 1 \end{bmatrix}, X = \begin{bmatrix} 0 & 1 \\ 1 & 0 \end{bmatrix}, Y = \begin{bmatrix} 0 & -i \\ i & 0 \end{bmatrix}, \text{ and } Z = \begin{bmatrix} 1 & 0 \\ 0 & -1 \end{bmatrix}.$$

They form a basis for all 2×2 complex matrices. Hence an arbitrary qubit operator (2×2 unitary matrix over $\mathcal{H}$) is a linear combinations of the mentioned matrices. The set
$$P_1 = \{\pm 1, \pm i\} \times \{I, X, Y, Z\}$$
is a multiplicative group of unitary matrices called the *Pauli group*. For a positive integer $n > 1$, an n-qubit *Pauli matrix* can be represented as
$$X^u Z^v = X^{u_1} Z^{v_1} \otimes X^{u_2} Z^{v_2} \otimes \cdots \otimes X^{u_n} Z^{v_n} \in \mathcal{H}^{2^n},$$
where $u = (u_1, u_2, \ldots, u_n), v = (v_1, v_2, \ldots, v_n) \in \mathbb{F}_2^n$. In particular, the matrix $X^u Z^v$ acts on the i-th qubit as $X^{u_i} Z^{v_i}$ for each $1 \leq i \leq n$. For example, $X^{(1,0,1,0)} Z^{(0,0,1,1)}$ is equivalent to the Pauli matrix $X \otimes I \otimes XZ \otimes Z$. Pauli matrices are also referred to as Pauli errors, as they are commonly considered as the foundational error model encompassing all potential errors capable of corrupting a quantum system. This is a consequence of the so-called error discretization that results from the syndrome extraction process, where arbitrary continuous errors discretize in such set of discrete errors [18]. The weight of an n-qubit Pauli matrix is the number of non-identity components in its tensor product. Analogously, the group generated by the n-dimensional Pauli matrices with the multiplicative factors $\{\pm 1, \pm i\}$, $P_n = \bigotimes^n P_1$, is called the *n-dimensional Pauli group*.

Recall that an $[\![n, k, d]\!]_2$ binary quantum code encodes k qubits of information into n physical qubits (a 2^k-dimensional subspace of $\mathcal{H}^{2^n}$) and is capable of correcting up to $\lfloor \frac{d-1}{2} \rfloor$ or fewer number of errors on the n physical qubits.

The following theorem describes the necessary conditions to construct a subclass of quantum codes known as *quantum stabilizer codes*, and also to determine their parameters. For any subgroup G of P_n, we denote the Pauli normalizer of G by $N(G)$, which is the set of Pauli matrices that commute with each element of G.

Theorem 4. *Let G be a commutative subgroup of P_n not containing a nontrivial multiple of the identity that is generated by a minimal set of $n - k$ Pauli matrices. Then, the common eigenspace with eigenvalue $+1$ of G is a $[\![n, k, d]\!]_2$ binary quantum code. If $k > 0$, then d is the minimum weight of error operators in $N(G) \setminus G$; if $k = 0$, then d is the minimum weight of error operators in $N(G) = G$.*

Proof. For the proof, see, for example, the discussion in Sect. 2.2 of [21] or Section 3.2 of [12].

A quantum code of Theorem 4 is called *degenerate* if it minimum distance d is greater than the minimum weight of error operators in $N(G)$. In general, directly constructing a quantum code or computing its parameters using the description provided in Theorem 4 is not practical. Instead, a more feasible approach to constructing stabilizer codes is through the Calderbank-Shor-Steane (CSS) construction, as detailed below.

Theorem 5. *[5, Theorem 9] Let $C_2 \subseteq C_1$ be binary linear codes of length n with dimensions k_2 and k_1, respectively. Then there exists an $[\![n, k_1 - k_2, d]\!]_2$ binary quantum stabilizer code, where*

$$d = \min\{d(C_1 \setminus C_2), d(C_2^\perp \setminus C_1^\perp)\}. \tag{3}$$

In particular, the stabilizer group of the quantum code described in Theorem 5 is generated by

$$\{X^u : u \in C_2\} \cup \{Z^v : v \in C_1^\perp\}.$$

Moreover, this quantum code is degenerate if $d > \min\{d(C_1), d(C_2^\perp)\}$. The quantum codes discussed in this paper are generated by applying the CSS construction to binary duadic and triadic codes. Our primary interest is in computing the parameters and identifying the degenerate codes constructed in this manner. Since Theorem 5 completely determines the parameters and degeneracy of quantum CSS codes, we never deal with the actual quantum code in the sense of Theorem 4.

A special case of CSS construction arises when the pair of binary codes satisfy $C_1 = C_2^\perp$ and both codes are cyclic. As discussed in [13,27], such quantum CSS codes support transversal Pauli operators, the $S = Z^{\frac{1}{2}}$ gate, and the Hadamard H gate (bitwise application of such operators maps the encoded Hilbert space onto itself). Indeed, many of the quantum CSS codes in this paper satisfy the mentioned property.

3 Existence of Duadic Codes and Quantum CSS Duadic Codes

In this section, we investigate the existence of binary duadic codes (Definition 2). Duadic codes with splittings satisfying $|X| = 1$ have already been discussed in the literature; see, for example, [1,9]. Our focus and new results lie in exploring the existence of splittings satisfying $|X| > 1$ and their corresponding binary CSS codes. This latter case has not been previously explored in the literature.

First note that the Euclidean dual-containing duadic codes are minimal among all Euclidean dual-containing cyclic codes in the sense of the following proposition. In particular, the odd-like (respectively even-like) duadic codes have the largest minimum distance among all dual-containing (respectively self-orthogonal) cyclic codes.

Proposition 1. *Let C be a binary dual-containing linear cyclic code of length n. Then*

$$C^\perp \subseteq D_1 \subseteq C_1 = D_1^\perp \subseteq C, \tag{4}$$

where D_1 and C_1 are even-like and odd-like duadic codes corresponding to a splitting that is given by μ_{-1}, respectively.

Proof. The proof is similar to that of [8, Proposition 3.1], thus, we only give a sketch of it. Let A be the defining set of C. By Theorem 1, we have $A \cap \mu_{-1}(A) = \emptyset$. Hence we can find a splitting of $\mathbb{Z}/n\mathbb{Z}$ in the form (X, S_1, S_2) that is given by μ_{-1}, where $A \subseteq S_1$, respectively. Let D_1 and C_1 be even-like and odd-like duadic codes with the defining sets $S_1 \cup X$ and S_1. Then $D_1 \subseteq C_1 = D_1^\perp$ and $C_1 \subseteq C$. Now the fact that C is dual-containing implies (4). □

According to the proposition above, among all quantum CSS codes constructed using a dual-containing cyclic code and its dual, the duadic pairs yield the largest minimum distance. Additionally, the existence of duadic codes relies on the existence of a splitting for $\mathbb{Z}/n\mathbb{Z}$. Furthermore, $\mu_b(\{0\}) = \{0\}$ for each multiplier μ_b. So in the next lemmas, we study when $a \in \mathbb{Z}/n\mathbb{Z}$ satisfies $Z(a) = Z(ba)$. The proofs are straightforward, but we state them for the sake of completeness. Recall that $\text{ord}_n(2) = |Z(1)|$ is the multiplicative order of 2 modulo n.

Lemma 1. *Let n and b be positive odd integers such that $\gcd(n, b) = 1$. Then $Z(s) = Z(bs)$ for each $s \in \mathbb{Z}/n\mathbb{Z}$ if and only if $n \mid 2^r - b$ for some $1 \leq r \leq \text{ord}_n(2)$.*

Proof. $\rightarrow$: The assumption implies that $Z(1) = Z(b)$ (choosing $s = 1$). Thus there exists $1 \leq r \leq \text{ord}_n(2)$ such that $2^r \equiv b \pmod{n}$, or equivalently $n \mid 2^r - b$.
$\leftarrow$: We have $2^r \equiv b \pmod{n}$. Now multiplying both sides by s gives the desired result. □

Therefore, if $n \mid 2^r - b$ for some integer $1 \leq r \leq \text{ord}_n(2)$, then there is no splitting of $\mathbb{Z}/n\mathbb{Z}$ that is given by μ_b. On the other hand, if $n \nmid 2^r - b$ for each integer $1 \leq r \leq \text{ord}_n(2)$, there may still exist some values $s \in \mathbb{Z}/n\mathbb{Z}$ for which the equality $Z(s) = Z(bs)$ holds. The trivial case occurs when $s = 0$. The following lemma classifies all other fixed cyclotomic cosets of $\mathbb{Z}/n\mathbb{Z}$ under the action of a multiplier μ_b.

Lemma 2. *Let n and b be positive odd integers such that $\gcd(n, b) = 1$, and $s \in \mathbb{Z}/n\mathbb{Z}$. Then $Z(s) = Z(bs)$ if and only if $\frac{n}{\gcd(n, 2^r - b)} \mid s$ for some $1 \leq r \leq \text{ord}_n(2)$.*

Proof. We have $Z(s) = Z(bs)$ if and only if $2^r s \equiv bs \pmod{n}$ or equivalently $n \mid s(2^r - b)$ for some $1 \leq r \leq \text{ord}_n(2)$. The latter is equivalent to $\frac{n}{\gcd(n, 2^r - b)} \mid s$. □

For positive integers n and b such that $\gcd(n, b) = 1$, we define

$$\text{Fix}_b(n) = \{s \in \mathbb{Z}/n\mathbb{Z} : \frac{n}{\gcd(n, 2^r - b)} \mid s \text{ for some } 1 \leq r \leq \text{ord}_n(2)\}. \quad (5)$$

Note that, by Lemma 2, the set $\text{Fix}_b(n)$ is a union of 2-cyclotomic cosets modulo n. Also, $Z(s) = Z(bs)$ if and only if $s \in \text{Fix}_b(n)$. Moreover, Lemma 1 implies that $\text{Fix}_b(n) = \mathbb{Z}/n\mathbb{Z}$ if and only if $n \mid 2^r - b$ for some $1 \leq r \leq \text{ord}_n(2)$. This leads to the following observation.

Theorem 6. *Let n be a positive odd integer.*
(1) Dual-containing binary duadic codes of length n exist if and only if $\mathrm{Fix}_{-1}(n) \subsetneq \mathbb{Z}/n\mathbb{Z}$.
(2) Each splitting of a duadic code that is given by μ_b is in the form (X, S_1, S_2), where $X = \mathrm{Fix}_b(n)$.
(3) $\mathrm{Fix}_b(n) = \mu_{-1}(\mathrm{Fix}_b(n))$.

Proof. First, using Lemma 1, one can argue that dual-containing binary duadic codes exist if and only if there exists a splitting of $\mathbb{Z}/n\mathbb{Z}$ that is given by μ_{-1} if and only if $\mathrm{Fix}_{-1}(n) \subsetneq \mathbb{Z}/n\mathbb{Z}$.

For the second part, note that $\mathrm{Fix}_b(n) \subseteq X$. Moreover, the definition of a splitting, Definition 1, implies that $X \subseteq \mathrm{Fix}_b(n)$. Thus, $X = \mathrm{Fix}_b(n)$.

The last part follows from (5) as for each $s \in \mathbb{Z}/n\mathbb{Z}$, we have $\frac{n}{\gcd(n, 2^r - b)} \mid s$ if and only if $\frac{n}{\gcd(n, 2^r - b)} \mid -s$. □

Now we use the above information to formally introduce quantum CSS duadic codes.

Theorem 7. *Let n and b be positive odd integers such that $\gcd(n, b) = 1$ and $(\mathrm{Fix}_b(n), S_1, S_2)$ is a splitting of $\mathbb{Z}/n\mathbb{Z}$ that is given by μ_b. Then there exists a binary quantum code with parameters $[\![n, |\mathrm{Fix}_b(n)|, d \geq \sqrt{d(C)}]\!]_2$, where $d(C)$ is the minimum distance of binary cyclic code with defining set $(\mathbb{Z}/n\mathbb{Z}) \setminus \mathrm{Fix}_b(n)$.*

Proof. Let D_1 and C_1 be binary even-like and odd-like duadic codes of length n with the defining sets $\mathrm{Fix}_b(n) \cup S_1$ and S_1, respectively. Then by Theorem 3 (2), we have $\dim(C_1) = \frac{n + |\mathrm{Fix}_b(n)|}{2}$ and $\dim(D_1) = \frac{n - |\mathrm{Fix}_b(n)|}{2}$. Moreover, Theorem 6 implies that $\mu_{-1}(\mathrm{Fix}_b(n)) = \mathrm{Fix}_b(n)$. So we have that $(\mathrm{Fix}_b(n), -S_1, -S_2)$ is also a splitting of $\mathbb{Z}/n\mathbb{Z}$ that is given by μ_b. Furthermore, $C_1^\perp$ and $D_1^\perp$ are binary even-like and odd-like duadic codes of length n with the defining sets $\mathrm{Fix}_b(n) \cup -S_2$ and $-S_2$, respectively (the codes D_1 and $C_1^\perp$, similarly C_1 and $D_1^\perp$, are permutation equivalent as μ_{-b} exchanges their defining sets). Now, Theorem 3 (5) implies that

$$d(C_1 \setminus D_1) \text{ and } d(D_1^\perp \setminus C_1^\perp) \geq \sqrt{d(C)}.$$

Applying the CSS construction (Theorem 5) to $D_1 \subseteq C_1$, proves the existence of an $[\![n, \mathrm{Fix}_b(n), d \geq \sqrt{d(C)}]\!]_2$. □

The mentioned theorem presents a generalization of the conventional definition of binary quantum CSS duadic codes as outlined in [1]. This is because when $X = \{0\}$, the code C in Theorem 7 has the minimum distance n (C is the code generated by all-one). This scenario gives the family of quantum CSS duadic codes that has parameters $[\![n, 1, d \geq \sqrt{n}]\!]_2$ as established in [1, Theorem 4].

Note also that in many cases minimum distance of the code C in Theorem 7 can be computed or bounded using the BCH bound. In particular, if s is the smallest non-zero element of $\mathrm{Fix}_b(n)$, then one can easily show that $[1, s-1] \subseteq$

$(\mathbb{Z}/n\mathbb{Z}) \setminus \text{Fix}_b(n)$. Therefore, applying the BCH lower bound implies that $d(C) \geq s$. For example, when $n = 21$, the code C has the defining set $(\mathbb{Z}/21\mathbb{Z}) \setminus \{0, 7, 14\}$ which contains the interval $[1, 6]$. Hence it has minimum distance $d(C) \geq 7$. In fact 7 is the minimum distance in this case.

4 Extended Splitting and Degenerate Codes

In this section, we study binary duadic codes that are constructed by extending the splittings of shorter-length binary duadic codes. This approach allows us to construct degenerate quantum duadic codes. We also provide lower and upper bounds, or even compute, the minimum odd-like distance of duadic codes constructed in this manner. To begin, we first define the extended splitting of duadic codes.

Definition 3. *Let n_1 and n_2 be positive odd integers, and $T = (T_0, T_1, T_2)$ and $U = (U_0, U_1, U_2)$ be two splittings of $\mathbb{Z}/n_1\mathbb{Z}$ and $\mathbb{Z}/n_2\mathbb{Z}$, respectively, that are given by μ_b such that $U_0 = \{0\}$. The set $S = (S_0, S_1, S_2)$, where $S_0 = \{in_2 : i \in T_0\}$ and*

$$S_k = \{in_2 : i \in T_k\} \cup \{i + jn_2 : i \in U_k, 0 \leq j \leq n_1 - 1\} \tag{6}$$

for $k = 1, 2$, is called the extended splitting *of T and U modulo $n_1 n_2$.*

When $T_0 = \{0\}$, it has been shown in [26] that the extended splitting of T and U is a splitting of $\mathbb{Z}/(n_1 n_2 \mathbb{Z})$ that is given by μ_b. The same conclusion is obtained in [8] for cyclic and constacyclic duadic codes over $\mathbb{F}_4$ that are given by μ_{-2}. A similar technique can be used to show that when $|T_0| > 1$, the set S is a splitting of $\mathbb{Z}/(n_1 n_2 \mathbb{Z})$ that is given by μ_b.

Note also that each splitting of $\mathbb{Z}/(n_1 n_2)\mathbb{Z}$ is not necessarily an extended splitting of two smaller length splittings. For instance, there is no splitting of $\mathbb{Z}/3\mathbb{Z}$ and $\mathbb{Z}/5\mathbb{Z}$; however, one can find a splitting for $\mathbb{Z}/15\mathbb{Z}$.

The next theorem connects codewords of even-like duadic codes of smaller length and its corresponding extended duadic code. Proof of the following theorem is similar to that of Theorem 3 in [26] and Theorem 4.8 in [8], and here we only provide a short proof.

Theorem 8. *Let C_S and C_U be even-like binary duadic cyclic codes of lengths $n_1 n_2$ and n_2 with defining sets $S_0 \cup S_1$ and $U_0 \cup U_1$, respectively, where (S_0, S_1, S_2) is an extended splitting of a length n_1 splitting and a length n_2 splitting in the form (U_0, U_1, U_2). If C_U has a vector of weight t, then C_S also has a vector of weight t. In particular,*

$$d(C_S) \leq d(C_U) < n_2.$$

Proof. Let δ be a primitive $(n_1 n_2)$-th root of unity such that the mentioned splittings are formed using the primitive roots δ, δ^{n_2}, and δ^{n_1}. First note that

$$n_1 S_i = \{(n_1 a) \mod n_1 n_2 : a \in S_i\} = \begin{cases} n_1 U_i & \text{if } i = 1, 2 \\ \{0\} & \text{if } i = 0. \end{cases}$$

Hence, if $c(x) \in C_U$ (or equivalently $c(\delta^{n_1 a}) = 0$ for each $a \in U_0 \cup U_1$), then $c'(x) = c(x^{n_1}) \in C_S$. This is because for any $b \in S_0 \cup S_1$, we have $c'(\delta^b) = c(\delta^{n_1 b})$ and as we showed above $n_1 b \in (n_1 S_0) \cup (n_1 S_1) = n_1 U_0 \cup n_1 U_1$ (note that $n_1 S_0 = n_1 U_0 = \{0\}$). Therefore, $c'(\delta^b) = 0$. Thus, $c'(x) \in C_S$ and it has the same weight as $c(x)$. The last part is an immediate consequence of the first one. □

Let T and U be two splittings of length n_1 and n_2 that are given by μ_a and μ_b, where $\gcd(n_1, n_2) = 1$. Then, using the Chinese remainder theorem, one can find $c \in \mathbb{Z}/(n_1 n_2)\mathbb{Z}$ such that $c \equiv a \pmod{n_1}$ and $c \equiv b \pmod{n_2}$. Thus the set S defined in Definition 3 is still an splitting that is given by μ_c. Therefore, the result of Theorem 8 holds in such cases too.

An application of the above theorem is in the construction of degenerate quantum CSS duadic codes with dimension greater than one.

Corollary 1. *Let n_1 and n_2 be positive odd integers, and let $T = (T_0, T_1, T_2)$ and $U = (U_0, U_1, U_2)$ be two splittings of length n_1 and n_2, respectively, that are either given by the same multiplier, or they have different multipliers and $\gcd(n_1, n_2) = 1$. If $\frac{n_1}{|T_0|} \geq n_2$, then there exists a degenerate binary quantum code with parameters $[\![n_1 n_2, |T_0|, d > n_2]\!]_2$.*

Proof. Let $S = (S_0, S_1, S_2)$ be the extended splitting of T and U that is given by μ_b for some $b \in \mathbb{Z}/(n_1 n_2)\mathbb{Z}$. Let C_S and C_U (respectively D_S and D_U) be even-like (respectively odd-like) binary duadic codes of lengths $n_1 n_2$ and n_2 with the defining sets $S_0 \cup S_1$ and $U_0 \cup U_1$ (respectively S_1 and U_1), respectively. Then by Theorem 8 we have $d(C_S) \leq d(C_U) \leq n_2$.

Moreover, the fact that $\frac{n_1}{|T_0|} \geq n_2$ implies that there exist integers $a < b \in T_0$ such that $b - a \geq n_2 - 1$ and $T_0 \cap \{a, a+1, \ldots, b\} = \emptyset$. Hence the the set S_0 has no element in the interval $[n_2 a, n_2 b]$ which has size n_2^2. Let C be the cyclic code of length $n_1 n_2$ and the defining set $A = (\mathbb{Z}/(n_1 n_2)\mathbb{Z}) \setminus S_0$. Then $[n_2 a, n_2 b] \subseteq A$, and the BCH bound (Theorem 1) implies that $d(C) > n_2^2$. Now, applying the result of Theorem 7 to the duadic codes corresponding to S (recall that $\mathrm{Fix}_b(n_1 n_2) = S_0$ and $|S_0| = |T_0|$) implies the existence of a binary quantum code Q with parameters $[\![n_1 n_2, |T_0|, d > n_2]\!]_2$, where Q is the CSS code of $C_S \subset D_S$. Moreover, Theorem 8 implies that $d(C_S) < n_2$ and since C_S and $D_S^\perp$ are permutation equivalent (see the proof of Theorem 7) we have $d(D_S^\perp) < n_2$. Therefore, Q is degenerate. □

A special case of above theorem is when $T_0 = \{0\}$ which is discussed in [1, Theorem 6] in more detail. A remaining problem is to compute or bound the minimum distance of such quantum codes using the odd-like distance of smaller duadic codes. In what follows we resolve this problem.

The next theorem gives a bound for the minimum odd-like distance of duadic codes that have been constructed using the extended splittings.

Theorem 9. *Let n_1 and n_2 be positive odd integers. Let C, C_1, and C_2 be odd-like binary duadic codes of lengths $n_1 n_2$, n_1, and n_2, respectively, where C is* constructed from an extended splitting of C_1 and C_2. Then

$$d_o(C_1)d(C_2) \le d_o(C) \le d_o(C_1)d_o(C_2).$$

In particular, if $d(C_2) = d_o(C_2)$, we have $d_o(C) = d_o(C_1)d_o(C_2)$.

Proof. The proof follows the logic of [8, Theorem 4.11], and we omit it here. □

This result provides an efficient way of computing or bounding the minimum odd-like distance of binary duadic codes of certain composite lengths. Note also that when the code C_2 in the above theorem is a QR code, then $d(C_2) = d_o(C_2)$ is odd [16, Theorem 6.6.22], so this bound gives the exact minimum odd-like distance of such codes.

Corollary 2. *Let C, C_1, and C_2 be as in Theorem 9 and C_2 be a QR code. Then $d_o(C) = d_o(C_1)d_o(C_2)$.*

For instance the binary odd-like QR code of length 7 which has parameters $[7, 4, 3]_2$ can be used to construct a binary quantum code with parameters $[[7, 1, 3]]_2$. This code has minimum even-like distance of 4. Now repeatedly extending the splittings and using the above distance bound, one get the family of $[[7^m, 1, 3^m]]_2$ quantum codes for each $m \ge 1$. Except for the case $m = 1$, the remaining codes are all degenerate, and their corresponding duadic codes always have minimum distance 4. This observation completes the discussion in Example 7 of [1] by computing the exact minimum distance of such codes.

We conclude this section by presenting another example and a list of quantum CSS duadic codes. The parameters of all small-length quantum codes discussed in this paper were computed using the computer algebra system Magma [3].

Example 1. One can find a splitting of $\mathbb{Z}/15\mathbb{Z}$ that is given by μ_{-1} in the form $T = (T_0 = Z(0) \cup Z(3) \cup Z(5), T_1 = Z(1), T_2 = Z(7))$. The quantum duadic code of length 15 corresponding to this splitting has parameters $[[15, 7, 3]]_2$. Moreover, there exists a splitting of $\mathbb{Z}/7\mathbb{Z}$ in the form $U = (U_0 = Z(0), U_1 = Z(1), U_2 = Z(3))$ that is given by μ_{-1}. The splitting U gives the $[[7, 1, 3]]_2$ quantum code discussed above. Extending the splittings T and U, one gets a splitting of $\mathbb{Z}/105\mathbb{Z}$ in the form

$$S = (S_0 = Z(0) \cup Z(21) \cup Z(35), S_1 = Z(1) \cup Z(7) \cup Z(9) \cup Z(11) \cup Z(15) \cup Z(25),$$
$$S_2 = Z(3) \cup Z(5) \cup Z(13) \cup Z(17) \cup Z(45) \cup Z(49)).$$

Now, Corollary 1 and Theorem 9 imply that the quantum duadic corresponding to the splitting S has parameters $[[105, 7, 9]]_2$ which is degenerate, and pure to 4. Further extending the splittings S and U, one gets a $[[7^2 \times 3 \times 5, 7, 3^3]]_2$ binary quantum code that is degenerate and pure to 4. Repeating this process one gets an infinite family of quantum codes with parameters $[[7^i \times 3 \times 5, 7, 3^{i+1}]]_2$ which is again degenerate and pure to 4.

Hence, one can generate many families of degenerate quantum codes with dimensions larger than one and arbitrarily large minimum distances using binary duadic codes.

Table 1 presents the parameters of some additional small-length quantum duadic codes with dimension greater than one. In the table, the coset leaders are those of the odd-like duadic codes and all the splitting are given by μ_{-1}. According to [13, Theorem 2.1], the weights of stabilizer elements (even-like duadic codes) are all divisible by 4. Therefore, as mentioned earlier, these codes support transversal Pauli operators, the $S = Z^{\frac{1}{2}}$ gate, and the Hadamard H gate.

It should be also mentioned that some of these codes offer more logical qubits compared to the best quantum (non-CSS) duadic codes constructed using duadic codes over $\mathbb{F}_4$ [8, Table 1]. For example, the quaternary duadic of length 45 yields a quantum code with parameters $[\![45, 9, 5]\!]_2$, whereas the binary duadic code of the same length results in a quantum code with parameters $[\![45, 13, 5]\!]_2$, providing four additional logical qubits.

Furthermore, each $[\![n, k, d]\!]_2$ quantum code listed in Table 1 with an odd minimum distance d can be used to generate a quantum code with parameters $[\![n+1, k-1, d+1]\!]_2$. This is achieved by employing the CSS construction on the extended code of odd-like duadic code and its dual.

Table 1. Small-length quantum duadic code of dimension larger than one.

Length	Coset Leaders	Parameters	Degenerate
15	1	$[\![15, 7, 3]\!]_2$	No
21	1, 3	$[\![21, 3, 5]\!]_2$	No
35	1, 5	$[\![35, 5, 6]\!]_2$	No
45	1, 3	$[\![45, 13, 5]\!]_2$	No
55	1	$[\![55, 15, 5]\!]_2$	No
85	1, 3, 7, 9	$[\![85, 21, 5]\!]_2$	No
91	1, 3, 9, 13	$[\![91, 13, 7]\!]_2$	No
93	1, 5, 7, 21, 33, 45	$[\![93, 3, 14]\!]_2$	No
95	1	$[\![95, 23, 5]\!]_2$	No
105	3, 5, 7, 11, 13, 15	$[\![105, 7, 12]\!]_2$	Yes
115	1, 5	$[\![115, 5, 14]\!]_2$	No

5 Quantum Triadic Codes

Triadic codes, introduced in [25], represent another infinite class of cyclic codes with many intriguing properties [4, 23]. In this section, we provide a brief overview of these codes before proceeding to construct a family of binary quantum codes using them.

Let n be a positive odd integer. The tuple $(X_\infty, X_0, X_1, X_2)$ is called a *3-splitting* of $\mathbb{Z}/n\mathbb{Z}$ if

- X_i is a union of 2-cyclotomic cosets modulo n for each $i = 0, 1, 2, \infty$.
- $(X_\infty, X_0, X_1, X_2)$ is a partition of $\mathbb{Z}/n\mathbb{Z}$.
- There exists a multiplier μ_b such that $\mu_b(X_0) = X_1$, $\mu_b(X_1) = X_2$, $\mu_b(X_2) = X_0$, and $\mu_b(Z(s)) = Z(s)$ for each $s \in X_\infty$.

Note that despite the duadic case, μ_b cannot be an involution, i.e., $b^2 \not\equiv 1 \pmod{n}$.

Definition 4. *Let $(X_\infty, X_0, X_1, X_2)$ be a 3-splitting of $\mathbb{Z}/n\mathbb{Z}$ that is given by μ_b. The binary linear cyclic codes C_1 and C_2 of length n with the defining sets X_i and $X_\infty \cup X_i \cup X_j$, where $i \neq j \in \{0, 1, 2\}$, are called a pair of odd-like and even-like binary triadic codes of length n.*

Some properties of triadic codes is summarized below.

Proposition 2. *Let C_1 and C_2 ($C_2 \subset C_1$) be a pair of odd-like and even-like binary triadic codes corresponding to a 3-splitting that is given by multiplier μ_b. Then*

1. *C_1 and $\mu_b(C_1)$ (similarly C_2 and $\mu_b(C_2)$) are a pair of permutation equivalent triadic codes.*
2. *$\dim(C_1) = \frac{2n + |X_\infty|}{3}$ and $\dim(C_2) = \frac{n - |X_\infty|}{3}$.*
3. *The codes $D_1 = C_2^\perp$ and $D_2 = C_1^\perp$ form a pair of odd-like and even-like triadic codes, and $d(C_1 \setminus C_2) = d(D_1 \setminus D_2)$.*

Proof. We only discuss the last part, as the proof for the other cases is straightforward. Let $(X_\infty, X_0, X_1, X_2)$ be a 3-splitting of $\mathbb{Z}/n\mathbb{Z}$ that is given by μ_b, and X_i and $X_\infty \cup X_i \cup X_j$ be the defining sets of C_1 and C_2 for some $i \neq j \in \{0, 1, 2\}$, respectively. A similar proof as in Theorem 6 shows that $\mu_{-1}(X_\infty) = X_\infty$. Thus $D_2 = C_1^\perp$ has the defining set

$$(\mathbb{Z}/n\mathbb{Z}) \setminus (-X_i) = X_\infty \cup -X_j \cup -X_k,$$

where $k \in \{0, 1, 2\} \setminus \{i, j\}$. Similarly $D_1 = C_2^\perp$ has the defining set

$$(\mathbb{Z}/n\mathbb{Z}) \setminus (X_\infty \cup -X_i \cup -X_j) = -X_k.$$

Therefore D_1 and D_2 are another pair of odd-like and even-like triadic codes (corresponding to the splitting $(X_\infty, -X_0, -X_1, -X_2)$).

For the second argument, note that we can find multiplier $b' \in \{b, b^2\}$ such that $\mu_{-b'}(X_i) = -X_k$ and $\mu_{-b'^2}(X_\infty \cup X_i \cup X_j) = X_\infty \cup -X_j \cup -X_k$. Thus the codes C_1 and D_1, respectively C_2 and D_2, are permutation equivalent and therefore have the same weight distribution. Now the facts that $C_2 \subset C_1$ and $D_2 \subset D_1$ imply that the sets $C_1 \setminus C_2$ and $D_1 \setminus D_2$ have the same weight distributions. This proves the last statement. □

For certain values of n (but not always), duadic codes of length n may not exist; however, triadic codes can be constructed. The following example illustrates this.

Example 2. Let $n = 43$. Then $\mathbb{Z}/n\mathbb{Z} = Z(0) \cup Z(1) \cup Z(3) \cup Z(7)$, where $|Z(1)| = |Z(3)| = |Z(7)|$. Moreover for any positive integer $1 \leq b \leq 42$, the multiplier μ_b either fixes all $Z(1), Z(3), Z(7)$ or fixes none of them. Therefore, there is no 2-splitting modulo 43. However, $(Z(0), Z(1), Z(3), Z(7))$ forms a 3-splitting that is given, for example, by μ_3.

In the next theorem, we discuss the structure of binary quantum triadic codes.

Theorem 10. *Let n be a positive odd integer, and $(X_\infty, X_0, X_1, X_2)$ be a 3-splitting of $\mathbb{Z}/n\mathbb{Z}$. Then there exists a binary quantum code with parameters $[\![n, \frac{n+2|X_\infty|}{3}, d]\!]_2$, where $d = d(C_1 \setminus C_2)$ and C_1 and C_2 are a pair of odd-like and even-like triadic codes, respectively.*

Proof. Let C_1 and C_2 be a pair of odd-like and even-like triadic codes, respectively. Then $C_2 \subset C_1$ and by Proposition 2 (2) we have $\dim(C_1) - \dim(C_2) = \frac{n+2|X_\infty|}{3}$. Moreover, Proposition 2 (3) implies that $d(C_1 \setminus C_2) = d(C_2^\perp \setminus C_1^\perp)$. Hence applying the quantum CSS construction to $C_2 \subset C_1$ implies the desired result. $\square$

Note that if p is a prime number such that $3 \mid p - 1$ and 2 is a cubic residue modulo p, then there exists a 3-splitting of $\mathbb{Z}/p\mathbb{Z}$ [17]. In this case, we have $X_\infty = \{0\}$, and the corresponding quantum triadic code has rate $\frac{n+2}{3n}$. The quantum triadic family of Theorem 10 has a non-zero (asymptotic) rate of $\geq \frac{1}{3}$. Let C_1 and C_2 be a pair of odd-like and even-like triadic codes of length n. Note also that when C_1 has an odd minimum distance, then $d(C_1 \setminus C_2) = d(C_1)$ (because C_2 only has even weight codewords). Thus one can find the cubic root minimum distance lower bound $d(C_1) \geq n^{\frac{1}{3}}$ using the discussion of [23, Theorem 16]. Such odd minimum distance triadic codes form a family of binary quantum triadic codes with a non-zero asymptotic rate and growing minimum distance.

Next we look at the smallest example of a quantum triadic code.

Example 3. Let $n = 31$. Then

$$(X_\infty = Z(0), X_0 = Z(1) \cup Z(15), X_1 = Z(3) \cup Z(7), X_2 = Z(5) \cup Z(15))$$

forms a splitting of $\mathbb{Z}/31\mathbb{Z}$ that is given by μ_5. Let C_1 and C_2 be a pair of binary odd-like and even-like triadic codes of length n with the defining sets X_0 and $X_\infty \cup X_0 \cup X_1$ respectively. The codes C_1 and C_2 have parameters $[31, 21, 5]_2$ and $[31, 10, 10]_2$. Hence applying the result of Theorem 10 gives a binary quantum code with parameters $[\![31, 11, 5]\!]_2$. This code is also a quantum BCH code and it is considered as a good candidate for quantum error correction [27].

Table 2. Small-length quantum triadic codes.

Length	Coset Leaders	Multiplier	Parameters
31	1, 3	μ_5	$[\![31, 11, 5]\!]_2$
43	1	μ_3	$[\![43, 15, 6]\!]_2$
93	1, 3, 9, 23	μ_5	$[\![93, 33, 7]\!]_2$
109	1	μ_3	$[\![109, 37, 10]\!]_2$
127	1, 5, 19, 27, 47, 63	μ_3	$[\![127, 43, 13]\!]_2$
129	1, 3, 19	μ_5	$[\![129, 45, 12]\!]_2$
155	1, 5, 23, 75	μ_3	$[\![155, 55, 10]\!]_2$

Table 2 presents the parameters of quantum triadic codes of small lengths. In the table, the coset leaders are those of the odd-like triadic codes.

Figure 1 also illustrates the minimum distance of quantum triadic codes relative to the code's length. The diagram indicates that all minimum distances exceed the cubic root of the code's length. We anticipate similar behavior for quantum triadic codes of greater lengths. Additionally, the codes depicted in black on the diagram exhibit a higher rate compared to the blue codes.

We have also made the Magma codes related to the splittings and constructions of cyclic, duadic, triadic, and related quantum codes available online at [6].

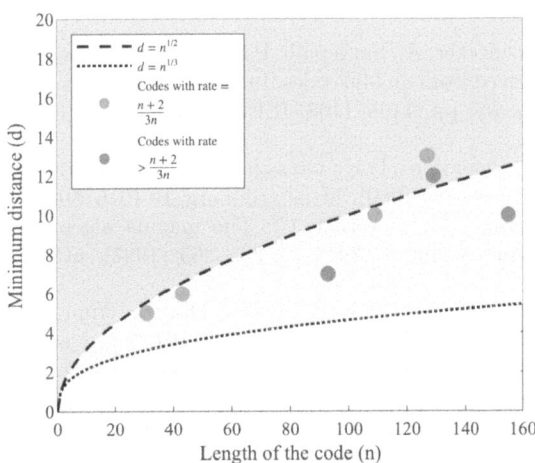

Fig. 1. Minimum distance of quantum triadic codes as a function of the code length.

6 Conclusion

To sum up, this study explored the construction of quantum CSS duadic codes with dimensions larger than one. We discussed a method for extending smaller splittings of quantum duadic codes to create larger, (degenerate) quantum duadic codes. Moreover, we introduced a technique for computing or bounding the minimum distances of duadic codes and their corresponding quantum codes constructed through the extended splittings. Additionally, we introduced quantum CSS triadic codes, a family capable of generating quantum codes with a rate of at least $\frac{1}{3}$.

Acknowledgments. The authors acknowledge support by the Spanish Ministry of Economy and Competitiveness through the MADDIE project (Grant No. PID2022-137099NB-C44), by the Spanish Ministry of Science and Innovation through the project "Few-qubit quantum hardware, algorithms and codes, on photonic and solid-state systems" (PLEC2021-008251), by the Ministry of Economic Affairs and Digital Transformation of the Spanish Government through the QUANTUM ENIA project call - Quantum Spain project, and by the European Union through the Recovery, Transformation and Resilience Plan - NextGenerationEU within the framework of the "Digital Spain 2026 Agenda".

Disclosure of Interests. The authors have no competing interests to declare that are relevant to the content of this article.

References

1. Aly, S.A., Klappenecker, A., Sarvepalli, P.K.: Remarkable degenerate quantum stabilizer codes derived from duadic codes. In: 2006 IEEE International Symposium on Information Theory, pp. 1105–1108. IEEE (2006). https://doi.org/10.1109/ISIT.2006.261955
2. Bose, R.C., Ray-Chaudhuri, D.K.: On a class of error correcting binary group codes. Inf. Control **3**(1), 68–79 (1960). https://doi.org/10.1016/S0019-9958(60)90287-4
3. Bosma, W., Cannon, J., Playoust, C.: The magma algebra system I: the user language. J. Symb. Comput. **24**(3–4), 235–265 (1997). https://doi.org/10.1006/jsco.1996.0125
4. Brualdi, R.A., Pless, V.S.: Polyadic codes. Discret. Appl. Math. **25**(1-2), 3–17 (1989). https://doi.org/10.1016/0166-218X(89)90042-5. Combinatorics and complexity (Chicago, IL, 1987)
5. Calderbank, A.R., Rains, E.M., Shor, P.W., Sloane, N.J.A.: Quantum error correction via codes over GF(4). IEEE Trans. Inform. Theory **44**(4), 1369–1387 (1998). https://doi.org/10.1109/18.681315
6. Dastbasteh, R.: Construction of cyclic and duadic codes, and related quantum CSS codes (2022). https://github.com/Dastbasteh/Cyclic-and-Duadic-codes.git
7. Dastbasteh, R.: New quantum codes, minimum distance bounds, and equivalence of codes. Ph.D. thesis, Simon Fraser University (2023). https://summit.sfu.ca/item/36338
8. Dastbasteh, R., Etxezarreta Martinez, J., Nemec, A., deMarti iOlius, A., Crespo Bofill, P.: Infinite class of quantum codes derived from duadic constacyclic codes. arXiv preprint arXiv:2312.06504 (2023)

9. Dastbasteh, R., Lisoněk, P.: New quantum codes from self-dual codes over F_4. Des. Codes Cryptogr. **92**(3), 787–801 (2024). https://doi.org/10.1007/s10623-023-01306-5
10. Ding, C.: Cyclic codes over finite fields. In: Huffman, W.C., Kim, J.L., Solé, P. (eds.) Concise Encyclopedia of Coding Theory. Chapman and Hall/CRC (2021). https://doi.org/10.1201/9781315147901
11. Gottesman, D.: Class of quantum error-correcting codes saturating the quantum Hamming bound. Phys. Rev. A **54**(3), 1862–1868 (1996). https://doi.org/10.1103/PhysRevA.54.1862
12. Grassl, M.: Algebraic quantum codes: linking quantum mechanics and discrete mathematics. Int. J. Comput. Math. Comput. Syst. Theory **6**(4), 243–259 (2021). https://doi.org/10.1080/23799927.2020.1850530
13. Grassl, M., Beth, T.: Cyclic quantum error-correcting codes and quantum shift registers. R. Soc. Lond. Proc. Ser. A Math. Phys. Eng. Sci. **456**(2003), 2689–2706 (2000). https://doi.org/10.1098/rspa.2000.0633
14. Guenda, K.: Quantum duadic and affine-invariant codes. Int. J. Quantum Inf. **7**(1), 373–384 (2009). https://doi.org/10.1142/S0219749909004979
15. Hocquenghem, A.: Codes correcteurs d'erreurs. Chiffres **2**, 147–156 (1959). https://cir.nii.ac.jp/crid/1573387450087403264
16. Huffman, W.C., Pless, V.: Fundamentals of Error-Correcting Codes. Cambridge University Press (2010). https://doi.org/10.1017/CBO9780511807077
17. Job, V.R.: M-adic residue codes. IEEE Trans. Inform. Theory **38**(2), 496–501 (1992). https://doi.org/10.1109/18.119710
18. Knill, E., Laflamme, R., Viola, L.: Theory of quantum error correction for general noise. Phys. Rev. Lett. **84**, 2525–2528 (2000). https://doi.org/10.1103/PhysRevLett.84.2525
19. Kuo, K.Y., Lai, C.Y.: Exploiting degeneracy in belief propagation decoding of quantum codes. NPJ Quantum Inf. **8**(1), 111 (2022). https://doi.org/10.1038/s41534-022-00623-2
20. Leon, J.S., Masley, J.M., Pless, V.: Duadic codes. IEEE Trans. Inform. Theory **30**(5), 709–714 (1984). https://doi.org/10.1109/TIT.1984.1056944
21. Panteleev, P., Kalachev, G.: Degenerate quantum LDPC codes with good finite length performance. Quantum **5**, 585 (2021). https://doi.org/10.22331/q-2021-11-22-585
22. Pless, V.: Q-codes. J. Combin. Theory Ser. A **43**(2), 258–276 (1986). https://doi.org/10.1016/0097-3165(86)90066-X
23. Pless, V.: Duadic codes and generalizations. In: Camion, P., Charpin, P., Harari, S. (eds.) Eurocode 1992. ICMS, vol. 339, pp. 3–15. Springer, Vienna (1993). https://doi.org/10.1007/978-3-7091-2786-5_1
24. Pless, V., Masley, J.M., Leon, J.S.: On weights in duadic codes. J. Combin. Theory Ser. A **44**(1), 6–21 (1987). https://doi.org/10.1016/0097-3165(87)90056-2
25. Pless, V., Rushanan, J.J.: Triadic codes. Linear Algebra Appl. **98**, 415–433 (1988). https://doi.org/10.1016/0024-3795(88)90174-7
26. Smid, M.: Duadic codes (corresp.). IEEE Trans. Inform. Theory **33**(3), 432–433 (1987). https://doi.org/10.1109/TIT.1987.1057300
27. Steane, A.M.: Efficient fault-tolerant quantum computing. Nature **399**(6732), 124–126 (1999). https://doi.org/10.1038/20127
28. Wilde, M.M., Hsieh, M.H., Babar, Z.: Entanglement-assisted quantum turbo codes. IEEE Trans. Inform. Theory **60**(2), 1203–1222 (2014). https://doi.org/10.1109/TIT.2013.2292052

Cryptography and Boolean Functions

Prescribing Traces of Primitive Elements in Finite Fields

Lucas Reis[✉] [iD]

Departamento de Matemática, Universidade Federal de Minas Gerais, Belo Horizonte, Minas Gerais 31270-901, Brazil
lucasreismat@gmail.com

Abstract. Let F be a finite field and let E be an n-degree extension of F. Given a family $\{F_1, \ldots, F_k\}$ of intermediate fields, we discuss the existence of primitive elements of E whose traces over the fields F_i are prescribed. In 2022, S. Ribas and the author studied this problem and, by employing a very standard approach, we provided asymptotic results under some mild restrictions on the intermediate fields. Moreover, we observed that such element can never exist if $[E : F_i] = 2$ for some $1 \leq i \leq k$ and the corresponding prescribed trace is zero. In this paper we show that, up to this genuine exception, such element exists if n is fixed and $\#F$ is large enough. In contrast to the ideas employed in our previous work, here our approach basically relies on showing that the corresponding set of elements in E with prescribed traces comprises an affine space with a generic algebraic property. The affine spaces satisfying this property were recently studied by the author, where it is shown that they present a good cancellation through multiplicative character sums (hence they contain a large number of primitive elements).

Keywords: primitive elements · field trace · normal elements

1 Introduction

Let q be a prime power and, for each positive integer n, let $\mathbb{F}_{q^n}$ be the finite field with q^n elements. It is well known that the multiplicative group $\mathbb{F}_{q^n}^* = \mathbb{F}_{q^n} \setminus \{0\}$ is cyclic and any generator of such group is a primitive element. Primitive elements play an important role in the theory of finite fields, mainly due to their applications such as the Discrete Logarithm Problem (e.g., the Diffie-Hellman key exchange [6]).

The existence of primitive elements with further specified properties has been extensively studied in the past few years: see [1,2,5,8,10]. A pioneer work in this issue is the Primitive Normal Basis Theorem, which states that for every prime power q and integer $n \geq 1$, there exists an element $\alpha \in \mathbb{F}_{q^n}$ that is primitive and normal over $\mathbb{F}_q$, i.e., the set $\{\alpha, \alpha^q, \ldots, \alpha^{q^{n-1}}\}$ is an $\mathbb{F}_q$-basis for $\mathbb{F}_{q^n}$ (regarded as an $\mathbb{F}_q$-vector space). The Primitive Normal Basis Theorem was proven by Lenstra and Schoof [9] in 1987 who needed a little help of computers, and a full computer-free proof of this result was later given by Cohen and Huczynska [4] in 2003.

In 1990, Cohen [3] explored the existence of primitive elements in finite field extensions whose trace (over the base field) is prescribed. More specifically if, for $\alpha \in \mathbb{F}_{q^n}$,

$$\mathrm{Tr}_{q^n/q}(\alpha) = \alpha + \alpha^q + \ldots + \alpha^{q^{n-1}} \in \mathbb{F}_q,$$

is the trace of α over $\mathbb{F}_q$, he showed that there exists a primitive element $\alpha \in \mathbb{F}_{q^n}$ such that $\mathrm{Tr}_{q^n/q}(\alpha) = a$ with the exception of the following genuine cases:

$$(q, n, a) = (q, 2, 0) \text{ or } (4, 3, 0).$$

In order to obtain this result he employs a character sum technique that is widely used nowadays. The approach relies on expressing the characteristic functions 1_P (of primitive elements) and 1_T (of elements with trace equals a) by means of characters of finite fields and proving that

$$\sum_{x \in \mathbb{F}_{q^n}} 1_P(x) \cdot 1_T(x) > 0.$$

In the case of this problem, there is a mix of multiplicative and additive character sums (1_P contributes with the multiplicative characters and 1_T with the additive ones), giving rise to some Gauss sums. By employing classical estimates on such Gauss sums, most of the work is over and the rest is completed by more elementary arguments. More recently, in [14] the authors explore the number of special elements (that cover the primitive elements) with prescribed trace.

Inspired by Cohen's work, in [13] the authors explore a generalization of this problem, considering the existence of primitive elements with prescribed traces over several intermediate extensions. In constrast to the case of just one trace, we cannot freely prescribe traces due to natural restrictions given by the transitivity of traces: if d divides n and t divides d, then

$$\mathrm{Tr}_{q^n/q^t}(\alpha) = \mathrm{Tr}_{q^d/q^t}(\mathrm{Tr}_{q^n/q^d}(\alpha)), \alpha \in \mathbb{F}_{q^n}.$$

For the sake of completeness, we provide one generic instance of this phenomenon. If p is the characteristic of $\mathbb{F}_q$, then one cannot have

$$\mathrm{Tr}_{q^{p(p+1)}/q^{p+1}}(\alpha) = 1 = \mathrm{Tr}_{q^{p(p+1)}/q^p}(\alpha).$$

In fact, the first equality implies $\mathrm{Tr}_{q^{p(p+1)}/q}(\alpha) = \mathrm{Tr}_{q^{p+1}/q}(1) = p + 1 = 1 \neq 0$, while the second one implies $\mathrm{Tr}_{q^{p(p+1)}/q}(\alpha) = \mathrm{Tr}_{q^p/q}(1) = p = 0$. Moreover, if one prescribes a trace over a specific extension, the traces of all its subextensions are already determined. So we can already assume that no intermediate extension is contained in another one. These observations are compiled in the following definition which is essentially taken from [13].

Definition 1. *Given integers $n, k \geq 2$, let $\Lambda_k(n)$ be the set of k-tuples $(d_1, \ldots, d_k)$, where $d_1 < d_2 < \ldots < d_k < n$ are divisors of n and no d_i divides* any d_j with $i \neq j$. Also, for $\mathbf{d} = (d_1, \ldots, d_k)$ and $\mathbf{a} = (a_1, \ldots, a_k) \in \mathbb{F}(\mathbf{d}) := \mathbb{F}_{q^{d_1}} \times \cdots \times \mathbb{F}_{q^{d_k}}$, we say that $\mathbf{a} \in \mathbb{F}(\mathbf{d})$ is $\mathbf{d}$-admissible if

$$\mathrm{Tr}_{q^{d_i}/q^{\gcd(d_i,d_j)}}(a_i) = \mathrm{Tr}_{q^{d_j}/q^{\gcd(d_i,d_j)}}(a_j),$$

for every $1 \leq i < j \leq k$.

Remark 1. According to Theorem 4.1 of [11], for $\mathbf{d} = (d_1, \ldots, d_k)$ and $\mathbf{a} = (a_1, \ldots, a_k) \in \mathbb{F}(\mathbf{d})$, there exists an element $\alpha \in \mathbb{F}_{q^n}$ such that $\mathrm{Tr}_{q^n/q^{d_i}}(\alpha) = a_i$ for every $1 \leq i \leq k$ if and only if the tuple $\mathbf{a}$ is $\mathbf{d}$-admissible. Moreover, in the affirmative case, the corresponding set of such α's has cardinality $q^{n-\lambda(\mathbf{d})}$, where

$$\lambda(\mathbf{d}) = d_1 + \ldots + d_k + \sum_{i=2}^{k} (-1)^{i+1} \sum_{1 \leq \ell_1 < \cdots < \ell_i \leq k} \gcd(d_{\ell_1}, \ldots, d_{\ell_i}).$$

Since the trace is an $\mathbb{F}_q$-linear map, it follows that the set of such α's is an $\mathbb{F}_q$-affine space of dimension $n - \lambda(\mathbf{d})$. Moreover, by Lemma 3.5 in [13], we have that $\lambda(\mathbf{d}) \leq n - \varphi(n)$, where $\varphi(n)$ is Euler's totient function at n. In particular, this affine space contains at least $q^{n-\lambda(\mathbf{d})} \geq q^{\varphi(n)} > 1$ elements. More recently, in [7] we discuss the existence of elements with prescribed traces in greater generality, covering many classes of finite Galois extensions of arbitrary fields.

In [13], the authors prove the following theorem.

Theorem 1. *Let $n, k \geq 2$ be integers, $\mathbf{d} = (d_1, \ldots, d_k) \in \Lambda_k(n)$ and let $\mathbf{a} = (a_1, \ldots, a_k) \in \mathbb{F}(\mathbf{d})$ be a $\mathbf{d}$-admissible tuple. Then there exists a primitive element $\alpha \in \mathbb{F}_{q^n}$ with prescribed traces $\mathrm{Tr}_{n/d_i}(\alpha) = a_i$ for every $1 \leq i \leq k$ provided that*

$$q^{n/2 - \lambda(\mathbf{d})} \geq W(q^n - 1), \tag{1}$$

where $W(t)$ denotes the number of squarefree divisors of t.

This theorem provides a nice range where we can prescribe traces of primitive elements but it is ineffective in some cases. For instance, if r is an odd prime and $n \equiv 0 \pmod{2r}$, $k = 2$ and $\mathbf{d} = (n/r, n/2)$, then

$$\frac{n}{2} - \lambda(\mathbf{d}) = -\frac{n}{2r},$$

and so Ineq. (1) is trivially false. Many other examples where $\lambda(\mathbf{d}) \geq n/2$ can be found in a similar way. The term $n/2$ in the inequality of Theorem 1 appears due to the Gauss sums that naturally arise in the approach employed in [13]. Similar obstructions also appears in this kind of problem when Weil's bound is employed (e.g., see [10]).

Among these obstructions in Theorem 1, there is the case of a primitive element with zero trace in quadratic extensions, which is a genuine exception and it was already noted in Cohen's work from 1990. We provide the proof here for the sake of completeness. If $\alpha \in \mathbb{F}_{Q^2}$ and

$$\mathrm{Tr}_{Q^2/Q}(\alpha) = \alpha + \alpha^Q = 0,$$

then either $\alpha = 0$ (which is not a primitive element) or $\alpha^{Q-1} = -1$. In the last case, we have $\alpha^{2(Q-1)} = 1$ and, in particular, α generates a group of order at most $2(Q-1)$. As $2(Q-1) < Q^2 - 1$, such an element cannot be a generator of $\mathbb{F}_{Q^2}^*$. Therefore, in the notation of Definition 1, no primitive element can appear for pairs $(\mathbf{d}, \mathbf{a})$ with $\mathbf{d} = (d_1, \ldots, d_k)$ and $\mathbf{a} = (a_1, \ldots, a_k)$ satisfying $(d_k, a_k) = (n/2, 0)$.

The main result of this paper entails an asymptotic general existence theorem on primitive elements with prescribed traces, providing a positive answer to the problem with the sole genuine exception of cases where there is a quadratic extension with zero prescribed trace. The theorem can be stated as follows.

Theorem 2. *Fix $n > 1$ an integer and $\mathbf{d} = (d_1, \ldots, d_k) \in \Lambda_k(n)$. Then there exists $c = c(n)$ with the following property: for every prime power $q > c$ and every $\mathbf{d}$-admissible tuple $\mathbf{a} = (a_1, \ldots, a_k) \in \mathbb{F}(\mathbf{d})$, there exists a primitive element $\alpha \in \mathbb{F}_{q^n}$ with prescribed traces $\mathrm{Tr}_{n/d_i}(\alpha) = a_i$ for every $1 \le i \le k$, with the sole genuine exception of the cases where n is even and the pair $(\mathbf{d}, \mathbf{a})$ satisfies $(d_k, a_k) = (n/2, 0)$.*

We end this section commenting on the proof of Theorem 2. In [12] we show that, if n is fixed and q is large enough, the $\mathbb{F}_q$-affine subspaces of $\mathbb{F}_{q^n}$ satisfying a generic algebraic condition contains a large number of primitive elements. This result can cover affine spaces of dimension smaller than $n/2$, which are usually not covered in these kind of problems due to estimates for Gauss sums and Weil's bound. In fact, a great improvement on the results of [10] (which essentially employs Weil's bound) was given in [12] as an application. For the proof of Theorem 2, most of our work concentrates on proving that the $\mathbb{F}_q$-affine space comprising the elements with prescribed traces as in Theorem 2 satisfies the algebraic conditions given in [12]. The latter is obtained by using the $\mathbb{F}_q[x]$-module structure of finite fields through normal bases, along with some basic number theoretical tools.

2 Preliminaries

Throughout this section, q is a fixed prime power, $n > 1$ is a positive integer and we write $\mathcal{A} = u + \mathcal{V} \subseteq \mathbb{F}_{q^n}$ for an $\mathbb{F}_q$-affine space, meaning that $\mathcal{V}$ is its underlying $\mathbb{F}_q$-vector space. We start with the following definition.

Definition 2. *An $\mathbb{F}_q$-affine space $\mathcal{A} = u + \mathcal{V} \subseteq \mathbb{F}_{q^n}$ has degree n if there exist $y \in \mathcal{A}$ and $z \in \mathcal{V}$ such that yz^{-1} has degree n over $\mathbb{F}_q$, i.e., $\mathbb{F}_q(yz^{-1}) = \mathbb{F}_{q^n}$.*

From Proposition 2.2 in [12], we have the following result.

Proposition 1. *Fix q a prime power and $1 < n < q$. If the $\mathbb{F}_q$-affine space $\mathcal{A} \subseteq \mathbb{F}_{q^n}$ of dimension $t \ge 1$ is not of degree n, then there exists a divisor $d < n$ of n and an element $\delta \in \mathcal{A}$ such that $\mathcal{A} \subseteq \delta \cdot \mathbb{F}_{q^d}$.*

From Proposition 4.1 in [12], we have the following existence result on primitive elements in affine spaces of degree n.

Theorem 3. *Let n be a positive integer. Then there exists a constant $c = c(n)$ such that for every $q > c$, the following holds: if $\mathcal{A} \subseteq \mathbb{F}_{q^n}$ is an $\mathbb{F}_q$-affine space of degree n, then there exists $\alpha \in \mathcal{A}$ that is a primitive element of $\mathbb{F}_{q^n}$.*

2.1 The $\mathbb{F}_q[x]$-Module Structure of $\mathbb{F}_{q^n}$

For a polynomial $f \in \mathbb{F}_q[x]$ with $f(x) = \sum_{i=0}^{m} a_i x^i$ and an element $\alpha \in \mathbb{F}_{q^n}$, set

$$f \circ \alpha = \sum_{i=0}^{m} a_i \alpha^{q^i} \in \mathbb{F}_{q^n}.$$

This defines an $\mathbb{F}_q[x]$-module structure on $\mathbb{F}_{q^n}$. In other words, for $f, g \in \mathbb{F}_q[x]$ and $\alpha, \beta \in \mathbb{F}_{q^n}$, the following hold:

(i) $f \circ (\alpha + \beta) = f \circ \alpha + f \circ \beta$
(ii) $(f + g) \circ \alpha = f \circ \alpha + g \circ \alpha$
(iii) $f \circ (g \circ \alpha) = fg \circ \alpha$
(iv) $1 \circ \alpha = \alpha$.

The proof of these identities are simple so we omit details. It is worth mentioning that this $\mathbb{F}_q[x]$-module structure naturally arises from the action of the cyclic infinite group $\mathrm{Gal}(\overline{\mathbb{F}}_q/\mathbb{F}_q)$ (generated by the Frobenius automorphism $\sigma : \alpha \mapsto \alpha^q$) on the subfield $\mathbb{F}_{q^n}$. From the identities (ii) and (iii) we conclude that, for each $\alpha \in \mathbb{F}_{q^n}$, the set $\mathcal{I}_\alpha = \{h \in \mathbb{F}_q[x] : h \circ \alpha = 0\}$ is an ideal of $\mathbb{F}_q[x]$. Since $(x^n - 1) \circ \alpha = \alpha^{q^n} - \alpha = 0$, we have $x^n - 1 \in \mathcal{I}_\alpha$ and then $\mathcal{I}_\alpha$ is nonzero. Since $\mathbb{F}_q[x]$ is a PID and $\mathbb{F}_q$ is a field, it follows that $\mathcal{I}_\alpha$ is generated by a polynomial $m_{\alpha,q}(x)$, which we can suppose to be monic.

Definition 3. *The polynomial $m_{\alpha,q}(x)$ is called the $\mathbb{F}_q$-order of $\alpha \in \mathbb{F}_{q^n}$.*

For instance, $m_{0,q}(x) = 1$ and, if $a \in \mathbb{F}_q^*$, then $m_{a,q}(x) = x - 1$. We have seen that $x^n - 1 \in \mathcal{I}_\alpha$ for every $\alpha \in \mathbb{F}_{q^n}$ and, in particular, we conclude that $m_{\alpha,q}(x)$ is always a divisor of $x^n - 1$. It is well known that there exists normal elements in every finite field extension. The following lemma compiles some useful basic results in the context of our discussion. For its proof, see Corollary 2.5 and Lemmas 2.3 and 2.6 in [11].

Lemma 1. *Let $\beta \in \mathbb{F}_{q^n}$ be normal over $\mathbb{F}_q$ and let $\alpha \in \mathbb{F}_{q^n}$. Then the following hold*

(i) $m_{\beta,q}(x) = x^n - 1$;
(ii) there exists a unique polynomial $h \in \mathbb{F}_q[x]$ of degree at most $n - 1$ such that $\alpha = h \circ \beta$;
(iii) If $g \in \mathbb{F}_q[x]$ and $\gamma = g \circ \alpha$, then $m_{\gamma,q}(x) = \frac{m_{\alpha,q}(x)}{\gcd(m_{\alpha,q}(x), g(x))}$.

In particular, normal elements are the ones of maximal $\mathbb{F}_q$-order and they work as generators of finite fields through this $\mathbb{F}_q[x]$-module structure. We end this section with an interpretation of the trace functions through the $\mathbb{F}_q[x]$-module structure of finite fields. If d is a divisor of n, the polynomial $f_{n,d}(x) = \frac{x^n-1}{x^d-1}$ satisfies $f_{n,d} \circ \alpha = \mathrm{Tr}_{q^n/q^d}(\alpha)$ for every $\alpha \in \mathbb{F}_{q^n}$. In particular, we directly obtain the following result.

Lemma 2. *If $\alpha \in \mathbb{F}_{q^n}$ and d is a divisor of n, then $\operatorname{Tr}_{q^n/q^d}(\alpha) = 0$ if and only if $m_{\alpha,q}(x)$ divides $\frac{x^n-1}{x^d-1}$.*

3 Proof of Theorem 2

Let $n, k \geq 2$ be integers, $\mathbf{d} = (d_1, \ldots, d_k) \in \Lambda_k(n)$ and let $\mathbf{a} = (a_1, \ldots, a_k) \in \mathbb{F}(\mathbf{d})$ be a $\mathbf{d}$-admissible tuple. Moreover, let $\mathcal{A} = \theta + \mathcal{V}$ be the corresponding $\mathbb{F}_q$-affine space comprising the elements $\alpha \in \mathbb{F}_{q^n}$ such that $\operatorname{Tr}_{q^n/q^{d_i}}(\alpha) = a_i$ for every $1 \leq i \leq k$ (whose existence is ensured by Remark 1). Since the trace is linear, it follows that

$$\mathcal{V} = \{y \in \mathbb{F}_{q^n} : \operatorname{Tr}_{q^n/q^{d_i}}(y) = 0, 1 \leq i \leq k\}. \tag{2}$$

Let $c = c(n)$ be as in Theorem 3, and take $q > \max\{c, n\}$. From the same theorem, it suffices to prove that $\mathcal{A}$ has degree n. We prove the latter by contradiction. Assume that $\mathcal{A}$ does not have degree n. Since $q > n$, Proposition 1 entails that there exists a divisor $d < n$ of n and an element $\delta \in \mathcal{A}$ such that $\mathcal{A} \subseteq \delta \cdot \mathbb{F}_{q^d}$. In particular, $\mathcal{V} = \{\alpha_1 - \alpha_2 : \alpha_1, \alpha_2 \in \mathcal{A}\} \subseteq \delta \cdot \mathbb{F}_{q^d}$. From Eq. (2) and the fact that $\operatorname{Tr}_{q^n/q^t}(z^q) = \operatorname{Tr}_{q^n/q^t}(z)^q$ for every $z \in \mathbb{F}_{q^n}$ and every divisor t of n, we see that $v^q \in \mathcal{V}$ for every $v \in \mathcal{V}$. In particular, $\mathcal{V} \subseteq \delta \cdot \mathbb{F}_{q^d} \cap \delta^q \cdot \mathbb{F}_{q^d}$. From Remark 1, we have that $\#\mathcal{V} = \#\mathcal{A} > 1$ and so $\delta \cdot \mathbb{F}_{q^d} \cap \delta^q \cdot \mathbb{F}_{q^d}$ has at least 2 elements. Therefore, there exist nonzero $v_1, v_2 \in \mathbb{F}_{q^d}$ such that $\delta v_1 = \delta^q v_2$, hence $\delta^{q-1} = v_1/v_2 \in \mathbb{F}_{q^d}$ and so

$$a := \delta^{q^d-1} \in \mathbb{F}_q.$$

In particular, the roots of $x^{q^d} - \delta^{q^d-1}x = x^{q^d} - ax$ are precisely the elements of $\delta \cdot \mathbb{F}_{q^d}$. Pick $\beta \in \mathbb{F}_{q^n}$ a normal element and set $F(x) = \operatorname{lcm}_{1 \leq i \leq k}\{x^{d_i} - 1\}$, $\Delta = F \circ \beta$. From Lemma 1, we have that

$$m_{\Delta,q}(x) = \frac{m_{\beta,q}(x)}{\gcd(m_{\beta,q}(x), F(x))} = \frac{x^n-1}{\gcd(x^n-1, F(x))} = \frac{x^n-1}{F(x)}.$$

Therefore, $m_{\Delta,q}(x)$ divides $\frac{x^n-1}{x^{d_i}-1}$ for every $1 \leq i \leq k$. From Lemma 2, we have that $\operatorname{Tr}_{q^n/q^{d_i}}(\Delta) = 0$ for every $1 \leq i \leq k$ and so $\Delta \in \mathcal{V}$. Since $\mathcal{V} \subseteq \delta \cdot \mathbb{F}_{q^d}$, we conclude that $\Delta^{q^d} - a\Delta = 0$. Since $a \in \mathbb{F}_q$, the latter can be rewritten as $(x^d - a) \circ \Delta = 0$. It follows by the definition that $m_{\Delta,q}(x) = \frac{x^n-1}{F(x)}$ divides $x^d - a$. In other words, $x^n - 1$ divides

$$(x^d - a)F(x) = (x^d - a) \cdot \operatorname{lcm}_{1 \leq i \leq k}\{(x^{d_i} - 1)\}.$$

We claim that $a = -1$, n is even and $d = \frac{n}{2}$. We split the proof of the claim into cases:

(a) n is not divisible by the characteristic p of $\mathbb{F}_q$. In this case, there exist primitive n-th roots of unity in some finite extension of $\mathbb{F}_q$ and any such element is a root of $x^n - 1$. Let ω be a primitive n-th root of unity. Since $d_i < n$ for every $1 \leq i \leq k$, the element ω is not a root of any $x^{d_i} - 1$. In particular, every primitive n-th root of unity is a root of the polynomial $x^d - a$. Therefore, ω and ω^{-1} are roots of $x^d - a$ and so $\omega^d = \omega^{-d} = a$. The latter implies that $a = a^{-1}$, i.e., $a = \pm 1$. However, as $d < n$, $a = \omega^d \neq 1$ and so $a = -1$. Hence $\omega^d = -1$ and then $\omega^{2d} = 1$, that is, n divides $2d$. As $d < n$ divides n, the latter implies that n is even and $d = \frac{n}{2}$. In particular, this case can never occur if the characteristic of $\mathbb{F}_q$ is 2.

(b) n is divisible by the characteristic p of $\mathbb{F}_q$. This case is slightly different to the previous one since we need to consider multiplicities of roots. Write $n = p^t \cdot u$, where $\gcd(u, p) = 1$ and $t \geq 1$. In particular, every primitive u-th root of unity is a root of $x^n - 1$ with multiplicity p^t. Therefore, such elements are also roots of $(x^d - a) \cdot \text{lcm}_{1 \leq i \leq k}\{(x^{d_i} - 1)\}$ with multiplicity at least p^t (as this polynomial is divisible by $x^n - 1$). Since $d_i < n$ is a divisor of n, either $\text{lcm}_{1 \leq i \leq k}\{x^{d_i} - 1\}$ does not vanish at any primitive u-th root of unity or it vanishes at all of them but with multiplicity at most p^{t-1} and, in the latter case, such polynomial divides $x^{\frac{n}{p}} - 1$. Therefore, we see that one of the following must occur:
- The polynomial $\text{lcm}_{1 \leq i \leq k}\{x^{d_i} - 1\}$ divides $x^{\frac{n}{p}} - 1$.
- Every primitive u-th root of unity is a root of $x^d - a$ with multiplicity at least p^t.

In the first case, we conclude that $x^d - a$ is divisible by $\frac{x^n - 1}{x^{n/p} - 1}$ and then $d \geq n - \frac{n}{p}$. Since $d < n$ divides n, we obtain $d \leq \frac{n}{2}$. Hence $p = 2$, n is even and $x^{\frac{n}{2}} + 1 = \frac{x^n - 1}{x^{n/2} - 1}$ divides $x^d - a$. As $d \leq \frac{n}{2}$, we conclude that $x^{\frac{n}{2}} + 1 = x^d - a$, that is, $d = \frac{n}{2}$ and $a = -1$.

In the second case, we conclude that $d = p^s \cdot d_0$ with $\gcd(d_0, p) = 1$ and $s \geq t$. Since d divides $n = p^t \cdot u$, it follows that d_0 divides u and $s \leq t$, hence $s = t$. Following the proof of case (a), we conclude that $a = -1$, u is even and $d_0 = \frac{u}{2}$. Hence $n = p^t u$ is even and $d = p^s d_0 = p^t \cdot \frac{u}{2} = \frac{n}{2}$.

All in all, we proved that n is even and $\delta \cdot \mathbb{F}_{q^d}$ equals the set $\mathcal{T}$ of elements $\alpha \in \mathbb{F}_{q^n}$ such that $\alpha^{q^{n/2}} + \alpha = 0$, i.e., $\text{Tr}_{q^n / q^{n/2}}(\alpha) = 0$. Recall that $\mathcal{A} \subseteq \delta \cdot \mathbb{F}_{q^d} = \mathcal{T}$. We now show that the inclusion $\mathcal{A} \subseteq \mathcal{T}$ is a contradiction with the following hypothesis in our main theorem: the pair $(\mathbf{d}, \mathbf{a})$ does not satisfy $(d_k, a_k) = (n/2, 0)$. In fact, as $d_1 < \cdots < d_k$ are proper divisors of n, then either $d_k = n/2$ or $d_i < n/2$ for every $1 \leq i \leq k$. If $d_k = n/2$, the hypothesis in our theorem implies that $a_k \neq 0$ and so every element $\alpha \in \mathcal{A}$ satisfies $\text{Tr}_{q^n/q^{n/2}}(\alpha) = a_k \neq 0$, a contradiction with $\mathcal{A} \subseteq \mathcal{T}$. Hence $d_k \neq n/2$ and so $d_i < n/2$ for every $1 \leq i < k$. However, recall that $x^n - 1$ must divide $(x^{n/2} + 1) \cdot \text{lcm}_{1 \leq i \leq k}\{(x^{d_i} - 1)\}$, and then $x^{n/2} - 1 = \frac{x^n - 1}{x^{n/2} + 1}$ must divide $\text{lcm}_{1 \leq i \leq k}\{x^{d_i} - 1\}$. By employing similar ideas to the ones used in items (a) and (b), the latter easily implies that $n/2$ divides one of the d_i's. But this is a contradiction with $d_i < n/2$ for every $1 \leq i \leq k$. The proof is complete.

Acknowledgments. This work was funded by CNPq (309844/2021-5), FAPEMIG (APQ-01712-23) and PRPq - UFMG (03/2024).

Disclosure of Interests. The author has no competing interests to declare that are relevant to the content of this article.

References

1. Booker, A.R., Cohen, S.D., Sutherland, N., Trudgian, T.: Primitive values of quadratic polynomials in a finite field. Math. Comput. **88**, 1903–1912 (2018)
2. Carvalho, C., Guardieiro, J.P., Neumann, V.G., Tizziotti, G.: On special pairs of primitive elements over a finite field. Finite Fields Their Appl. **73**, 101839 (2021)
3. Cohen, S.D.: Primitive elements and polynomials with arbitrary trace. Discret. Math. **83**, 1–7 (1990)
4. Cohen, S.D., Huczynska, S.: The primitive normal basis theorem - without a computer. J. Lond. Math. Soc. **67**, 41–56 (2003)
5. Cohen, S.D., Oliveira e Silva, T., Trudgian, T.: On consecutive primitive elements in a finite field. Bull. Lond. Math. Soc. **47**, 418–426 (2015)
6. Diffie, W., Hellman, M.: New directions in cryptography. IEEE Trans. Inf. Theory **22**, 644–654 (1976)
7. Greither, C., Reis, L.: On linearly Chinese field extensions. Commun. Algebra **49**, 1884–1894 (2020)
8. Kapetanakis, G., Reis, L.: Variations of the primitive normal basis theorem. De. Codes Cryptogr. **87**, 1459–1480 (2018)
9. Lenstra, H.W., Schoof, R.J.: Primitive normal bases for finite fields. Math. Comput. **48**, 217–217 (1987)
10. Reis, L.: Existence results on k-normal elements over finite fields. Revista matemática iberoamericana **35**, 805–822 (2019)
11. Reis, L.: Counting solutions of special linear equations over finite fields. Finite Fields Their Appl. **68**, 101759 (2020)
12. Reis, L.: Character sums over affine spaces and applications. Finite Fields Their Appl. **83**, 102067 (2022)
13. Reis, L., Ribas, S.: Generators of finite fields with prescribed traces. J. Aust. Math. Soc. **112**, 355–366 (2021)
14. Tuxanidy, A., Wang, Q.: On the number of n-free elements with prescribed trace. J. Number Theory **160**, 536–565 (2016)

On Cryptographic Properties of a Class of Power Permutations in Odd Characteristic

Mohit Pal[✉]

Department of Informatics, University of Bergen, PB 7803, 5020 Bergen, Norway
mohit.pal@uib.no

Abstract. Recently, interest in bijective functions over finite fields of odd characteristic with good cryptographic properties increased as many cryptographic primitives have been proposed in the literature which operate on prime field $\mathbb{F}_p$ for some large prime p. Here, we consider the boomerang uniformity and the algebraic degree of a class of differentially 4-uniform power permutations over finite fields of odd characteristic. We also determine the compositional inverse of this class of power permutations and compute the algebraic degree of its compositional inverse.

Keywords: Finite fields · Differential uniformity · Boomerang uniformity

1 Introduction

Permutation polynomials over finite fields are fundamental objects due to their applications in cryptography, coding theory, combinatorial design, and related topics. For example, in symmetric key cryptography, permutation polynomials have been used frequently in the design of block ciphers. More precisely, permutation polynomials are used in the substitution box which is the only non-linear part of a block cipher. In order to be a good candidate for the substitution box of a block cipher, permutation polynomials should possess good cryptographic properties to resist various kinds of cryptographic attacks. For instance, a low differential uniformity [17] to resist the differential attacks [1], a low boomerang uniformity [3,5] to thwart a generalization of differential attacks called boomerang attacks [19], and a high algebraic degree to resist the higher-order differential attacks [13], among others. In this paper, we shall consider the compositional inverse, boomerang uniformity and algebraic degree of a class of differentially 4-uniform power permutations over finite fields of odd characteristic.

Let $\mathbb{F}_q$ be the finite field with $q = p^n$ elements, where p is an odd prime and n is a positive integer. We denote by $\mathbb{F}_q^*$ the multiplicative cyclic group of nonzero elements of $\mathbb{F}_q$ and by $\mathbb{F}_q[X]$ the ring of polynomials in indeterminate X and coefficients in $\mathbb{F}_q$. It is well-known, due to Lagranges' interpolation formula, that any function $f : \mathbb{F}_q \to \mathbb{F}_q$ can be uniquely expressed by a polynomial $f(X) \in \mathbb{F}_q[X]$ of degree $\leq q - 1$. Therefore, in what follows, we shall use the term function and polynomial for f, interchangeably. In addition to the degree of this univariate polynomial there is another notion of degree associated to it called the algebraic degree of f, denoted by $deg(f)$. For any power map $f(X) = X^d$, let $d = \sum_{i=0}^{n-1} c_i p^i$ be the base p representation of the exponent d. Then the algebraic degree of f is given by the sum $\sum_{i=0}^{n-1} c_i$. For an arbitrary f, its algebraic degree is the maximum of the algebraic degrees of its monomials. A polynomial $f(X) \in \mathbb{F}_q[X]$ is called a permutation polynomial (PP) if the induced mapping $c \mapsto f(c)$ permutes the elements of $\mathbb{F}_q$.

For any function $f : \mathbb{F}_q \to \mathbb{F}_q$ and $a, b \in \mathbb{F}_q$, the difference distribution table (DDT) entry at point (a, b), denoted by $\Delta_f(a, b)$, is defined as $\Delta_f(a, b) = |\{X \in \mathbb{F}_q \mid f(X + a) - f(X) = b\}|$. The differential uniformity of the function f, denoted by Δ_f, is then defined as the maximum of $\Delta_f(a, b)$, where $a \in \mathbb{F}_q^*$ and $b \in \mathbb{F}_q$. When $\Delta_f = 1$ then f is called perfect nonlinear (also known as planar) function and when $\Delta_f = 2$ then f is called almost perfect nonlinear (APN) function. In the particular case when $f(X) = X^d$ for some positive integer d then it is easy to see that $\Delta_f(a, b) = \Delta_f\left(1, \frac{b}{a^d}\right)$, for all $a \in \mathbb{F}_q^*$ and $b \in \mathbb{F}_q$. Therefore, for determining differential properties of a power map f, it is sufficient to consider DDT entries with $a = 1$. In 2010, Blondeau et al. [2] introduced another notion associated to the differential properties of power maps called the differential spectrum. For any power map $f(X) = X^d$ and $0 \le i \le \Delta_f$, let $\omega_i = |\{b \in \mathbb{F}_q \mid \Delta_f(1, b) = i\}|$ then the differential spectrum of f, denoted by DS_f, is defined as $DS_f = \{\omega_i > 0 \mid 0 \le i \le \Delta_f\}$.

Here, we consider the class of power maps $f(X) = X^{\frac{q+3}{2}} \in \mathbb{F}_q[X]$. In [9], Helleseth and Sandberg considered the differential uniformity of this class of power maps and showed that its differential uniformity Δ_f is given by

$$\Delta_f \le \begin{cases} 1 & \text{if } p = 3 \text{ and } n \text{ is even,} \\ 3 & \text{if } p \ne 3 \text{ and } p^n \equiv 1 \pmod 4, \\ 4 & \text{otherwise.} \end{cases}$$

The differential spectrum of this power map in all the above mentioned cases is known. When $p = 3$ and n is even then this function is linear equivalent to the planar function $X^{\frac{3^{n-1}+1}{2}}$ over $\mathbb{F}_{3^n}$ and hence its differential spectrum is trivial. The differential spectrum in the case when $p = 3$ and n is odd is determined in [12]. In the case when $p > 3$ and $p^n \equiv 1 \pmod 4$, the differential spectrum is given in [11]. Very recently, the differential spectrum of this power map in the remaining case $q \equiv 3 \pmod 4$ has been determined in [21]. It is easy to see that when $q \equiv 3 \pmod 4$ then f is a permutation.

The boomerang attack [19], introduced by Wagner in 1999, is a cryptanalysis technique against block ciphers which may be thought of as an extension of the differential attack [1]. The boomerang attack exploits the dependency between two differentials, one for the upper part of the cipher and one for the lower part. Analogous to the DDT entries concerning differential attack, Cid et al. [5] introduced the notion of boomerang connectivity table (BCT), that permits to simplify this analysis. In [5], the BCT entries were defined for permutation functions in even characteristic and the knowledge of the inverse of the permutation was required to compute the BCT entries. In 2019, Li et al. [14] gave an equivalent technique to compute the BCT, which does not require the compositional inverse of the permutation polynomial $f(X)$ at all. For any $a, b \in \mathbb{F}_q$, the BCT entry of the function f at point (a, b), denoted by $\mathcal{B}_f(a, b)$, is the number of solutions $(X, Y) \in \mathbb{F}_q \times \mathbb{F}_q$ of the following system of equations

$$\begin{cases} f(X) - f(Y) = b, \\ f(X + a) - f(Y + a) = b. \end{cases}$$

The boomerang uniformity [3] of the function f, denoted by $\mathcal{B}_f$, is then defined as the maximum of $\mathcal{B}_f(a, b)$, where $a, b \in \mathbb{F}_q^*$. In the particular case when $f(X) = X^d$ for some positive integer d then $\mathcal{B}_f(a, b) = \mathcal{B}_f\left(1, \frac{b}{a^d}\right)$. Thus, for the power maps it is sufficient to consider BCT entries with $a = 1$. The boomerang uniformity of power maps over finite fields of even characteristic has been studied a lot in recent years (see [4,6–8,14,16,22,24] and references therein) whereas there are very few results in the case of odd characteristic (see Table 1). Also, one may note that among all the power maps in Table 1, there are only two permutations, namely, the inverse map X^{p^n-2} and the particular case $p = 3$ of the class of power permutations $X^{\frac{p^n+3}{2}}$ over $\mathbb{F}_{p^n}$ with $p^n \equiv 3 \pmod 4$. In this paper, we shall consider the boomerang uniformity of the power permutation $X^{\frac{p^n+3}{2}}$, where $p^n \equiv 3 \pmod 4$ for all $p > 3$ and show that its boomerang uniformity is ≤ 23. Moreover, we also obtain the compositional inverse of this power permutation. We also determine the algebraic degrees of this permutation and its compositional inverse.

The paper is organised in the following way. Section 2 will be devoted to the boomerang uniformity of the power permutation $f(X) = X^{\frac{q+3}{2}}$, $q \equiv 3 \pmod 4$. We first give a list of all the power maps with known boomerang uniformity over finite fields of odd characteristic in Table 1 and then we recall the proof for the differential uniformity of f as we need some ingredients of this proof in determining an upper bound for the boomerang uniformity of f. In Sect. 3, we first determine the compositional inverse, denoted by f^{-1}, of the power permutation f. Moreover, we compute algebraic degrees of f and f^{-1}. Finally, we summarize the paper with some future directions in Sect. 4.

2 Boomerang Uniformity of f

In this section, we first give a brief survey of the power maps with known boomerang uniformity over finite fields of odd characteristic. In [18], Stănică gave a complete description for the BCT (and its generalization known as c-BCT) entries of a power map X^d over $\mathbb{F}_q$ in terms of double Weil sums. Since explicit evaluation of Weil sums of arbitrary functions over finite fields is a quite difficult problem, alternate techniques were needed for the computation of the boomerang uniformity. To the best of our knowledge, the list of monomials over finite fields of odd characteristic with known boomerang uniformity is listed in Table 1. For a recent progress on boomerang uniformity of functions over finite fields, the reader may refer to a nice survey by Mesnager et al. [15].

Table 1. Monomials X^d over $\mathbb{F}_{p^n}$ with known boomerang uniformity.

	p	d	Condition	$\mathcal{B}_f$	Is PP?	References
C_1	3	$\frac{p^n+3}{2}$	n odd	3	Yes	[12]
C_2	$p > 2$	$p^m - 1$	$n = 2m, p \not\equiv 2 \pmod{3}$	2	No	[20]
C_3	$p > 2$	$\frac{(p^m+3)(p^m-1)}{2}$	$n = 2m$	2	No	[20]
C_4	$p > 2$	$\frac{p^n-3}{2}$	$p^n \equiv 3 \pmod{4}$	≤ 6	No	[23]
C_5	$p > 2$	$p^n - 2$	any n	≤ 5	Yes	[12]
C_6	$p > 2$	$k(p^m - 1)$	$n = 2m,\ \gcd(k, p^m + 1) = 1$	2	No	[10]
C_7	$p > 3$	$\frac{p^n+3}{2}$	$p^n \equiv 3 \pmod{4}$	≤ 23	Yes	This paper

In the past few years, many cryptographic primitives such as homomorphic encryption (HE), multi-party computation (MPC) and zero-knowledge (ZK) protocols have been proposed in the literature which operates on prime field $\mathbb{F}_p$ for some large prime p. Therefore, finding invertible functions with good cryptographic properties over $\mathbb{F}_{p^n}$ for large primes p is an intriguing problem. It is easy to observe, from the Table 1, that the inverse function X^{p^n-2} is the only invertible power map with arbitrary p whose boomerang uniformity is known. It is worth mentioning here that the low differential uniformity of a function does not imply a low boomerang uniformity. For instance, over binary fields, Calderini and Villa [4] showed that the boomerang uniformity of the differentially 4-uniform Bracken-Leander cubic function is ≤ 24. In this paper, we consider the boomerang uniformity of the class of power permutations C_7 for $p > 3$. The differential uniformity [9] and the differential spectrum [21] of this power map is known. We are including the proof of the differential uniformity here as it will help us in simplyfying certain cases in the computation of the boomerang uniformity of this power map. The following proposition gives the differential uniformity of the power permutation $f(X) = X^{\frac{q+3}{2}}$, where $q \equiv 3 \pmod{4}$.

Proposition 1. *Let $q \equiv 3 \pmod{4}$ then the differential uniformity of the power map $f(X) = X^{\frac{q+3}{2}} \in \mathbb{F}_q[X]$ is 4.*

Proof. Recall that the DDT entry of the function f at b, denoted by $\Delta_f(1, b)$, is given by the number of solutions $X \in \mathbb{F}_q$ of the equation $D_f(X) = b$, where

$$D_f(X) = \left(X + \frac{1}{2}\right)^{\frac{q+3}{2}} - \left(X - \frac{1}{2}\right)^{\frac{q+3}{2}}$$
$$= \left(X + \frac{1}{2}\right)^2 \chi\left(X + \frac{1}{2}\right) - \left(X - \frac{1}{2}\right)^2 \chi\left(X - \frac{1}{2}\right)$$
$$= \left(X^2 + \frac{1}{4}\right)\left(\chi\left(X + \frac{1}{2}\right) - \chi\left(X - \frac{1}{2}\right)\right) + X\left(\chi\left(X + \frac{1}{2}\right) + \chi\left(X - \frac{1}{2}\right)\right), \quad (1)$$

and $\chi: \mathbb{F}_q \to \{0, 1, -1\}$ is the quadratic character of the finite field $\mathbb{F}_q$ defined as follows:

$$\chi(a) = \begin{cases} 0 & \text{if } a = 0; \\ 1 & \text{if } X^2 = a \text{ has a solution } X \in \mathbb{F}_q^*; \\ -1 & \text{if } X^2 = a \text{ has no solution } X \in \mathbb{F}_q^*. \end{cases}$$

It is straightforward to see that if $X \in \left\{\frac{1}{2}, -\frac{1}{2}\right\}$ then $b = 1$. Thus, when $X \notin \left\{\frac{1}{2}, -\frac{1}{2}\right\}$ then $\chi\left(X + \frac{1}{2}\right), \chi\left(X + \frac{1}{2}\right) \in \{1, -1\}$ and we shall consider the following four cases:

Case 1. Let $\chi\left(X + \frac{1}{2}\right) = 1 = \chi\left(X - \frac{1}{2}\right)$. In this case we have $X = \frac{b}{2}$. Notice that this solution will also be a solution of Eq. (1) if and only if

$$\chi\left(\frac{b+1}{2}\right) = 1 = \chi\left(\frac{b-1}{2}\right).$$

Case 2. Let $\chi\left(X + \frac{1}{2}\right) = -1 = \chi\left(X - \frac{1}{2}\right)$. In this case we have $X = -\frac{b}{2}$. Again, this solution will also be a solution of Eq. (1) if and only if

$$\chi\left(\frac{-b+1}{2}\right) = -1 = \chi\left(\frac{-b-1}{2}\right) \iff \chi\left(\frac{b-1}{2}\right) = 1 = \chi\left(\frac{b+1}{2}\right).$$

Case 3. Let $\chi\left(X + \frac{1}{2}\right) = 1$ and $\chi\left(X - \frac{1}{2}\right) = -1$. In this case we have $X^2 = \frac{2b-1}{4}$. It is easy to observe that we have two solutions of Eq. (1) from this subcase if and only if

$$\chi(2b-1) = 1, \quad \chi\left(\frac{\sqrt{2b-1}+1}{2}\right) = 1 \text{ and } \chi\left(\frac{\sqrt{2b-1}-1}{2}\right) = -1.$$

The above conditions are equivalent to the following conditions

$$\chi(2b-1) = 1, \quad \chi\left(\frac{\sqrt{2b-1}+1}{2}\right) = 1 \text{ and } \chi\left(\frac{b-1}{2}\right) = -1. \tag{2}$$

Case 4. Let $\chi\left(X + \frac{1}{2}\right) = -1$ and $\chi\left(X - \frac{1}{2}\right) = 1$. In this case we have $X^2 = \frac{-2b-1}{4}$. Again, we have two solutions of Eq. (1) from this subcase if and only if

$$\chi(-2b-1) = 1, \quad \chi\left(\frac{\sqrt{-2b-1}+1}{2}\right) = -1 \text{ and } \chi\left(\frac{\sqrt{-2b-1}-1}{2}\right) = 1,$$

which is equivalent to the following conditions

$$\chi(-2b-1) = 1, \quad \chi\left(\frac{\sqrt{-2b-1}+1}{2}\right) = -1 \text{ and } \chi\left(\frac{b+1}{2}\right) = 1. \tag{3}$$

One may note that if X is a solution of Eq. (1) then so is $-X$. Therefore, Eq. (1) always has even number of solutions except when $X = -X \implies X = 0$ is a solution of Eq. (1). This happens when

$$b = \left(\frac{1}{2}\right)^{\frac{q+3}{2}} - \left(-\frac{1}{2}\right)^{\frac{q+3}{2}} = 2 \cdot \left(\frac{1}{2}\right)^{\frac{q+3}{2}} = \frac{\chi(2)}{2} = \begin{cases} \frac{1}{2} & \text{if } \chi(2) = 1, \\ -\frac{1}{2} & \text{if } \chi(2) = -1. \end{cases}$$

Let $\chi(2) = 1$ then $b = \dfrac{1}{2}$. In this case we do not have solutions from Case 1 and Case 2 as $\chi\left(\dfrac{b-1}{2}\right) = \chi\left(-\dfrac{1}{4}\right) \neq 1$. From Case 3, we have $X = 0$ as a solution and we do not have solutions from Case 4 as $\chi(2) = 1$.

Let $\chi(2) = -1$ then $b = -\dfrac{1}{2}$. We consider two cases, namely, $\chi(3) = 1$ and $\chi(3) = -1$. When $\chi(3)=1$ then we do not have solutions from Case 1 and Case 2. We have two solutions $X = \pm\dfrac{\sqrt{-2}}{2}$ from Case 3 if $\chi\left(\dfrac{\sqrt{-2}+1}{2}\right) = 1$. We get a solution $X = 0$ from Case 4. When $\chi(3) = -1$. We get one solution from each of the Case 1 and Case 2, no solution from Case 3, and a solution $X = 0$ from Case 4. From here, we conclude that

$$\Delta_f\left(1, \dfrac{\chi(2)}{2}\right) = \begin{cases} 3 & \text{if } \chi(2) = -1 = \chi(3), \\ 3 & \text{if } \chi(2) = -1, \chi(3) = 1 \text{ and } \chi\left(\dfrac{\sqrt{-2}+1}{2}\right) = 1, \\ 1 & \text{otherwise.} \end{cases}$$

Similarly, when $b = 1$ then $X = \pm\dfrac{1}{2}$ are two solutions of Eq. (1). We do not have solutions from Case 1, Case 2 and Case 3. We have two solutions from Case 4 if and only if

$$\chi(3) = -1, \ \chi\left(\dfrac{\sqrt{-3}+1}{2}\right) = -1.$$

We summarize above discussion in Table 2. This completes the proof.

Table 2. Number of solutions from different cases related to Proposition 1.

$b \in \mathbb{F}_q \setminus \{1, \frac{\chi(2)}{2}\}$	$\chi\left(\frac{b+1}{2}\right)$	$\chi\left(\frac{b-1}{2}\right)$	Case 1	Case 2	Case 3	Case 4
$b = -1$	0	-1	0	0	at most 2	0
$b \notin \{-1, 1, \frac{\chi(2)}{2}\}$	1	1	1	1	0	at most 2
$b \notin \{-1, 1, \frac{\chi(2)}{2}\}$	1	-1	0	0	at most 2	at most 2
$b \notin \{-1, 1, \frac{\chi(2)}{2}\}$	-1	1	0	0	0	0
$b \notin \{-1, 1, \frac{\chi(2)}{2}\}$	-1	-1	0	0	0	at most 2

The following theorem gives an upper bound for the boomerang uniformity of the power permutation $X^{\frac{q+3}{2}}$, where $q \equiv 3 \pmod{4}$.

Theorem 1. *Let $q \equiv 3 \pmod{4}$ then the boomerang uniformity of the power permutation $f(X) = X^{\frac{q+3}{2}}$ is ≤ 23.*

Proof. Recall that the boomerang connectivity table entry of the power permutation $f(X) = X^{\frac{q+3}{2}}$ at point $(1, b)$ is given by the number of solutions $(X, Y) \in \mathbb{F}_q \times \mathbb{F}_q$ of the following system of equations:

$$\begin{cases} X^{\frac{q+3}{2}} - Y^{\frac{q+3}{2}} = b, \\ (X+1)^{\frac{q+3}{2}} - (Y+1)^{\frac{q+3}{2}} = b. \end{cases}$$

By substituting $X \mapsto X - \frac{1}{2}, Y \mapsto Y - \frac{1}{2}$ and replacing second equation of the above system by the difference of second equation and the first equation, we have

$$\begin{cases} \left(X - \frac{1}{2}\right)^{\frac{q+3}{2}} - \left(Y - \frac{1}{2}\right)^{\frac{q+3}{2}} = b, \\ \left(X + \frac{1}{2}\right)^{\frac{q+3}{2}} - \left(X - \frac{1}{2}\right)^{\frac{q+3}{2}} = \left(Y + \frac{1}{2}\right)^{\frac{q+3}{2}} - \left(Y - \frac{1}{2}\right)^{\frac{q+3}{2}}, \end{cases} \quad (4)$$

$$\iff \begin{cases} \left(X - \frac{1}{2}\right)^2 \chi\left(X - \frac{1}{2}\right) - \left(Y - \frac{1}{2}\right)^2 \chi\left(Y - \frac{1}{2}\right) = b, \\ D_f(X) = D_f(Y). \end{cases}$$

Now, we shall consider different cases depending on the values of $\chi\left(X + \frac{1}{2}\right)$ and $\chi\left(X - \frac{1}{2}\right)$.

Case 1. Let $X = \frac{1}{2}$. In this case we have

$$\begin{cases} \left(Y - \frac{1}{2}\right)^{\frac{q+3}{2}} = -b, \\ \left(Y + \frac{1}{2}\right)^{\frac{q+3}{2}} - \left(Y - \frac{1}{2}\right)^{\frac{q+3}{2}} = 1. \end{cases}$$

From Proposition 1, we know that $Y = \pm\frac{1}{2}$ are solutions of the second equation of the above system and if $\chi(3) = -1 = \chi\left(\frac{\sqrt{-3}+1}{2}\right)$ then we have two more solutions $Y = \pm\frac{\sqrt{-3}}{2}$. It is easy to observe that $Y = \frac{1}{2}$ can not be a solution of the above system as $b \neq 0$. When $Y = -\frac{1}{2}$ then from the first equation of the system, we have $b = 1$. When $Y = \frac{\sqrt{-3}}{2}$ then from the first equation of the above system we have

$$b = -\left(\frac{\sqrt{-3}-1}{2}\right)^{\frac{q+3}{2}} = -\chi\left(\frac{\sqrt{-3}-1}{2}\right)\left(\frac{\sqrt{-3}-1}{2}\right)^2 = \frac{\sqrt{-3}+1}{2}.$$

Similarly, when $Y = -\frac{\sqrt{-3}}{2}$ then from the first equation of the above system we have

$$b = \left(\frac{\sqrt{-3}+1}{2}\right)^{\frac{q+3}{2}} = \chi\left(\frac{\sqrt{-3}+1}{2}\right)\left(\frac{\sqrt{-3}+1}{2}\right)^2 = -\frac{\sqrt{-3}-1}{2}.$$

Case 2. Let $X = -\frac{1}{2}$. In this case we have

$$\begin{cases} \left(Y - \frac{1}{2}\right)^{\frac{q+3}{2}} = -(b+1), \\ \left(Y + \frac{1}{2}\right)^{\frac{q+3}{2}} - \left(Y - \frac{1}{2}\right)^{\frac{q+3}{2}} = 1. \end{cases}$$

Again, from Proposition 1, we know that $Y = \pm\dfrac{1}{2}$ are solutions of the second equation of the above system and if $\chi(3) = -1 = \chi\left(\dfrac{\sqrt{-3}+1}{2}\right)$ then we have two solutions $Y = \pm\dfrac{\sqrt{-3}}{2}$. It is easy to observe that $Y = -\dfrac{1}{2}$ can not be a solution of the above system as $b \neq 0$. When $Y = \dfrac{1}{2}$ then from the first equation of the system, we have $b = -1$. When $Y = \dfrac{\sqrt{-3}}{2}$ then from the first equation of the above system we have

$$b = -\left(\dfrac{\sqrt{-3}-1}{2}\right)^{\frac{q+3}{2}} - 1 = -\chi\left(\dfrac{\sqrt{-3}-1}{2}\right)\left(\dfrac{\sqrt{-3}-1}{2}\right)^2 - 1 = \dfrac{\sqrt{-3}-1}{2}.$$

Similarly, when $Y = -\dfrac{\sqrt{-3}}{2}$ then from the first equation of the above system we have

$$b = \left(\dfrac{\sqrt{-3}+1}{2}\right)^{\frac{q+3}{2}} - 1 = \chi\left(\dfrac{\sqrt{-3}+1}{2}\right)\left(\dfrac{\sqrt{-3}+1}{2}\right)^2 - 1 = -\dfrac{\sqrt{-3}+1}{2}.$$

When $X \notin \left\{\dfrac{1}{2}, -\dfrac{1}{2}\right\}$ then $\chi\left(X + \dfrac{1}{2}\right), \chi\left(X + \dfrac{1}{2}\right) \in \{1, -1\}$ and we shall consider the following four more cases:

Case 3. Let $\chi\left(X + \dfrac{1}{2}\right) = 1 = \chi\left(X - \dfrac{1}{2}\right)$. Then, from system (4), we have

$$\begin{cases} (X - \frac{1}{2})^2 - (Y - \frac{1}{2})^2 \chi (Y - \frac{1}{2}) = b, \\ D_f(Y) = 2X. \end{cases} \tag{5}$$

Notice that, when $Y = \pm\dfrac{1}{2}$ then from the second equation of the above system, we have $X = \dfrac{1}{2}$ and hence we do not have solutions of Eq. (4) in this case. Now, when $Y \neq \pm\dfrac{1}{2}$ then we consider following four subcases:

Subcase 3.1. Let $\chi\left(Y + \dfrac{1}{2}\right) = 1 = \chi\left(Y - \dfrac{1}{2}\right)$. In this case, from the second equation of system (4), we have $X = Y$. Since $b \in \mathbb{F}_q^*$, we do not have solutions of Eq. (5) from this subcase.

Subcase 3.2. Let $\chi\left(Y + \dfrac{1}{2}\right) = -1 = \chi\left(Y - \dfrac{1}{2}\right)$. In this case system (5) becomes

$$\begin{cases} (X - \frac{1}{2})^2 + (Y - \frac{1}{2})^2 = b, \\ X = -Y, \end{cases} \Longleftrightarrow \begin{cases} Y^2 + \frac{1}{4} = \frac{b}{2}, \\ X = -Y. \end{cases}$$

Thus, we can have at most 2 solutions of Eq. (4) from this subcase only if

$$1 = \chi\left(Y^2 - \frac{1}{4}\right) = \chi\left(\frac{b-1}{2}\right).$$

Subcase 3.3. Let $\chi\left(Y + \frac{1}{2}\right) = 1$ and $\chi\left(Y - \frac{1}{2}\right) = -1$. Then, from system (5), we have

$$\begin{cases} \left(X - \frac{1}{2}\right)^2 + \left(Y - \frac{1}{2}\right)^2 = b, \\ Y^2 + \frac{1}{4} = X. \end{cases}$$

Notice that

$$-1 = \chi\left(Y + \frac{1}{2}\right) \cdot \chi\left(Y - \frac{1}{2}\right) = \chi\left(Y^2 - \frac{1}{4}\right) = \chi\left(X - \frac{1}{2}\right) = 1,$$

which is a contradiction. Therefore, we do not have solutions of Eq. (4) from this subcase.

Subcase 3.4. Let $\chi\left(Y + \frac{1}{2}\right) = -1$ and $\chi\left(Y - \frac{1}{2}\right) = 1$. In this case system (5) becomes

$$\begin{cases} \left(X - \frac{1}{2}\right)^2 - \left(Y - \frac{1}{2}\right)^2 = b, \\ Y^2 + \frac{1}{4} = -X, \end{cases} \iff \begin{cases} X^2 + Y + \frac{1}{4} - b = 0, \\ X + Y^2 + \frac{1}{4} = 0, \end{cases}$$

Thus, we can have at most 4 solutions of Eq. (4) from this subcase.

Case 4. Let $\chi\left(X + \frac{1}{2}\right) = -1 = \chi\left(X - \frac{1}{2}\right)$. Then from system (4), we have

$$\begin{cases} -\left(X - \frac{1}{2}\right)^2 - \left(Y - \frac{1}{2}\right)^2 \chi\left(Y - \frac{1}{2}\right) = b, \\ D_f(Y) = -2X. \end{cases} \quad (6)$$

Again, when $Y = \pm\frac{1}{2}$ then from the second equation of the above system, we have $X = -\frac{1}{2}$ and hence we do not have solutions of Eq. (4) in this case. Now, when $Y \neq \pm\frac{1}{2}$, we consider following four subcases.

Subcase 4.1. Let $\chi\left(Y + \frac{1}{2}\right) = 1 = \chi\left(Y - \frac{1}{2}\right)$. In this case system (6) becomes

$$\begin{cases} \left(X - \frac{1}{2}\right)^2 + \left(Y - \frac{1}{2}\right)^2 = -b, \\ X = -Y, \end{cases} \iff \begin{cases} Y^2 + \frac{1}{4} = -\frac{b}{2}, \\ X = -Y. \end{cases}$$

Thus, we can have at most 2 solutions of Eq. (4) from this subcase only if

$$1 = \chi\left(Y^2 - \frac{1}{4}\right) = \chi\left(\frac{-b-1}{2}\right) = -\chi\left(\frac{b+1}{2}\right).$$

Subcase 4.2. Let $\chi\left(Y + \frac{1}{2}\right) = -1 = \chi\left(Y - \frac{1}{2}\right)$. Then, from the second equation of the system (6), we have $X = Y$ and since $b \in \mathbb{F}_q^*$, we do not have solutions of Eq. (4) from this subcase.

Subcase 4.3. Let $\chi\left(Y + \frac{1}{2}\right) = 1$ and $\chi\left(Y - \frac{1}{2}\right) = -1$. Then system (6) becomes
$$\begin{cases} -\left(X - \frac{1}{2}\right)^2 + \left(Y - \frac{1}{2}\right)^2 = b, \\ Y^2 + \frac{1}{4} = -X. \end{cases}$$

Notice that
$$-1 = \chi\left(Y + \frac{1}{2}\right) \cdot \chi\left(Y - \frac{1}{2}\right) = \chi\left(Y^2 - \frac{1}{4}\right) = -\chi\left(X + \frac{1}{2}\right) = 1,$$
which is a contradiction. Therefore, we do not have solutions of Eq. (4) from this subcase.

Subcase 4.4. Let $\chi\left(Y + \frac{1}{2}\right) = -1$ and $\chi\left(Y - \frac{1}{2}\right) = 1$. In this case system (6) reduces to
$$\begin{cases} \left(X - \frac{1}{2}\right)^2 + \left(Y - \frac{1}{2}\right)^2 = -b, \\ -X + Y^2 + \frac{1}{4} = 0, \end{cases} \iff \begin{cases} X^2 - Y + \frac{1}{4} + b = 0, \\ -X + Y^2 + \frac{1}{4} = 0. \end{cases}$$

Therefore, we can have at most 4 solutions of Eq. (4) from this subcase.

Case 5. Let $\chi\left(X + \frac{1}{2}\right) = 1$ and $\chi\left(X - \frac{1}{2}\right) = -1$. Then, from system (4), we have
$$\begin{cases} -\left(X - \frac{1}{2}\right)^2 - \left(Y - \frac{1}{2}\right)^2 \chi\left(Y - \frac{1}{2}\right) = b, \\ 2 \cdot \left(X^2 + \frac{1}{4}\right) = D_f(Y). \end{cases} \quad (7)$$

It is easy to see that when $Y \in \left\{\frac{1}{2}, -\frac{1}{2}\right\}$ then $X \in \left\{\frac{1}{2}, -\frac{1}{2}\right\}$. Therefore, we not have solutions of system (4) in these cases. Now, $Y \notin \left\{\frac{1}{2}, -\frac{1}{2}\right\}$ then we consider following four subcases.

Subcase 5.1. Let $\chi\left(Y + \frac{1}{2}\right) = 1 = \chi\left(Y - \frac{1}{2}\right)$. Then system (7) reduces to
$$\begin{cases} \left(X - \frac{1}{2}\right)^2 + \left(Y - \frac{1}{2}\right)^2 = -b, \\ X^2 + \frac{1}{4} = Y. \end{cases}$$

Notice that
$$-1 = \chi\left(X + \frac{1}{2}\right) \cdot \chi\left(X - \frac{1}{2}\right) = \chi\left(X^2 - \frac{1}{4}\right) = \chi\left(Y - \frac{1}{2}\right) = 1,$$

which is a contradiction. Therefore, we do not have solutions of Eq. (4) from this subcase.

Subcase 5.2. Let $\chi\left(Y+\frac{1}{2}\right) = -1 = \chi\left(Y-\frac{1}{2}\right)$. Then, from system (7), we have
$$\begin{cases} -\left(X-\frac{1}{2}\right)^2 + \left(Y-\frac{1}{2}\right)^2 = b, \\ X^2 + \frac{1}{4} = -Y \end{cases}$$

Again, one may note that
$$-1 = \chi\left(X+\frac{1}{2}\right)\cdot\chi\left(X-\frac{1}{2}\right) = \chi\left(X^2-\frac{1}{4}\right) = -\chi\left(Y+\frac{1}{2}\right) = 1,$$

which is a contradiction. Therefore, we do not have solutions of Eq. (4) from this subcase.

Subcase 5.3. Let $\chi\left(Y+\frac{1}{2}\right) = 1$ and $\chi\left(Y-\frac{1}{2}\right) = -1$. In this case system (7) becomes
$$\begin{cases} -\left(X-\frac{1}{2}\right)^2 + \left(Y-\frac{1}{2}\right)^2 = b, \\ X^2 = Y^2. \end{cases}$$

When $X = Y$ then we have $b = 0$, which is a contradiction. When $X = -Y$ then from the first equation, we have
$$-\left(Y+\frac{1}{2}\right)^2 + \left(Y-\frac{1}{2}\right)^2 = b \implies Y = -\frac{b}{2}.$$

It is straightforward to see that this solution will also be a solution of Eq. (4) if and only if
$$\chi\left(\frac{b+1}{2}\right) = 1 \text{ and } \chi\left(\frac{b-1}{2}\right) = -1.$$

Subcase 5.4. Let $\chi\left(Y+\frac{1}{2}\right) = -1$ and $\chi\left(Y-\frac{1}{2}\right) = 1$. Then system (7) reduces to
$$\begin{cases} -\left(X-\frac{1}{2}\right)^2 - \left(Y-\frac{1}{2}\right)^2 = b, \\ \left(X^2+\frac{1}{4}\right) = -\left(Y^2+\frac{1}{4}\right), \end{cases} \iff \begin{cases} X+Y = b, \\ X^2 + Y^2 + \frac{1}{2} = 0. \end{cases}$$

Thus, we can have at most 2 solutions of Eq. (4) from this subcase.

Case 6. Let $\chi\left(X+\frac{1}{2}\right) = -1$ and $\chi\left(X-\frac{1}{2}\right) = 1$. In this case system (4) becomes
$$\begin{cases} \left(X-\frac{1}{2}\right)^2 - \left(Y-\frac{1}{2}\right)^2 \chi\left(Y-\frac{1}{2}\right) = b, \\ -2\cdot\left(X^2+\frac{1}{4}\right) = D_f(Y). \end{cases} \tag{8}$$

It is straightforward to see that when $Y = \pm\frac{1}{2}$ then form the second equation of the above system, we have $X^2 = -\frac{3}{4}$. Now if $\chi(3) = 1$ then we do not have solutions of system (4) with $Y = \pm\frac{1}{2}$. Assume that $\chi(3) = -1$ then $Y = \pm\frac{1}{2}$ gives $X = -\left(b \pm \frac{1}{2}\right)$. When $Y \neq \pm\frac{1}{2}$ then we consider the following four subcases:

Subcase 6.1. Let $\chi\left(Y + \frac{1}{2}\right) = 1 = \chi\left(Y - \frac{1}{2}\right)$. Then, system (8) reduces to

$$\begin{cases} (X - \frac{1}{2})^2 - (Y - \frac{1}{2})^2 = b, \\ X^2 + Y + \frac{1}{4} = 0, \end{cases} \iff \begin{cases} X + Y^2 + b + \frac{1}{4} = 0, \\ X^2 + Y + \frac{1}{4} = 0. \end{cases}$$

Thus, we can have at most 4 solutions of Eq. (4) from this subcase.

Subcase 6.2. Let $\chi\left(Y + \frac{1}{2}\right) = -1 = \chi\left(Y - \frac{1}{2}\right)$. In this case system (8) becomes

$$\begin{cases} (X - \frac{1}{2})^2 + (Y - \frac{1}{2})^2 = b, \\ X^2 - Y + \frac{1}{4} = 0, \end{cases} \iff \begin{cases} -X + Y^2 - b + \frac{1}{4} = 0, \\ X^2 - Y + \frac{1}{4} = 0, \end{cases}$$

Again, we can have at most 4 solutions of Eq. (4) from this subcase.

Subcase 6.3. Let $\chi\left(Y + \frac{1}{2}\right) = 1$ and $\chi\left(Y - \frac{1}{2}\right) = -1$. Then, from system (8), we have

$$\begin{cases} (X - \frac{1}{2})^2 + (Y - \frac{1}{2})^2 = b, \\ X^2 + Y^2 + \frac{1}{2} = 0, \end{cases} \iff \begin{cases} X + Y = -b, \\ X^2 + Y^2 + \frac{1}{2} = 0. \end{cases}$$

Thus, we can have at most 2 solutions of Eq. (4) from this subcase.

Subcase 6.4. Let $\chi\left(Y + \frac{1}{2}\right) = -1$ and $\chi\left(Y - \frac{1}{2}\right) = 1$. In this case system (8) reduces to

$$\begin{cases} (X - \frac{1}{2})^2 - (Y - \frac{1}{2})^2 = b, \\ X^2 = Y^2. \end{cases}$$

When $X = Y$ then we have $b = 0$, which is a contradiction. When $X = -Y$ then from the first equation, we have

$$\left(Y + \frac{1}{2}\right)^2 - \left(Y - \frac{1}{2}\right)^2 = b \implies Y = \frac{b}{2}.$$

Also, notice that this solution will also be a solution of Eq. (4) if and only if

$$\chi\left(\frac{b+1}{2}\right) = -1 \text{ and } \chi\left(\frac{b-1}{2}\right) = 1.$$

Table 3. Number of solutions from different subcases related to Theorem 1.

Subcase	System of equations	Conditions	Max no. of solutions
3.2	$Y^2 + \frac{1}{4} = \frac{b}{2}$, $X = -Y$.	$\chi\left(\frac{b-1}{2}\right) = 1$	2
3.4	$X^2 + Y - b + \frac{1}{4} = 0$, $X + Y^2 + \frac{1}{4} = 0$,	none	4
4.1	$Y^2 + \frac{1}{4} = -\frac{b}{2}$, $X = -Y$.	$\chi\left(\frac{b+1}{2}\right) = -1$	2
4.4	$X^2 + Y + b + \frac{1}{4} = 0$, $-X + Y^2 + \frac{1}{4} = 0$,	none	4
5.3	$Y = -\frac{b}{2}$, $X = -Y$.	$\chi\left(\frac{b-1}{2}\right) = -1$ and $\chi\left(\frac{b+1}{2}\right) = 1$	1
5.4	$X^2 + Y^2 + \frac{1}{2} = 0$, $X + Y = b$.	none	2
6.1	$X + Y^2 + b + \frac{1}{4} = 0$, $X^2 + Y + \frac{1}{4} = 0$.	none	4
6.2	$-X + Y^2 - b + \frac{1}{4} = 0$, $X^2 - Y + \frac{1}{4} = 0$.	none	4
6.3	$X^2 + Y^2 + \frac{1}{2} = 0$, $X + Y = -b$.	none	2
6.4	$Y = \frac{b}{2}$, $X = -Y$.	$\chi\left(\frac{b+1}{2}\right) = -1$ and $\chi\left(\frac{b-1}{2}\right) = 1$	1

We shall now summarize the above discussion. We do not have solutions of Eq. (4) from Subcase 3.1, Subcase 3.3, Subcase 4.2, Subcase 4.3, Subcase 5.1 and Subcase 5.2. We list maximum number of solutions of system (4) from remaining subcases in the Table 3.

3 Inverse and Algebraic Degree of f

In this section, we shall consider the algebraic degree of the power permutation $f(X) = X^{\frac{p^n+3}{2}}$, where $p^n \equiv 3 \pmod 4$. We then obtain the compositional inverse of f and determine its algebraic degree. The following proposition gives the algebraic degree of the power permutation f.

Proposition 2. *Let $p > 3$ such that $p^n \equiv 3 \pmod 4$. Then the algebraic degree of the power permutation $f(X) = X^{\frac{p^n+3}{2}}$ is $\frac{n(p-1)}{2} + 2$.*

Proof. Notice that

$$\frac{p^n + 3}{2} = \frac{p^n - 1}{2} + 2 = \frac{p-1}{2}(p^{n-1} + \cdots + p + 1) + 2.$$

Therefore, $deg(f) = \frac{n(p-1)}{2} + 2$.

The following theorem gives the compositional inverse of f.

Theorem 2. *Let $q \equiv 3 \pmod 4$ then the compositional inverse of the power permutation $f(X) = X^{\frac{q+3}{2}}$ is given by*

$$f^{-1}(X) = \begin{cases} X^{\frac{q+1}{4}} & \text{if } p \equiv 3 \pmod 8, \\ X^{\frac{3q-1}{4}} & \text{if } p \equiv 7 \pmod 8. \end{cases}$$

Proof. It is straightforward to see that $f(X) = 0 \iff X = 0$. Let $X, Y \in \mathbb{F}_q^*$ then

$$f(X) = Y \iff X^2 \cdot \chi(X) = Y \iff \begin{cases} X^2 = Y & \text{if } \chi(X) = 1, \\ -X^2 = Y & \text{if } \chi(X) = -1. \end{cases} \quad (9)$$

From above it is easy to see that $\chi(X) = \chi(Y)$. Now we shall consider the following two cases:

Case 1. Let $\chi(X) = \chi(Y) = 1$. Then, from Eq. (9), we have

$$X^2 = Y$$
$$\iff X = \pm\sqrt{Y}$$
$$\iff X = \begin{cases} \sqrt{Y} & \text{if } \chi(\sqrt{Y}) = 1, \\ -\sqrt{Y} & \text{if } \chi(\sqrt{Y}) = -1, \end{cases} \quad (10)$$
$$\iff X = \chi(\sqrt{Y}) \cdot \sqrt{Y}.$$

Case 2. Let $\chi(X) = \chi(Y) = -1$. In this case, we have

$$X^2 = -Y$$
$$\iff X = \pm\sqrt{-Y}$$
$$\iff X = \begin{cases} -\sqrt{-Y} & \text{if } \chi(\sqrt{-Y}) = 1, \\ \sqrt{-Y} & \text{if } \chi(\sqrt{-Y}) = -1, \end{cases} \quad (11)$$
$$\iff X = -\chi(\sqrt{-Y}) \cdot \sqrt{-Y}.$$

From the above discussion, we may deduce that

$$X = \chi(Y) \cdot \chi(\sqrt{Y \cdot \chi(Y)}) \cdot \sqrt{Y \cdot \chi(Y)} = Y^{\frac{q-1}{2}} \cdot Y^{\frac{q+1}{4} \cdot \frac{q-1}{2}} \cdot Y^{\frac{q+1}{4}}.$$

We further simplify this expression by dividing the case $q \equiv 3 \pmod 4$ into two cases, namely, $q \equiv 3 \pmod 8$ and $q \equiv 7 \pmod 8$. When $q \equiv 3 \pmod 8$ then the product of the first two terms is equal to 1 as

$$\frac{q+1}{4} \cdot \frac{q-1}{2} + \frac{q-1}{2} = \frac{q+5}{4} \cdot \frac{q-1}{2} = \frac{q+5}{8} \cdot (q-1) \equiv 0 \pmod{q-1}.$$

Thus, in this case $f^{-1}(X) = X^{\frac{q+1}{4}}$. When $q \equiv 7 \pmod{8}$ then the middle term becomes 1 and in this case we have $f^{-1}(X) = X^{\frac{3q-1}{4}}$. Thus,

$$f^{-1}(X) = \begin{cases} X^{\frac{q+1}{4}} & \text{if } p \equiv 3 \pmod 8, \\ X^{\frac{3q-1}{4}} & \text{if } p \equiv 7 \pmod 8. \end{cases}$$

This completes the proof.

The following theorem gives the algebraic degree of f^{-1}.

Theorem 3. *Let $p \equiv 3 \pmod 4$ and $n = 2m + 1$ for some non-negative integer m then the algebraic degree of the inverse of the power permutation $f(X) = X^{\frac{p^n+3}{2}}$ is*

$$\begin{cases} m(p-1) + \frac{p+1}{4} & \text{if } p \equiv 3 \pmod 8, \\ m(p-1) + \frac{3p-1}{4} & \text{if } p \equiv 7 \pmod 8. \end{cases}$$

Proof. From Theorem 2, we know that when $p \equiv 3 \pmod 8$ then the inverse of f is given by $X^{\frac{q+1}{4}}$. It is easy to observe that the p-adic expansion of $\frac{q+1}{4}$ is given by

$$\frac{q+1}{4} = \sum_{i=0}^{2m} c_i p^i,$$

where

$$c_i = \begin{cases} \frac{p+1}{4} & \text{if } i = 0, \\ \frac{p-3}{4} & \text{if } i > 0 \text{ such that } i \equiv 0 \pmod 2, \\ \frac{3p-1}{4} & \text{if } i > 0 \text{ such that } i \equiv 1 \pmod 2. \end{cases}$$

Thus, the algebraic degree is given by

$$m \cdot \frac{p-3}{4} + m \cdot \frac{3p-1}{4} + \frac{p+1}{4} = m(p-1) + \frac{p+1}{4}.$$

When $p \equiv 7 \pmod 8$ then the inverse of f is given by $X^{\frac{3q-1}{4}}$. One may note that the p-adic expansion of $\frac{3q-1}{4}$ is given by

$$\frac{3q-1}{4} = \sum_{i=0}^{2m} c_i p^i,$$

where

$$c_i = \begin{cases} \frac{3p-1}{4} & \text{if } i > 0 \text{ such that } i \equiv 0 \pmod 2, \\ \frac{p-3}{4} & \text{if } i > 0 \text{ such that } i \equiv 1 \pmod 2. \end{cases}$$

Thus, the algebraic degree is given by

$$(m+1) \cdot \frac{3p-1}{4} + m \cdot \frac{p-3}{4} = m(p-1) + \frac{3p-1}{4}.$$

This completes the proof.

4 Conclusion

In this paper, we considered two important cryptographic properties of a class of differentially 4-uniform power permutations over finite fields of odd characteristic, namely, the boomerang uniformity and the algebraic degree. We gave an upper bound for the boomerang uniformity of this class of power maps and determined its algebraic degree. Moreover, we obtained the compositional inverse of this class of power permutations and determined the algebraic degree of the inverse. It would be interesting to determine the nonlinearity of this class of power permutations.

Acknowledgements. The research of Mohit Pal is supported by the Research Council of Norway under Grant No. 314395. The author would like to sincerely thank Hridesh Kumar for his careful reading of the initial draft and for his many valuable discussions.

References

1. Biham, E., Shamir, A.: Differential cryptanalysis of DES-like cryptosystems. J. Cryptol. **4**(1), 3–72 (1991)
2. Blondeau, C., Canteaut, A., Charpin, P.: Differential properties of power functions. Int. J. Inf. Coding Theory **1**(2), 149–170 (2010)
3. Boura, C., Canteaut, A.: On the boomerang uniformity of cryptographic s-boxes. IACR Trans. Symmetric Cryptol. **3**, 290–310 (2018)
4. Calderini, M., Villa, I.: On the boomerang uniformity of some permutation polynomials. Cryptogr. Commun. **12**(6), 1161–1178 (2020). https://doi.org/10.1007/s12095-020-00439-x
5. Cid, C., Huang, T., Peyrin, T., Sasaki, Yu., Song, L.: Boomerang connectivity table: a new cryptanalysis tool. In: Nielsen, J.B., Rijmen, V. (eds.) EUROCRYPT 2018. LNCS, vol. 10821, pp. 683–714. Springer, Cham (2018). https://doi.org/10.1007/978-3-319-78375-8_22
6. Garg, K., Hasan, S.U., Stănică, P.: Boomerang uniformity of some classes of functions over finite fields. Discret. Appl. Math. **343**, 166–179 (2024)
7. Hasan, S.U., Pal, M., Stănică, P.: Boomerang uniformity of a class of power maps. Des. Codes Crypt. **89**(11), 2627–2636 (2021). https://doi.org/10.1007/s10623-021-00944-x
8. Hasan, S.U., Pal, M., Stănică, P.: The binary gold function and its c-boomerang connectivity table. Cryptogr. Commun. **14**(6), 1257–1280 (2022)
9. Helleseth, T., Sandberg, D.: Some power mappings with low differential uniformity. Appl. Algebra Eng. Commun. Comput. **8**, 363–370 (1997)
10. Hu, Z., Li, N., Xu, L., Zeng, X., Tang, X.: The differential spectrum and boomerang spectrum of a class of locally-APN functions. Des. Codes Crypt. **91**(5), 1695–1711 (2023)
11. Jiang, S., Kangquan, L., Yubo, L., Longjiang, Q.: Differential spectrum of a class of power functions. J. Cryptol. Res. **9**(3), 484–495 (2022)
12. Jiang, S., Li, K., Li, Y., Qu, L.: Differential and boomerang spectrums of some power permutations. Cryptogr. Commun. **14**(2), 371–393 (2022)
13. Lai, X.: Higher order derivatives and differential cryptanalysis. In: Communications and Cryptography: Two Sides of One Tapestry, pp. 227–233 (1994)

14. Li, K., Qu, L., Sun, B., Li, C.: New results about the boomerang uniformity of permutation polynomials. IEEE Trans. Inf. Theory **65**(11), 7542–7553 (2019)
15. Mesnager, S., Mandal, B., Msahli, M.: Survey on recent trends towards generalized differential and boomerang uniformities. Cryptogr. Commun. **14**, 691–735 (2022)
16. Mesnager, S., Özbudak, F.: Boomerang uniformity of power permutations and algebraic curves over $\mathbb{F}_{2^n}$. Adv. Geom. **23**(1), 107–134 (2023)
17. Nyberg, K.: Differentially uniform mappings for cryptography. In: Helleseth, T. (ed.) EUROCRYPT 1993. LNCS, vol. 765, pp. 55–64. Springer, Heidelberg (1994). https://doi.org/10.1007/3-540-48285-7_6
18. Stănică, P.: Using double Weil sums in finding the c-boomerang connectivity table for monomial functions on finite fields. Appl. Algebra Eng. Commun. Comput. **34**(4), 581–602 (2023)
19. Wagner, D.: The boomerang attack. In: Knudsen, L. (ed.) FSE 1999. LNCS, vol. 1636, pp. 156–170. Springer, Heidelberg (1999). https://doi.org/10.1007/3-540-48519-8_12
20. Yan, H., Li, Z., Song, Z., Feng, R.: Two classes of power mappings with boomerang uniformity 2. Adv. Math. Commun. **16**(4), 1111–1120 (2022)
21. Yan, H., Mesnager, S., Tan, X.: The complete differential spectrum of a class of power permutations over odd characteristic finite fields. IEEE Trans. Inf. Theory **69**(11) (2023)
22. Yan, H., Zhang, Z., Li, Z.: Boomerang spectrum of a class of power functions. In: 2022 10th International Workshop on Signal Design and Its Applications in Communications (IWSDA), pp. 1–4. IEEE (2022)
23. Yan, H., Zhang, Z., Zhou, Z.: A class of power mappings with low boomerang uniformity. In: International Workshop on the Arithmetic of Finite Fields, pp. 288–297. Springer, Cham (2022)
24. Zha, Z., Hu, L.: The boomerang uniformity of power permutations x^{2^k-1} over $\mathbb{F}_{2^n}$. In: 2019 Ninth International Workshop on Signal Design and its Applications in Communications (IWSDA), pp. 1–4. IEEE (2019)

Generating Gaussian Pseudorandom Noise with Binary Sequences

Francisco-Javier Soto[1](✉)[iD], Ana I. Gómez[1][iD], and Domingo Gómez-Pérez[2][iD]

[1] Universidad Rey Juan Carlos, Madrid, Spain
franciscojavier.soto@urjc.es
[2] Facultad de Ciencias, Cantabria, Spain

Abstract. Gaussian random number generators attract a widespread interest due to their applications in several fields. Important requirements include easy implementation, tail accuracy, and, finally, a flat spectrum. In this work, we study the applicability of uniform pseudorandom binary generators in combination with the Central Limit Theorem to propose an easy to implement, efficient and flexible algorithm that leverages the properties of the pseudorandom binary generator used as an input, specially with respect to the correlation measure of higher order, to guarantee the quality of the generated samples. Our main result provides a relationship between the pseudorandomness of the input and the statistical moments of the output. We propose a design based on the combination of pseudonoise sequences commonly used on wireless communications with known hardware implementation, which can generate sequences with guaranteed statistical distribution properties sufficient for many real life applications and simple machinery. Initial computer simulations on this construction show promising results in the quality of the output and the computational resources in terms of required memory and complexity.

Keywords: Pseudorandom number generator · Gold code · m-sequence · Central Limit Theorem · Gaussian distribution

1 Introduction

The performance of many industrial applications depends on the simulation of random events, for example biological simulations, communication channels measurement or electronic instrument calibration. Although one of the most commonly demanded statistical distribution is the Gaussian, the principal body of work is focused in the uniform distribution case. There are well-known results about construction of pseudorandom numbers with uniform distribution [14], that have been proposed as the basis to generate any other distribution. In a general setting, some methods rely on the rejection method and variations of it, such as acceptance-rejection, composition rejection, rejection with squeeze, etc. Others are based on the inversion method, or use special properties of the normal distribution like the Box-Muller algorithm that is widely used in practice due to its improved performance [11]. However, there are downsides which limit the applicability of this algorithm: it requires computation of elementary functions including the logarithm and square root as well as trigonometric functions which are costly to implement with logic gates. Moreover, the tail accuracy is directly dependent on the implementation [16].

The previous limitations show just a few of the found technical difficulties to implement a Gaussian Random Number generator (GRNG) over hardware. This is of maximum importance for many different real world applications where memory, computation time and throughput are constrained. In this case hardware-related parameters (e.g., number of logical gates, buffers, circuit layout,...) have to be minimized and it is recommended to reuse circuitry and functions, when possible. Different hardware-based techniques have been compared based on the amount of required hardware resources, the statistical precision and the tail accuracy [1].

A straightforward implementation of sums of truly random samples tends to the Gaussian population due to the well known Central Limit Theorem (CLT), although the resulting distribution will approximate poorly in the tail to the Gaussian distribution. The ease of implementation of this approach has fostered the proposal of CLT-based GRNGs with improved statistical properties and tail accuracy such as CLT correction, CLT inversion and multi hat methods [16]. Still, there is room for improvement on the performance and speed by limiting the statistical accuracy properties, aiming at achieving a minimalist solution that maintains an acceptable flexibility and ease of implementation.

In this work, we study the theoretical framework to apply the CLT to a sum of pseudorandom binary sequences with good correlation properties. We note that this approach provides a worse approximation compared to uniform numbers for the same number of terms (see Irwin-Hall distribution of the sums of numbers in the interval $(0,1)$ [17]). Still, it is widely seen as an acceptable trade off because the former allows to work at bit level. We note that also there are already efficient implementation for most pseudorandom sequences because of the widespread use in wireless communications.

This work focus on "simple" GRNG, which requires controlling moments up to third and fourth order, i.e., skewness and kurtosis, that have been studied for m-sequences [23]. The family of m-sequences is a good candidate due to the easy implementation by Linear Feedback Shift Registers (LFSRs). However m-sequences and several derived families show peaks on the higher order correlations [2,8], that will affect the quality of the output of any CLT based GRNG as we will prove in Sect. 2.

Increasing the number of LFSRs and, at the same time, using their states as part of the input can reduce previous limitations. Recent works [3–5,13] improve previous GRNGs architectures by using this idea and validate the results through computer experiments. This heuristic approach seems to also have nice properties, such that reduced latency and allowing parallelism. This study is the first step to fill this gap between theory and practice.

The outline of this work is the following: Sect. 2 describes the main results regarding the moments of Gaussian pseudorandom sequences constructed from the combination of binary sequences, focusing on Gold codes as a proof of concept. Section 3 provides computational experiments and Sect. 4 concludes the article with a discussion on the parameters, a comparison with other LFSR-based GRNGs and some open problems.

2 Gaussian Random Number Generation from Binary Sequences

This section bounds the moments of Gaussian pseudorandom sequences by the correlation measure of the binary sequences used to generate them.

We denote the binary sequences of period N as $s(i) \in \{-1, 1\}$ or simply s when possible. Also, for the reader's convenience, we recall the definition of *combined correlation measure of order k* for the periodic case [10].

Definition 1. *Given a binary sequence s of period N, $\theta_k(s, N)$ is the combined correlation measure of order k, defined as*

$$\theta_k(s, N) = \max_{L,D,T} \left| \sum_{i=1}^{T} s(L \cdot i + d_1) \cdots s(L \cdot i + d_k) \right|, \tag{1}$$

where $D = (d_1, \ldots, d_k)$ with $0 \leq d_1 < \cdots < d_k < N$, the sum on i run such all values $L \cdot i + d_1, \ldots, L \cdot i + d_k \in \{1, \ldots, N\}$ and $T \leq N$.

The combined correlation measure of order k is a powerful measure for asserting the pseudorandomness of a binary sequence, which calculates the correlation over arithmetic subsequences.

This is the discrete version of the product moments that have been characterized for continuous Gaussian random variables [20,21]. For a continuous Gaussian random variable with zero mean, the product moments of odd order k must be exactly zero [21, Corollary 2]. This also holds for every order k when the random variables are different [21, Remark 5]. In particular, we will prove that if the correlation measure of the binary sequence s is well-bounded and M is much smaller than T the generated sequences satisfy these properties.

Theorem 1. *Let $s(i)$ be a binary sequences of period N, M a positive integer with $M \ll N$, and k a non-negative integer. Then the following holds,*

$$\frac{1}{T} \sum_{i=1}^{T} \left(\sum_{n=1}^{M} s(i+n) \right)^k \leq (M(k-1))^{k/2} + \frac{M^k \max_{1 \leq r \leq k} \theta_r(s, N)}{T}, \tag{2}$$

where if k is odd, the first term on the right of the inequality disappears.

Proof. We follow the argument of Davenport and Erdös [6, Lemma 3]. First, we start from the expansion of the left side of Eq. (2).

$$\sum_{i=1}^{T}\left(\sum_{n=1}^{M} s(i+n)\right)^{k} =$$

$$\sum_{d_1=1}^{M} \cdots \sum_{d_k=1}^{M} \sum_{i=1}^{T} s(i+d_1) \cdots s(i+d_k) \leq$$

$$\sum_{d_1=1}^{M} \cdots \sum_{d_k=1}^{M} \left|\sum_{i=1}^{T} s(i+d_1) \cdots s(i+d_k)\right|, \quad (3)$$

then we can bound the inner sum depending on the integers $d_1, \ldots, d_k$. If each value of $d_1, \ldots, d_k$ is repeated an even number of times, we bound the inner sum by T. That number of choices for $d_1, \ldots, d_k$ is less than $(M(k-1))^{k/2}$.

When not all $d_1, \ldots, d_k$ appear an even number of times, we can remove repetitions, giving a different subset $d'_1, ..., d'_r$ in the inner sum

$$\left|\sum_{i=1}^{T} s(i+d'_1) \cdots s(i+d'_r)\right| \leq \theta_r(s, N). \quad (4)$$

The number of such cases is bounded by M^k, therefore the contribution in Eq. (3) is less than

$$M^k \max_{1 \leq r \leq k} \theta_r(s, N).$$

This finishes the proof. □

We define the following sequence S, whose elements approximate a Gaussian distribution by the CLT:

$$S(i) = \left(M^{-1/2}\left(\sum_{n=1}^{M} s(n+iM)\right)\right). \quad (5)$$

This definition aims to minimize the dependence between consecutive terms of the generated Gaussian sequence by taking sums of blocks of M terms without reusing them. We emphasize that for every $1 \leq M \leq N$ similar properties that for Eq. (2) must hold, where there are overlaps between each consecutive sum of M terms.

Theorem 2. *Let S be the sequence defined in the Eq. (5), M a non-negative integer with $M \ll T$. Let us take k positive integers $0 \leq d_1 < \ldots < d_k < T - M$. Under these conditions and for every M such that $0 \leq M \leq T$, we have*

$$\left|\frac{1}{T}\sum_{i=1}^{T} S(i+d_1) \cdots S(i+d_k)\right| \leq (k-1)^{k/2} + \frac{M^{k/2} \max_{1 \leq r \leq k} \theta_r(s, N)}{T},$$

where the first term on the right part of the inequality disappears for k odd.

Proof. The proof is similar to Theorem 1. First,

$$\left|\sum_{i=1}^{T} S(i+d_1)\cdots S(i+d_k)\right| =$$

$$\frac{1}{M^{k/2}}\left|\sum_{n_1=1}^{M}\cdots\sum_{n_k=1}^{M}\sum_{i=1}^{T}\left(\prod_{j=1}^{k} s(Mi+Md_j+n_j)\right)\right| \leq$$

$$\frac{1}{M^{k/2}}\sum_{n_1=1}^{M}\cdots\sum_{n_k=1}^{M}\left|\sum_{i=1}^{T}\left(\prod_{j=1}^{k} s(Mi+Md_j+n_j)\right)\right|.$$

Now, we expand the inner sum,

$$\left|\sum_{i=1}^{T}\left(\prod_{j=1}^{k} s(Md_j+iM+n_j)\right)\right| =$$

$$\left|\sum_{i=1}^{T} s(Mi+Md_1+n_1)\cdots s(Mi+Md_k+n_k)\right|.$$

The number of possible ways to take the integers $n_1,\ldots,n_k$ in the set $\{1,\ldots,M\}$ is M^k. Thus if k is odd we can bound this internal sum by $M^k \max_{1\leq r\leq k} \theta_r(s,N)$.

If k is even, it can happen that $Md_1+n_1,\ldots,Md_k+n_k$ appears an even number of times each value, therefore making the sum equal to T. We can calculate the number of times that this happens as in Theorem 1. This finishes the proof. □

2.1 Correlation Properties of Gold Codes

There are several families of binary sequences with good bounds for the combined correlation measure for many values of k. We will focus on sequences generated by two LFSRs such as Gold codes as a proof of concept, due to its simple implementation by two LFSRs and an XOR gate [19].

For a finite field of characteristic 2, denoted by $\mathbb{F}_q$ with $q = 2^n$, the *trace function* is defined as the following map, $\text{Tr}: \mathbb{F}_q \to \mathbb{F}_2$,

$$\text{Tr}(x) = \sum_{j=1}^{n} x^{2^j}.$$

Definition 2. *Let $\alpha \in \mathbb{F}_q$ be a primitive element and let $f(x) = x + x^{2^r+1}$ where $(r,n)=1$ be a polynomial in the $\mathbb{F}_q$. A Gold code is the following binary sequence s with period $q-1$,*

$$s(i) = \psi(f(\alpha^i)), \tag{6}$$

where $\psi(x) = (-1)^{Tr(x)}$, i.e., it is an additive character defined by the trace function.

Instead of the usual binary LFSRs architecture, we must convert the binary $\{0,1\}$ sequence to $\{-1,1\}$ by Eq. (6).

Now we compile previous results from [2,7,8] in Theorem 3. These results on the correlation measure of Gold codes hold for any n.

Theorem 3. *Let $\mathbb{F}_q$ be a finite field where $q = 2^n$, $q-1$ is a Mersenne prime and s is defined in Eq. (6). Then, for every k such that $1 \leq k \leq 4$*

$$\theta_k(s, q-1) \leq 9n(2^r+1)2^{n/2}.$$

Also, there is a full peak at $k = 5$.

3 Computational Experiments

In this section we compare different GRNGs based on the CLT and the Tausworthe model.

Let $f(x) = \sum_{i=0}^{n} b_i x^i$ be a polynomial with coefficients in $\mathbb{F}_2$ such that $b_0 = b_n = 1$ and $f(x)$ is a primitive polynomial of the finite field $\mathbb{F}_q$, i.e., the minimal polynomial of a primitive element α of $\mathbb{F}_q$. We recall the concept of maximum length LFSR for the reader's convenience.

Definition 3. *Using the preceding notation, a maximum length LFSR of n registers, consists of an initial state $\mathbf{e_0} \in \mathbb{F}_2^n$, a transition function, $T: \mathbb{F}_2^n \to \mathbb{F}_2^n$, such that*

$$T(e_1, ..., e_n) = \left(e_2, ..., e_n, \sum_{i=1}^{n} e_i b_{i-1}\right)$$

where $b_0, \ldots, b_n$ are the coefficients of $f(x)$ and an output function defined by

$$\text{out}\left(T^j(\mathbf{e_0})\right) = \text{out}(e_{1+j}, ..., e_{n+j}) = e_{1+j} \quad \text{for every } j \geq 0,$$

with

$$T^0(\mathbf{e_0}) = \mathbf{e_0} = (e_1, ..., e_n).$$

The Tausworthe model is a frequently used model to construct pseudorandom numbers with uniform distribution. Each state e_i defines a real number in the interval $(0, 1)$ in binary notation by the following application:

$$\varphi: \quad \begin{matrix} \mathbb{F}_2^B \mapsto & [0,1) \\ (e_1, \ldots, e_B) \mapsto & \sum_{i=1}^{B} e_i 2^{-i}. \end{matrix} \quad (7)$$

Different states can be also combined or partially taken by the application depending on the bit depth B, which is the number of registers used to define the uniform pseudorandom numbers.

We consider the following binary sequences for the experiments which are suitable for practical applications. Our experiments takes a random initial state, although the results do not depend on the initial values. First, we consider an m-sequence whose characteristic polynomial is $x^{89} + x^{38} + 1$. Second, we consider a Gold code with $r = 1$ as in Definition 2 with characteristic polynomials $f_1(x) = x^{89} + x^{38} + 1$ and $f_2(x) = x^{89} + x^{72} + x^{55} + x^{38} + 1$.

The parameters for Eq. (5) are $M = 256$ and the Tausworthe model has 32 bit depth, i.e., taking eight sums of consecutive 32-bit numbers. We have measured the first four moments with a sample size of $T = 10^5$ for both models. The results are summarized in Table 1.

These tests have been considered in wireless communications, where signals are interfered by delayed versions with different shifts. This is studied through the product moments and the polyspectrum [9,22]. The triple product moments normalized by the sample size T of each considered configuration is shown in Fig. 2 in a window 100×100.

Table 1. Statistical moments of order k binary model by Eq. (5) and Tausworthe model by Eq. (7) for binary sequences m-sequence and Gold code with Mersenne prime period with $M = 256$.

Order k	m-sequence $S(i)$	Gold code $S(i)$	m-sequence Tausworthe	Gold code Tausworthe
1	0.0037	−0.0012	0.0003	0.0018
2	1.0043	1.0011	1.0033	0.9992
3	0.3609	0.0049	0.0031	0.0022
4	3.2182	3.0061	2.8849	2.8421

Results in Table 1 show that the Gold code outperforms the m-sequence used as binary sequence. The m-sequences show deviations from the expected value in the third and fourth moments. In the case of the GRNGs based on the Tausworthe model there is non significant difference between using the Gold code and the m-sequence. The fourth moment in the Tausworthe model is far from the expected value 3 because of the smaller number of sums, which affects the behavior in the tails of the distribution.

The bit depth and the continuous approximation to the Gaussian distribution is better for the Tausworthe model as seen in Fig. 1. This agrees with previous results of similar GRNG constructions in the literature using m-sequences [3, 13]. Only in the binary sequence method using an m-sequence shows a clearly asymmetry due to the presence of peaks on the correlation.

Figure 2 displays patterns in the bispectrum depending on type of binary sequence, independently of the model. The Gold code behaves as expected, close to a Gaussian pseudonoise [21, Corollary 2] while the m-sequence show peaks, where the highest peaks appears in the case of the binary sequence model.

Furthermore, we note that the bound of Theorem 2 normalized by the period $2^{89} - 1$ can be interpreted as the average value of the results obtained for samples of the size T used. For the binary sequence model using the Gold code it gives us a value close to 10^{-6} for $k = 3$ by the Theorem 3 while for the m-sequence we get a value much larger than zero [15, Equation (7d)] which partly explains the observed behaviors.

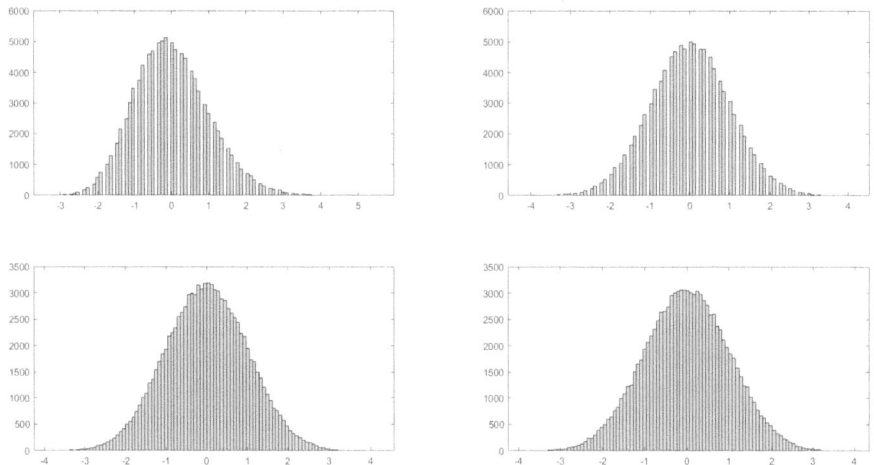

Fig. 1. Comparison of histograms with 100 bins generated by GRNGs using the m-sequence (left) and the Gold code (right). The histograms for binary sequence model are shown on the top row, while the Tauworthe model is shown on the bottom row.

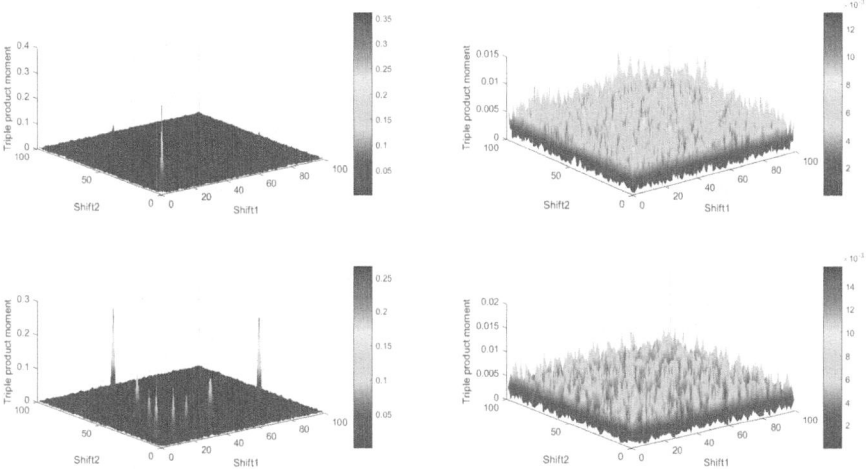

Fig. 2. Comparison of the triple product moments (absolute value) generated by GRNGs using the m-sequence (left) and the Gold code (right). The calculations for the binary sequence model are shown on the top row, while the Tauworthe model is shown on the bottom row.

4 Conclusions and Future Work

We have provided bounds for the higher-order product moments of Gaussian sources constructed from binary sequences. These bounds depend on the combined correlation measure of order k of the chosen binary sequences. Although these bounds can be improved for special sequences, they are sufficient for real applications and provides the missing link between good binary sequences and Gaussian Random Number Generator.

In this work we consider Gold codes with a Mersenne prime period, that provides a simple implementation and do not show peaks in the third and fourth correlation measures. It is interesting to characterize other lengths such that the Gold codes presents good properties. However, the length being a Mersenne prime is not a big restriction for practical purposes. The Mersenne prime period in the experiments has been chosen to be the smallest possible for its application. Larger values should perform better, according to our results. Regarding computational resources, Gold codes guarantee better statistical properties at the cost of doubling the size of the state and the number of XOR gates by four with respect to m-sequences. Initial computer experiments for other sequences with known implementations such as small Kasami and a Gold code with non-prime period has been done. However, the results varies from the distribution of the peaks on the correlation and further research is needed.

Computational results show that Gold codes offer better statistical and pseudorandom properties regardless of the used model that will depend on the practical application. In the case of the GRNG using an m-sequence with the Tausworthe model, we notice that both the moments and the histogram do not show significant deficiencies, but Fig. 2 shows well-localized peaks, contradicting what is expected for a Gaussian random variable. This negative phenomenon in the case of the m-sequence agrees with the existing literature where the distribution of moments is known under the name of distribution of weights of subsequences of m-sequences [12,15]. We leave it as an open problem to study how the peaks affect other desirable properties for a GRNG. A first step is to analyze the influence on Gaussian multivariate properties such as orthogonal invariance, i.e., spherical symmetry that may be of special interest in Monte-Carlo and Quasi Monte-Carlo algorithms or sampling in global optimization. Also, we propose to further study the Tausworthe model. The standard method to evaluate the Tausworthe model via character sums [18, Theorem 3.12] provides no information due to the presence of peaks in the correlation of order $k = 5$ for both sequences.

Acknowledgment. The authors want to thank Andrew Tirkel for pointing the problem and useful discussions. Also, we would like to thank the reviewers for their time, comments and suggestions.

Domingo Gómez-Pérez is partially supported by Research Project "PROTOCOLOS SEGUROS EN REDES DESCENTRALIZADAS.(AYUDA FINANCIADA CONTRATO PROGRAMA GOB CANTABRIA - UC)". In addition, Francisco-Javier Soto acknowledges support from the "PREDOCT2022-006" of Universidad Rey Juan Carlos.

References

1. Alimohammad, A., Fard, S.F., Cockburn, B.F., Schlegel, C.: A compact and accurate gaussian variate generator. IEEE Trans. Very Large Scale Integr. (VLSI) Syst. **16**(5), 517–527 (2008)
2. Chen, Z., Gómez, A.I., Gómez-Pérez, D., Tirkel, A.: Correlation measure, linear complexity and maximum order complexity for families of binary sequences.

Finite Fields Their Appl. **78**(nil), 101977 (2022). https://doi.org/10.1016/j.ffa.2021.101977
3. Condo, C., Gross, W.: Pseudo-random gaussian distribution through optimised LFSR permutations. Electron. Lett. **51**(25), 2098–2100 (2015)
4. Cotrina, G., Peinado, A., Ortiz, A.: Gaussian pseudorandom number generator based on cyclic rotations of linear feedback shift registers. Sensors **20**(7), 2103 (2020)
5. Cotrina, G., Peinado, A., Ortiz, A.: Gaussian pseudorandom number generator using linear feedback shift registers in extended fields. Mathematics **9**(5), 556 (2021)
6. Davenport, H., Erdos, P.: The distribution of quadratic and higher residues. Publ. Math. Debrecen **2**(3–4), 252–265 (1952)
7. Folláth, J.: Construction of pseudorandom binary sequences using additive characters over gf (2 k). Period. Math. Hung. **57**(1), 73–81 (2008)
8. Gómez, A.I., Gomez-Perez, D., Tirkel, A.: Correlation measure of binary sequence families with trace representation. In: Mesnager, S., Zhou, Z. (eds.) Arithmetic of Finite Fields, pp. 313–319. Springer, Cham (2023)
9. Green, D.R.: The utility of higher-order statistics in gaussian noise suppression. Ph.D. thesis, Naval Postgraduate School (2003)
10. Gyarmati, K.: On a family of pseudorandom binary sequences. Period. Math. Hung. **49**, 45–63 (2004)
11. Hörmann, W., Leydold, J., Derflinger, G.: Automatic Nonuniform Random Variate Generation. Springer, Cham (2004)
12. Jordan, H.F., Wood, D.C.: On the distribution of sums of successive bits of shift-register sequences. IEEE Trans. Comput. **100**(4), 400–408 (1973)
13. Kang, M.: FPGA implementation of gaussian-distributed pseudo-random number generator. In: 6th International Conference on Digital Content, Multimedia Technology and its Applications, pp. 11–13. IEEE (2010)
14. Kuipers, L., Niederreiter, H.: Uniform Distribution of Sequences. Courier Corporation (2012)
15. Lindholm, J.: An analysis of the pseudo-randomness properties of subsequences of long m-sequences. IEEE Trans. Inf. Theory **14**(4), 569–576 (1968)
16. Malik, J.S., Hemani, A.: Gaussian random number generation: a survey on hardware architectures. ACM Comput. Surv. (CSUR) **49**(3), 1–37 (2016)
17. Marengo, J.E., Farnsworth, D.L., Stefanic, L.: A geometric derivation of the Irwin-hall distribution. Int. J. Math. Math. Sci. **2017** (2017)
18. Niederreiter, H.: Random number generation and quasi-Monte Carlo methods. SIAM (1992)
19. Sarwate, D.V., Pursley, M.B.: Crosscorrelation properties of pseudorandom and related sequences. Proc. IEEE **68**(5), 593–619 (1980)
20. Song, I.: A proof of the explicit formula for product moments of multivariate gaussian random variables. arXiv preprint arXiv:1705.00163 (2017)
21. Song, I., Lee, S.: Explicit formulae for product moments of multivariate gaussian random variables. Stat. Probab. Lett. **100**, 27–34 (2015)
22. Tugnait, J.K.: Detection of non-gaussian signals using integrated polyspectrum. IEEE Trans. Signal Process. **42**(11), 3137–3149 (1994)
23. Wainberg, S., Wolf, J.: Subsequences of pseudorandom sequences. IEEE Trans. Commun. Technol. **18**(5), 606–612 (1970)

An FPGA Accelerated Search Method for Maximum Period NLFSRs

Amund Askeland[1,2]

[1] University of Bergen, Bergen, Norway
amund.askeland@uib.no
[2] Nasjonal Sikkerhetsmyndighet, Oslo, Norway

Abstract. Maximum period nonlinear feedback shift registers (NLFSRs) are promising building blocks for stream ciphers and pseudorandom number generators. Unfortunately, many fundamental problems related to NLFSRs remain open, and in particular it is not known how to construct simple ones whose periods are of maximum length. In this paper, we describe a search method for finding maximum period NLFSRs. The method is based on using an accelerator implemented on a field programmable gate array (FPGA) to test NLFSR periods, and an initial pruning step that checks for short cycles. We use this method to build build a dataset with complete lists of maximum-period NLFSRs of certain forms up to a length of 32. We also release the source code for our FPGA implementation together with the dataset.

Keywords: Feedback shift registers · NLFSR · FPGA

1 Introduction

Feedback shift registers (FSRs) are simple constructions made from a few memory elements and a feedback function. When chosen properly, the feedback function can make the shift register cycle through all, or all but one, of its possible states before repeating itself. This is referred to as maximum period feedback shift registers and has many applications including in stream ciphers and pseudorandom number generators. The main strengths of these constructions are their simplicity and speed, especially in hardware implementations where they can be made with very few resources.

We divide feedback shift registers into linear feedback shift registers (LFSRs) and nonlinear feedback shift registers (NLFSRs), depending on the feedback function that is used. LFSRs are fairly well studied, and in particular, it is known that maximum period LFSRs correspond to primitive polynomials over the binary field $\mathbb{F}_2$. Maximum period LFSRs are useful building blocks in stream ciphers, where they for example can be used to guarantee long periods of the keystream as in grain-like ciphers [1,9,10]. Constructions relying on LFSRs, such as filter generators, can however have security issues related to the linearity of the feedback functions [5,14]. This motivates the use of NLFSRs, either as replacements for or in conjunction with LFSRs. We do not have efficient methods for generating maximum period NLFSRs, and the theory behind them is more complex and less understood. Golomb's fundamental book on shift register sequences from 1967 [8] contains much of the available knowledge on this subject. Since then there have been some advancements, like Mykkeltveit's proof of Golomb's conjecture for the maximum number of cycles an NLFSR can decompose its states into [12]. Still, fairly little is known about the cycle structure of NLFSRs.

Finding maximum period NLFSRs might be useful in developing the theory on the subject, as well as potentially having direct use cases in for example stream ciphers. In order to find maximum period NLFSRs we can apply brute-force methods. This process is computationally heavy and requires both powerful hardware and efficient methodology, especially if one wants to find large NLFSRs. There is a historical analog to this with LFSRs, as in the 1970s computers were used to find and build tables of irreducible polynomials [11]. Today the amount of computing power needed to build such tables is trivial, but finding large maximum period NLFSRs remains challenging.

There have been several efforts to find large maximum period NLFSRs using brute-force methods. In 2012, Dubrova published a list of maximum period NLFSRs for three types of quadratic feedback functions, which contains 578 NLFSRs with lengths between 4 and 25 [7]. Others have used more specialized hardware in order to speed up the process and find larger maximum-period NLFSRs. In [15] the authors use FPGAs to generate random feedback functions and test their period, resulting in them finding some 27-bit maximum period NLFSRs. In [13] a similar approach and performance is achieved, but the experiments are run longer to find some 29-bit NLFSRs. In [3], the authors use GPUs in order to find quadratic NLFSRs up to a length of 28. Other work, like [6] has found larger maximum period NLFSRs by starting from a maximum period LFSR and adding some high-degree nonlinear terms.

In this paper, we set out to find larger maximum period NLFSRs than what has been done before and to build an extensive dataset that might be useful in developing theory for NLFSRs. We do this by building an FPGA-based accelerator for testing the period of NLFSRs that can offload this task from a computer. Specifically, we focus on feedback functions with a low algebraic degree and a low number of terms. Our contributions can be summarized as follows: 1. We provide a large dataset containing more than 21000 maximum period NLFSRs for 10 different forms of feedback functions. 2. We find several 32-bit maximum period NLFSRs with simple feedback functions. 3. We release the source code for our FPGA implementation, lowering the barrier for others who want to find large maximum period NLFSRs.

2 Preliminaries

A feedback shift register is a collection of n storage elements referred to as stages whose values are in the binary field, $\mathbb{F}_2$. The state, i.e. the collective value of the stages of the FSR, is a vector in the n-dimensional vector space $\mathbb{F}_2^n$. The feedback function of an FSR is a Boolean function from $\mathbb{F}_2^n$ to $\mathbb{F}_2$. For an FSR with state $(x_0, x_1, \ldots, x_{n-1})$, the next state is determined from the current state by a shifting operation and the feedback function. The shifting is such that the new value of x_0 is the previous value of x_1, and so on. The new value of x_{n-1} is determined by the feedback function, $f(x_0, x_1, \ldots, x_{n-1})$ such that the new state becomes $(x_1, x_2, \ldots, x_{n-1}, f(x_0, x_1, \ldots, x_{n-1}))$. This configuration, where the feedback is only applied to the end stage of the FSR, is referred to as the Fibonacci configuration. Figure 1 shows an example of an 8-bit feedback shift register with the feedback function $f(x_0, x_1, \ldots, x_{n-1}) = x_0 + x_1 + x_6 + x_1 \cdot x_2$.

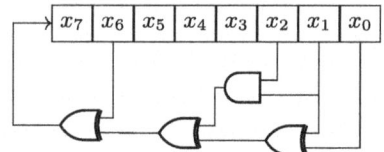

Fig. 1. Example of an NLFSR

Feedback shift registers can be represented by directed graphs, where the 2^n possible states are nodes, and directional edges connect the states according to the feedback function. A feedback shift register is guaranteed to have a branchless graph, i.e. every state has only one successor and predecessor, if its feedback function is on the form shown in Eq. 1, where g does not depend on x_0 [8]. For the remainder of this paper, by shift register, we implicitly mean branchless feedback shift registers.

$$f(x_0, x_1, \ldots, x_{n-1}) = x_0 + g(x_1, x_2, \ldots, x_{n-1}) \tag{1}$$

As a shift register progresses through its states, there will come a point where a state will be repeated. The longest period we can have is a path that visits all the states once, i.e. a Hamiltonian path. The number of such shift registers is $2^{2^{n-1}-n}$ as found by de Brujin [4]. Since the number of branchless feedback functions is $2^{2^{n-1}}$, the probability of a random branchless feedback function resulting in a maximum period shift register is 2^{-n}. A feedback function that does not use all n stages as inputs will decompose the space of states into an even number of cycles [8]. This means that shift registers with a single cycle of length 2^n have to use all n stages in the feedback function. Shift registers that instead result in two cycles of lengths $2^n - 1$ and 1 do not have this requirement, and can have quite simple feedback functions.

There are only two possible cycles of length 1, namely the cycles stuck at the all-zero or all-one state. A shift register resulting in one cycle with the all-zero state, and one cycle with the remaining $2^n - 1$ states can have its two cycles joined together by adding the term $\overline{x}_1 \cdot \overline{x}_2 \cdot \ldots \cdot \overline{x}_{n-1}$ to its feedback function (where $\overline{x}$ represents the logical negation of x) resulting in a single cycle of length

2^n. The same operation can be used to remove the all-zero state from the full cycle again. Similarly, if we add a constant (+1) term to a feedback function with one cycle of period $2^n - 1$ and a cycle involving a single state, we will change the single state cycle between the all-zero and all-one state. Consequently, there are $2^{2^{n-1}-n}$ shift registers of each of these three types, and we can trivially convert one to another. This also serves as a justification for referring to feedback shift registers with a cycle of period $2^n - 1$ as maximum period shift registers, although they are missing one state. When writing feedback functions in algebraic normal form (ANF), the feedback functions resulting in a cycle with only the all-zero state are somehow nicer, as they neither involve a constant term nor have to use all the stages. We will therefore focus on this type of feedback function.

For any feedback shift register, there is a reciprocal shift register that has the same number of cycles with the same periods, but where the corresponding sequence are in reverse order. The feedback function of this reciprocal shift register is given by Eq. 2 [8], where the inputs to g are in reverse order compared to in Eq. 1. We can say that FSRs come in pairs, if we have a feedback function that results in a maximum period shift register, we can trivially find another one corresponding to the reverse sequence.

$$f_r(x_0, x_1, \ldots, x_{n-1}) = x_0 + g(x_{n-1}, x_{n-2}, \ldots, x_1) \qquad (2)$$

3 Search Method

In order to test the period of an NLFSR, one can pick an initial value and shift the register until this initial value recurs. The number of shifts will then be the period of the cycle involving the initial state. In a search for maximum period NLFSRs the most time-consuming task will be this period-testing, especially for large n. In his book, Golomb finds that the average number of shifts until a state is repeated is about 2^{n-1} for simple feedback functions with random tap positions and all-one as the initial value [8]. In this work, we will speed up the search process by building an FPGA accelerator that performs the period testing of many NLFSRs in parallel. Other parts of the search process, like generating NLFSR candidates that we want to test, will be done in software. This allows us to get the benefits of software, such as simpler and more agile development, while letting the FPGA do the heavy lifting computation-wise. The software part of our search process will be used to generate candidate NLFSRs that are sent to the FPGA for period testing. Some of the theory from the preliminary section will be useful in speeding up the process. For instance, one can halve the search space by not testing the reciprocal feedback functions given by Eq. 2. We will also use software to perform an initial pruning step to discard NLFSRs with short cycles before testing the remaining candidates with the FPGA.

In our search, we will focus on feedback functions that have a low number of linear terms and few, low-degree, nonlinear terms in their ANF representation. This allows for a manageable search space and exhaustive searches to find all maximum period NLFSRs of a form. Focusing on ANF simplifies analysis and aligns with previous work, facilitating comparison. Although simple ANFs typically lead to low complexity in hardware implementations, an alternative approach could be to directly search for functions that can be implemented with minimal gate count or propagation delay.

3.1 Failing Fast Using Necklaces

A randomly chosen NLFSR can, and is likely to, have some very short cycles. When determining if a given NLFSR has a maximum period cycle, a sensible first step is to check if it has a short cycle since testing for short cycles can be done very quickly. For instance, there is only one possible cycle of period 2, consisting of the two states with alternating ones and zeroes. In order to test for the presence of this cycle we only need to load one of these two states into the NLFSR, clock the register twice, and check for a repetition.

Definition 1. *A binary necklace of length l is an equivalence class of binary strings of length l under cyclic shifts. A necklace is aperiodic if the strings it contains are aperiodic, i.e. no two cyclic shifts of the necklace are equal.*

Definition 2. *A binary Lyndon word is a binary string that is strictly smaller in lexicographic order than any of its cyclic shifts.*

More generally, in order to check for the presence of cycles with period l, we can use binary aperiodic necklaces to form initial values and clock the register l times. The number of such necklaces with length l can be counted using Witt's formula as given in Eq. 3 [16], where μ is the Möbius function. When forming the initial values, we use Lyndon words as representatives for the necklaces and concatenate them to fill the length of the register. Using Lyndon words results in initial values that are not equivalent under cyclic shifts, as well as not being covered by initial values for shorter cycles.

$$M(l) = \frac{1}{l} \sum_{d|l} \mu\left(\frac{l}{d}\right) 2^d \qquad (3)$$

This method was tested experimentally by enumerating all NLFSRs of length $n = 22$ with feedback functions of the form $x_0 + x_a + x_b + x_c + x_d + x_e \cdot x_f$, and testing them for short cycles using binary Lyndon words as initial values. We did not include the initial values that correspond to cycles of length 1, since an even number of terms in the feedback functions guarantees that the all-one state is not followed by itself and the all-zero state is known to be a cycle of length 1.

Table 1 shows the percentage of the NLFSRs that had cycles of length L, as well as the cumulative result for cycles of length $\leq L$. In the rightmost column, we can see the percentage of feedback functions that we can discard after checking for short cycles up to a length of L. We find similar results for other values of n, for example, 77.5% out of the feedback functions result in cycles with period ≤ 12 for $n = 32$ and the same type of feedback functions.

As we check for the presence of larger cycles, the number of initial values to test grows quickly, and the number of shifts that need to be computed grows even faster. In practice, we find that testing for short cycles up to a length of 12 can be quite effective as an initial pruning step, as it allows us to discard about 78% of the feedback functions at an early stage without being very computationally heavy. Testing for the presence of larger cycles becomes more time-consuming as the number of Lyndon words becomes quite big and the number of discarded functions rises more slowly.

Table 1. Percentage of NLFSRs with small cycles for $n = 22$

L	$M(L)$	NLFSRs with cycles of length L (%)	$\sum_{l=2}^{L} M(l)$	NLFSRs with cycles of length $1 < l \leq L$ (%)
2	1	26.4	1	26.4
3	2	25.2	3	45.0
4	3	22.7	6	57.2
5	6	11.6	12	62.2
6	9	11.3	21	64.3
7	18	8.9	39	67.4
8	30	11.0	69	72.1
9	56	8.2	125	74.4
10	99	5.6	224	75.6
11	186	6.2	410	77.1
12	335	8.0	745	78.4
13	630	6.0	1375	79.8
14	1161	6.5	2536	80.6
15	2182	6.2	4718	81.7
16	4080	5.7	8798	82.9

3.2 FPGA Accelerator

When implemented in FPGAs, the feedback calculation and the state update of an NLFSR can be done in a single clock cycle, and few resources are required. This means that testing the period of NLFSRs in FPGAs can be done quite effectively and with many tests in parallel. Next, we will describe the modules we design for testing periods of NLFSR cycles, and how these are used in the full design. The source code for the FPGA accelerator is written in Verilog and includes cocotb testbenches written in Python. The source code is open-source and made available for download [2].

NLFSR Tester Module. Our NLFSR tester module is a hardware module that takes in a feedback function configuration and a start signal. When started, the module will run until the period of the shift register cycle has been determined, and raise a success signal if the period was $2^n - 1$. The success signal is held high until the feedback function register has been read. When the module is not performing a test, and the success signal is low, the idle signal is held high to indicate that the module is ready for a new test. The module consists of two reciprocal shift registers, one running "backward" and one running "forward", producing each other's reverse sequence. The feedback functions of these two registers come from the same configuration, and are related by Eq. 2. The reason for using two registers is that by running the tester module until the contents of the two registers match, we find the period of a cycle in half the number of shifts compared to a single register. Note that the contents will only match for even-length cycles. In order for this to also work for odd-length cycles, we keep an extra bit in one of the registers and we have two comparison functions. We guarantee that the tested shift registers are branchless by forcing the feedback functions to be of the form in Eq. 1, the input configuration controls $g(x_1, x_2, \ldots, x_{n-1})$ and x_0 is hard-wired to be part of the feedback function.

A simplified block diagram of the tester module is shown in Fig. 2. The NLFSR tester module is parameterized so that before synthesis one specifies the shift register length, the number of nonlinear terms, and the maximal order of the nonlinear terms.

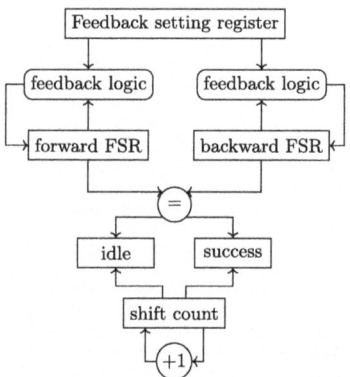

Fig. 2. Simplified Block Diagram of the Tester Module

Top Level Design. The goal of the FPGA accelerator is to reach a high speed in terms of how many feedback functions we can test per unit of time. In our design, the speed will come from having a high number of NLFSR tester modules working in parallel. With many parallel testers, the main design challenge will be the distribution of data and control signals to and from each tester, while at the same time avoiding high-fanouts and deep combinatorial paths that could limit the speed of the system. In our design, this distribution is realized with a tree structure, where the leaf nodes are the tester modules from Fig. 2. For the branch nodes, we design a distributor module that has the same interface as the testers, such that they are indifferent to whether they distribute signals to testers or other distributors. These distributor modules receive a feedback function configuration and select one idle child to pass it on to. Whenever a child's success signal is raised, the successful feedback functions will be passed up the tree. All signals passing through the distributor modules are held by registers in order to keep the combinatorial paths short. Before synthesis, the distributor modules are parameterized with a maximum branch number and the number of tester modules that they should have as descendants.

The entire tree is built by instantiating a single distributor at the top level, which recursively generates other distributors until the desired number of leaf nodes is met. Figure 3 shows a simplified block diagram of the top-level design, where we can see the tree structure used for distribution. In the figure, we have drawn two branches per distributor for simplicity. The number of branches per node will be a tradeoff between timing-related metrics such as fanout and logical depth against resource usage in the FPGA. In practice, we have found a branch number of 5 to be more suitable. The distributor at the root of the recursive tree is connected to two first-in-first-out (FIFO) buffers. One buffer holds feedback function settings that are to be tested, and one holds settings that yield maximum periods. These FIFOs also serve clock domain crossings between a fast clock domain and a slow clock domain. This is done to ease the timing requirements for the parts of the design that do not limit the general speed of the system.

At the other end of the FIFOs, there is an interface module with a serial interface that can be controlled by a PC. This module receives feedback functions to be tested and responds to various commands such as reset, status request, or successful feedback function readouts.

Accelerator Performance. The test rate of the FPGA accelerator, in terms of the average number of NLFSRs tested per second, can be estimated by Eq. 4. Here, N_T is the number of tester modules and f is the frequency of the fast clock domain in the FPGA implementation. In this equation, we have used 2^{n-1} as the average cycle length of random NLFSRs, and since we iterate through the states both forward and backward, the average number of cycles to test one NLFSR becomes 2^{n-2}. In our experiments, the estimation from Eq. 4 aligns well with the observed test rate. Our experiments utilized a Digilent Genesys 2, which is a midrange FPGA board with a Kintex-7 FPGA. The number of NLFSR tester modules that can fit on the FPGA varies depending on the length of the shift register and the form of the feedback function. For $n = 32$ and feedback functions with one quadratic term, we can fit 1300 NLFSR tester modules onto the FPGA, utilizing 156015 LUTs and 196311 flip flops, while operating at 350 MHz.

$$F(n) = \frac{N_T \cdot f}{2^{n-2}} \tag{4}$$

4 Experimental Results

We perform a set of experiments where we use the described search method to find complete lists of all feedback functions for certain combinations of shift reg-

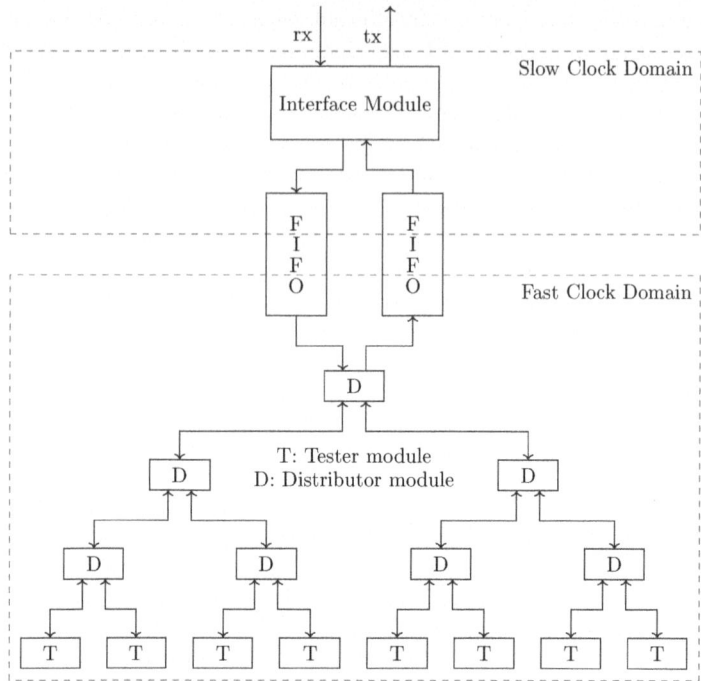

Fig. 3. High-Level Block Diagram of the FPGA Implementation

ister lengths, n, and forms of feedback functions. In our tests, we use software to generate candidate functions and to perform the initial pruning step to discard NLFSRs with cycles of period ≤ 12 from Sect. 3.1. A stream of candidate functions is then sent to the FPGA which sends back a filtered stream with only the maximum period feedback functions. Although the software component performs a non-trivial amount of work, the speed of the process is still purely limited by the period testing performed by the FPGA, especially for large n.

4.1 Dataset

We build a dataset that contains a complete list of the maximum period NLFSRs for lengths $10 \leq n \leq 32$ for 10 different forms of feedback functions. The types of feedback functions that we search for are functions whose ANF has 3, 5, 7, or 9 linear terms alongside one quadratic term, as well as functions with 2, 4, or 6 linear terms combined with two quadratic terms. Additionally, we consider functions with 3, 5, or 7 linear terms and one cubic term. We use a shorthand notation representing the number of terms in ascending order. For example "7,0,1" denotes functions with 7 linear terms and one cubic term. In accordance with Eq. 2, a reciprocal NLFSR can be trivially derived for every NLFSR found. In our dataset, we include only one of the two, specifically opting for the one with the lowest lexicographical order. Table 2 shows the number of maximum period NLFSR in the dataset for each of the different tested forms and shift register lengths. Combinations of forms and n where we have not performed the search are marked with a dash. In total, the dataset contains 21071 maximum period NLFSRs. The full dataset is available for download at [2].

Table 2. Number of Maximum Period NLFSRs for Different Forms

Form	n																						
	10	11	12	13	14	15	16	17	18	19	20	21	22	23	24	25	26	27	28	29	30	31	32
3,1	7	3	3	2	1	1	1	0	0	1	0	0	0	0	0	0	0	0	0	0	0	0	0
5,1	15	25	18	14	14	10	8	7	5	4	2	4	3	2	1	0	0	0	0	0	0	0	0
7,1	17	13	32	32	38	35	42	35	22	23	24	17	10	8	10	7	3	5	1	3	1	2	0
9,1	2	7	15	14	38	44	51	75	63	74	62	66	52	45	26	24	41	15	20	15	6	7	10
2,2	28	22	15	19	10	8	6	6	2	3	0	0	2	1	0	0	0	0	0	0	0	0	0
4,2	297	314	342	337	279	236	203	187	128	87	63	65	43	28	15	18	7	7	3	3	2	2	0
6,2	421	584	890	1067	1273	1330	1331	1329	1190	1006	892	757	604	448	391	279	211	148	120	–	–	–	–
3,0,1	13	15	17	7	11	4	3	1	3	3	2	0	1	0	0	1	0	0	0	0	0	0	0
5,0,1	54	49	69	84	76	66	52	39	32	19	28	17	15	9	6	1	3	3	3	1	0	0	0
7,0,1	53	58	96	148	160	169	199	178	156	150	121	136	104	75	63	49	17	20	25	10	7	7	1

4.2 32-Bit NLFSRs

As can be seen in Table 2 we have found 10 32-bit NLFSR with form "9,1" and 1 with form "7,0,1". The feedback functions for these are listed below. To the best of our knowledge, these are the largest maximum period NLFSRs with feedback functions of degree 2 or 3 publicly known. 32-bit NLFSRs are of special interest since their state can be represented by a whole number of bytes and they can be fairly efficiently implemented on 32-bit architecture systems. These functions could potentially be used as permutations in $\mathbb{F}_{2^{32}}$. We note that f_{32_2} does not use any of the 9 upper bits of the state, and thus one can trivially make an unrolled implementation that advances the state 9 bits at the time.

$$f_{32_1} = x_0 + x_4 + x_{10} + x_{12} + x_{17} + x_{22} + x_{25} + x_{26} + x_{31} + x_7 \cdot x_{10}$$
$$f_{32_2} = x_0 + x_1 + x_5 + x_6 + x_9 + x_{13} + x_{20} + x_{21} + x_{22} + x_{11} \cdot x_{15}$$
$$f_{32_3} = x_0 + x_4 + x_5 + x_{11} + x_{16} + x_{20} + x_{23} + x_{28} + x_{31} + x_3 \cdot x_{17}$$
$$f_{32_4} = x_0 + x_1 + x_3 + x_{16} + x_{26} + x_{27} + x_{28} + x_{29} + x_{31} + x_{11} \cdot x_{17}$$
$$f_{32_5} = x_0 + x_2 + x_6 + x_{10} + x_{15} + x_{17} + x_{18} + x_{29} + x_{30} + x_3 \cdot x_{19}$$
$$f_{32_6} = x_0 + x_8 + x_9 + x_{11} + x_{12} + x_{22} + x_{27} + x_{29} + x_{31} + x_8 \cdot x_{20}$$
$$f_{32_7} = x_0 + x_8 + x_{11} + x_{15} + x_{18} + x_{19} + x_{22} + x_{30} + x_{31} + x_4 \cdot x_{26}$$
$$f_{32_8} = x_0 + x_2 + x_3 + x_5 + x_{17} + x_{19} + x_{21} + x_{23} + x_{27} + x_6 \cdot x_{26}$$
$$f_{32_9} = x_0 + x_3 + x_{16} + x_{20} + x_{22} + x_{23} + x_{26} + x_{27} + x_{28} + x_1 \cdot x_{29}$$
$$f_{32_{10}} = x_0 + x_1 + x_8 + x_{10} + x_{15} + x_{19} + x_{22} + x_{26} + x_{30} + x_3 \cdot x_{29}$$
$$f_{32_{11}} = x_0 + x_2 + x_{10} + x_{18} + x_{19} + x_{22} + x_{30} + x_7 \cdot x_{19} \cdot x_{20}$$

5 Conclusion

In this paper, we have described the design of an FPGA-based accelerator for testing the period of NLFSRs. We have also described an initial pruning step based on using binary Lyndon words to quickly determine if an NLFSR contains very short cycles. The combination of these two was used to perform complete searches for maximum period NLFSRs with some simple feedback functions up to a length of 32. A dataset with the found NLFSRs and the source code for the FPGA implementation is made available for access.

Disclosure of Interest. The authors have no competing interests to declare that are relevant to the content of this article.

References

1. Armknecht, F., Mikhalev, V.: On lightweight stream ciphers with shorter internal states. Cryptology ePrint Archive, Paper 2015/131 (2015). https://eprint.iacr.org/2015/131
2. Askeland, A.: An FPGA accelerator for NLFSR period testing (2024). https://github.com/amundas/NLFSR
3. Augustynowicz, P., Kanciak, K.: Scalable method of searching for full-period nonlinear feedback shift registers with gpgpu. New list of maximum period nlfsrs. Int. J. Electron. Telecommun. **64**(2) (2018). https://doi.org/10.24425/119365, http://journals.pan.pl/Content/103846/PDF/24_1288-4312-1-PB.pdf
4. de Bruijn, N.: A combinatorial problem. Proceedings of the Section of Sciences of the Koninklijke Nederlandse Akademie van Wetenschappen te Amsterdam **49**(7), 758–764 (1946)
5. Canteaut, A.: Open problems related to algebraic attacks on stream ciphers. In: Ytrehus, Ø. (ed.) Coding and Cryptography, pp. 120–134. Springer, Berlin, Heidelberg (2006)
6. Dabrowski, P., Labuzek, G., Rachwalik, T., Szmidt, J.: Searching for nonlinear feedback shift registers with parallel computing. Inf. Process. Lett. **114**(5), 268–272 (2014)
7. Dubrova, E.: A list of maximum period NLFSRs. Cryptology ePrint Archive, Report 2012/166 (2012). https://eprint.iacr.org/2012/166
8. Golomb, S.W.: Shift Register Sequences. WORLD SCIENTIFIC, 3rd revised edn. (2017). https://doi.org/10.1142/9361, https://www.worldscientific.com/doi/abs/10.1142/9361
9. Hell, M., Johansson, T., Meier, W.: Grain: a stream cipher for constrained environments. Int. J. Wire. Mob. Comput. **2**(1), 86–93 (2007). https://doi.org/10.1504/IJWMC.2007.013798
10. Mikhalev, V., Armknecht, F., Müller, C.: On ciphers that continuously access the non-volatile key. IACR Trans. Symmetric Cryptol. **2016**(2), 52–79 (2017). https://doi.org/10.13154/tosc.v2016.i2.52-79, https://tosc.iacr.org/index.php/ToSC/article/view/565
11. Mossige, S.: Table of irreducible polynomials over GF[2] of degrees 10 through 20. Math. Comput. **26**(120), 1007–1009 (1972). http://www.jstor.org/stable/2005888

12. Mykkeltveit, J.: A proof of golomb's conjecture for the de bruijn graph. J. Comb. Theory Ser. B **13**, 40–45 (1972). https://api.semanticscholar.org/CorpusID: 42589667
13. Poluyanenko, N.: Development of the search method for non-linear shift registers using hardware, implemented on field programmable gate arrays. EUREKA: Phys. Eng. (1), 53–60 (2017). https://doi.org/10.21303/2461-4262.2017.00271, https://journal.eu-jr.eu/engineering/article/view/271
14. Preneel, B.: A survey of recent developments in cryptographic algorithms for smart cards. Comput. Netw. **51**(9), 2223–2233 (2007). https://doi.org/10.1016/j.comnet.2007.01.008, https://www.sciencedirect.com/science/article/pii/S1389128607000217, (1) Advances in Smart Cards and (2) Topics in Wireless Broadband Systems
15. Rachwalik, T., Szmidt, J., Wicik, R., Zablocki, J.: Generation of nonlinear feedback shift registers with special-purpose hardware. Cryptology ePrint Archive, Report 2012/314 (2012). https://eprint.iacr.org/2012/314
16. Witt, E.: Treue darstellung liescher ringe. Journal f ür die reine undangewandte Mathematik **1937**(177), 152–160 (1937). https://doi.org/10.1515/crll.1937.177.152

On Fat Linearized Polynomials

Olga Polverino, Paolo Santonastaso, and Ferdinando Zullo(✉)

Dipartimento di Matematica e Fisica, Università degli Studi della Campania "Luigi Vanvitelli", Viale Lincoln, 5, 81100 Caserta, Italy
{olga.polverino,paolo.santonastaso,ferdinando.zullo}@unicampania.it

Abstract. Fat polynomials were recently introduced as a generalization of scattered polynomials. These polynomials define linear sets on the projective line with r points of weight greater than one and also rank-metric codes with a certain number of matrices of lower rank. In this paper, we explore 1- and 2-fat polynomials, providing examples and some classification results.

Keywords: linearized polynomial · scattered polynomial · linear set · rank-metric code

1 Introduction

Linearized polynomials have been widely used to construct and to explore many objects, such as rank-metric codes, PN-functions and linear sets. An $\mathbb{F}_q$-**linearized polynomial** (or **q-polynomial**) in $\mathbb{F}_{q^n}[x]$ is a polynomial of the form

$$f(x) = \sum_{i=0}^{t} a_i x^{q^i}.$$

If $a_t \neq 0$ then t is called the q-**degree** of f. It is easy to see that any $\mathbb{F}_q$-linearized polynomial defines an $\mathbb{F}_q$-linear map of $\mathbb{F}_{q^n}$. The set of $\mathbb{F}_q$-linearized polynomials is denoted by $\mathcal{L}_{n,q}[x]$ and it forms an $\mathbb{F}_q$-algebra. We mainly consider the quotient algebra $\tilde{\mathcal{L}}_{n,q}[x] = \mathcal{L}_{n,q}[x]/(x^{q^n} - x)$, which is isomorphic to the algebra of all the $\mathbb{F}_q$-linear maps of $\mathbb{F}_{q^n}$. Elements of this algebra can also be identified with the polynomials of q-degree at most $n-1$. Recently, in connection with some combinatorial objects and optimal codes in the rank-metric, the notion of a scattered polynomial was introduced in [21]. An $\mathbb{F}_q$-linearized polynomial $f(x) \in \tilde{\mathcal{L}}_{n,q}[x]$ is said to be **scattered** if

$$\dim_{\mathbb{F}_q}(\ker(f(x) - mx)) \leq 1,$$

for every $m \in \mathbb{F}_{q^n}$. If an $\mathbb{F}_q$-linearized polynomial has the property that

$$\dim_{\mathbb{F}_q}(\ker(f(x) - mx)) \leq 1,$$

for every m in $\mathbb{F}_{q^n}$ except one, denoted $\overline{m}$, and if

$$i = \dim_{\mathbb{F}_q}(\ker(f(x) - \overline{m}x)),$$

then f is called an i-**club** polynomial. Extending these two notions, other generalizations have been proposed in [2,4,16] and investigated in [1,3,5,14].

In this direction, the notion of r-**fat polynomial** was recently introduced in [2, Definition 1.2], which can be seen as an extension of the definition of scattered polynomials.

Definition 1. *An $\mathbb{F}_q$-linearized polynomial $f(x) \in \tilde{\mathcal{L}}_{n,q}[x]$ is called r-**fat** if there exist exactly r elements $m_1, \ldots, m_r \in \mathbb{F}_{q^n}$ such that*

$$w_i = \dim_{\mathbb{F}_q}(\ker(f(x) - m_i x))) > 1 \qquad (1)$$

for any $i \in \{1, \ldots, r\}$. We say that $f(x)$ has maximum weight M if

$$\max\{\dim_{\mathbb{F}_q}(\ker(f(x) - m_i x)) : i \in \{1, \ldots, r\}\} = M.$$

*We say that f is an **exceptional** r-fat polynomial if it is an r-fat polynomial over infinitely many extensions of $\mathbb{F}_{q^n}$.*

The condition of being r-fat polynomial is equivalent to requiring that the linear set on the projective line defined by the function $f(x)$ has exactly r points of weight greater than one and M corresponds to the maximum weight of the points in such a linear set; as we will see later. As an example, scattered polynomials are 0-fat polynomials, and an i-club polynomial is a 1-fat polynomial with maximum weight i. Our aim is to give some classification results, which relies on the following notion of equivalence based on the action of $\Gamma L(2, q^n)$ on the graphs of the polynomials. Here, $\Gamma L(2, q^n)$ denotes the general semilinear group of order 2 over $\mathbb{F}_{q^n}$. The **graph** of the function f is defined as

$$\mathcal{G}_f = \{(x, f(x)) : x \in \mathbb{F}_{q^n}\} \subseteq \mathbb{F}_{q^n}^2.$$

So, two polynomials f and g in $\tilde{\mathcal{L}}_{n,q}[x]$ are said to be **equivalent** if there exists $\varphi \in \Gamma L(2, q^n)$ mapping the graph of f onto the graph of g, that is such that $\varphi(\mathcal{G}_f) = \mathcal{G}_g$ where $\mathcal{G}_f$ and $\mathcal{G}_g$ are respectively the graphs of f and g. The most general form of function equivalence that is known is the CCZ-equivalence, introduced by Carlet, Charpin and Zinoviev in [8]. For the case of interest of this paper, we say that two polynomials f and g in $\tilde{\mathcal{L}}_{n,q}[x]$ are said to be **CCZ-equivalent** if there exists an affine map φ of $\mathbb{F}_{q^n}^2$ such that $\varphi(\mathcal{G}_f) = \mathcal{G}_g$. Clearly, if f and g are equivalent, then they are also CCZ-equivalent. In this paper we will consider the equivalence based on the action of $\Gamma L(2, q^n)$ because of its connection with rank-metric codes and also because it turns out to be not interesting in this context, as we will prove later.

The problem of the classification of 0-fat polynomials has been deeply studied during the last years, and some results on the classification of exceptional 0-fat polynomials have been obtained in a series of papers; cf. [1,3,4,14]. On the other hand, in [2, Theorem 3.9], the authors prove that there is no exceptional r-fat polynomial in $\mathbb{F}_{q^n}[x]$ when $r \geq 1$. Therefore, the problem of finding examples of r-fat polynomials or classification results strongly depends on the field extension on which the polynomials are defined. The most challenging examples to be found or to classify are those for which r does not depend on q and n. In the next we will describe examples of 1- and 2-fat polynomials and we derive some properties. We will analyze the equivalence between a special class of 2-fat polynomials and we will pay attention to the case of polynomials over $\mathbb{F}_{q^5}$. Finally, we will describe consequences on rank-metric codes.

2 Connection with Linear Sets

Let V be a 2-dimensional $\mathbb{F}_{q^n}$-vector space and let $\Lambda = \mathrm{PG}(V, \mathbb{F}_{q^n}) = \mathrm{PG}(1, q^n)$ be the projective line defined by V over $\mathbb{F}_{q^n}$. The points of Λ correspond to the one-dimensional $\mathbb{F}_{q^n}$-subspace of V, that is the subspaces of the form $\langle \mathbf{u} \rangle_{\mathbb{F}_{q^n}}$ for some $\mathbf{u} \in V \setminus \{0\}$. Let $U \neq \{0\}$ be an $\mathbb{F}_q$-subspace of V, then the set

$$L_U = \{ \langle \mathbf{u} \rangle_{\mathbb{F}_{q^n}} : \mathbf{u} \in U \setminus \{0\} \}$$

is said to be an $\mathbb{F}_q$-**linear set** of rank $\dim_{\mathbb{F}_q}(U)$. The rank of L_U will also be denoted by $\mathrm{Rank}(L_U)$. The **weight** of a point $P = \langle \mathbf{v} \rangle_{\mathbb{F}_{q^n}} \in \Lambda$ in L_U is defined as $\dim_{\mathbb{F}_q}(U \cap \langle \mathbf{v} \rangle_{\mathbb{F}_{q^n}})$.

Denote by N_i the number of points of Λ having weight $i \in \{0, \ldots, k\}$ in L_U. The N_i's satisfy the following relations.

Proposition 1. *[19] Let L_U be an $\mathbb{F}_q$-linear set of rank k. Then*

$$|L_U| = N_1 + \ldots + N_k, \tag{2}$$

$$N_1 + N_2(q+1) + \ldots + N_k(q^{k-1} + \ldots + q + 1) = q^{k-1} + \ldots + q + 1, \tag{3}$$

and

$$|L_U| \leq \frac{q^k - 1}{q - 1}. \tag{4}$$

Linear sets where all its points have weight equal to 1 are called **scattered**, originally introduced in [7]. An i-**club of rank** k in $\mathrm{PG}(1, q^n)$ is an $\mathbb{F}_q$-linear set of rank k in $\mathrm{PG}(1, q^n)$ such that one point has weight i and all the others have weight one. If $k = n$, then we will simply call it an i-club.

When $\dim_{\mathbb{F}_q}(U) = n$, there exists a q-polynomial $f(x) \in \mathbb{F}_{q^n}[x]$ such that U is mapped by an element of $\mathrm{GL}(2, q^n)$ into

$$U_f = \{(x, f(x)) : x \in \mathbb{F}_{q^n}\}.$$

Hence L_U is mapped by the action of this matrix on the points of L_U into

$$L_f = \{ \langle (x, f(x)) \rangle_{\mathbb{F}_{q^n}} : x \in \mathbb{F}_{q^n}^* \}.$$

Note that the polynomial $f(x)$ is scattered if and only if L_f is a scattered $\mathbb{F}_q$-linear set. The above connection can be extended to r-fat polynomials.

Proposition 2. *Let $f(x) \in \tilde{\mathcal{L}}_{n,q}[x]$. Then $f(x)$ is an r-fat linearized polynomial with maximum weight M if and only if L_f has exactly r points with weight greater than one and $M = \max\{w_{L_U}(P) \colon P \in L_U\}$.*

Proof. Observe that $\langle (0,1) \rangle_{\mathbb{F}_{q^n}} \notin L_f$ and all the other points of $\mathrm{PG}(1, q^n)$ are of the form $\langle (1, m) \rangle_{\mathbb{F}_{q^n}}$ for $m \in \mathbb{F}_{q^n}$. For any $m \in \mathbb{F}_{q^n}$ we have that

$$w_{L_f}(\langle (1,m) \rangle_{\mathbb{F}_{q^n}}) = \dim_{\mathbb{F}_q}(U_f \cap \langle (1,m) \rangle_{\mathbb{F}_{q^n}}) = \dim_{\mathbb{F}_q}(\ker(f(x) - mx)). \quad (5)$$

Therefore, L_f has exactly r points of weight greater than one if and only if there exist r distinct elements $m_1, \ldots, m_r \in \mathbb{F}_{q^n}$ such that

$$\dim_{\mathbb{F}_q}(\ker(f(x) - m_i x))) > 1.$$

Clearly, using again (5) we have that $M = \max\{w_{L_U}(P) \colon P \in L_U\}$.

From now on we will assume that $n \geq 3$ as when $n = 2$ all nonzero $\mathbb{F}_q$-linearized polynomials not proportional to x are 0-fat polynomials.

3 Equivalence of Linearized Polynomials

The equivalence given on linearized polynomials based on the action of the semilinear group on the graph of linearized polynomial f preserve the following properties:

- the r-fat property;
- the set $\{\dim_{\mathbb{F}_q}(\ker(f(x) - mx)) \colon m \in \mathbb{F}_{q^n}\}$;
- the size of the value set of $f(x)/x$;
- the equivalence of the related rank-metric codes (see last section);

see e.g. [9].

As already noted in the Introduction, this notion is different from that of CCZ-equivalence. In this section we will show that the CCZ-equivalence for linearized polynomials is not useful as they all turns out to be CCZ equivalent to x and we point out the properties that the equivalence preserves.

First we show that a linearized polynomial is equivalent/CCZ-equivalent to an invertible linearized polynomial.

Lemma 1. *Let $f(x) \in \tilde{\mathcal{L}}_{n,q}[x]$ be a nonzero polynomial. Then f is equivalent to an invertible polynomial in $\tilde{\mathcal{L}}_{n,q}[x]$.*

Proof. Consider the following set of polynomials

$$\mathcal{A} = \{f(x) + mx \colon m \in \mathbb{F}_{q^n}\}.$$

Note that all the polynomials in $\mathcal{A}$ are equivalent to f. Suppose that in $\mathcal{A}$ all the polynomials are non-invertible, i.e.

$$\dim_{\mathbb{F}_q}(\ker(f(x) - mx)) > 0, \quad (6)$$

for any $m \in \mathbb{F}_{q^n}$. Observe that

$$w_{L_{U_f}}(\langle(1,m)\rangle_{\mathbb{F}_{q^n}}) = \dim_{\mathbb{F}_q}(U_f \cap \langle(1,m)\rangle_{\mathbb{F}_{q^n}}) = \dim_{\mathbb{F}_q}(\ker(f(x) - mx)),$$

for any m. Hence by (6) we have

$$w_{L_{U_f}}(\langle(1,m)\rangle_{\mathbb{F}_{q^n}}) > 0,$$

for any m, which implies that $L_{U_f} = \mathrm{PG}(1, q^n)$, a contradiction to (4). Therefore, there exists at least one invertible polynomial in $\mathcal{A}$.

We can now prove that all nonzero linearized polynomials are CCZ-equivalent to x.

Theorem 1. *Any nonzero linearized polynomial in $\tilde{\mathcal{L}}_{n,q}[x]$ is CCZ-equivalent to x.*

Proof. Let $f \in \tilde{\mathcal{L}}_{n,q}[x]$ be a nonzero polynomial. By Lemma (1) we may assume that f is invertible. Consider the map $\varphi \colon (a,b) \in \mathbb{F}_{q^n}^2 \mapsto (a, f^{-1}(b)) \in \mathbb{F}_{q^n}^2$, then we have that

$$\varphi(\mathcal{G}_f) = \mathcal{G}_x.$$

Therefore, because of Theorem 1 and because of the fact that the equivalence from the semilinear group preserve the fat property, in this paper we will only consider the notion of equivalence relying on the action of the semilinear group.

4 1-fat Polynomials

In [9, Problem 4] the authors asked to classify and to find examples of q-polynomials $f(x) \in \tilde{\mathcal{L}}_{n,q}[x]$ such that in the multiset $\{\{f(\alpha)/\alpha \colon \alpha \in \mathbb{F}_{q^n}^*\}\}$ there is a unique element which is represented more than $q - 1$ times, namely $q^i - 1$ times for some $i \in \{1, \ldots, n-1\}$. These polynomials correspond to 1-fat polynomials with weight i as L_f is an i-club.

In what follows we will describe the known examples of 1-fat polynomials arising from the known examples of i-clubs in $\mathrm{PG}(1, q^n)$.

Example of 1-fat polynomial in $\mathbb{F}_{q^n}$ with $M = n - 1$

The trace function of $\mathbb{F}_{q^n}$ over $\mathbb{F}_q$ defined as $\mathrm{Tr}_{q^n/q}(x) = x + x^q + \ldots + x^{q^{n-1}}$ is a trivial example of 1-fat polynomial. It can be proved that all the 1-fat polynomials in $\tilde{\mathcal{L}}_{n,q}[x]$ with maximum weight $n - 1$ are equivalent to $\mathrm{Tr}_{q^n/q}(x)$; see e.g. [10, Theorem 3.7].

When the degree n of the extension field is not prime, the trace example can be extended in the following way.

Examples of 1-fat polynomials in $\mathbb{F}_{q^n}$ with $M = n - m$ where $m \mid n$

Let $n = \ell m$, $\gcd(s,m) = 1$ and $\sigma \colon x \in \mathbb{F}_{q^n} \mapsto x^{q^s} \in \mathbb{F}_{q^n}$. Then

$$T(x) = \left(\mathrm{Tr}_{q^{\ell m}/q^m} \circ \sigma\right)(x) \in \tilde{\mathcal{L}}_{n,q}[x] \tag{7}$$

is a 1-fat polynomial with maximum weight $n - m$; see [11, Theorem 3.3].

Examples of 1-fat polynomials in $\mathbb{F}_{q^n}$ with $M = n - 2$

In [17, Theorem 5.1] an $(n-2)$-club of type L_f is described, hence by Proposition 2 the corresponding polynomial f is a 1-fat polynomial in $\mathbb{F}_{q^n}$ with maximum weight $n-2$. Precisely, let $\lambda \in \mathbb{F}_{q^n}^*$ such that $\{1, \lambda, \ldots, \lambda^{n-1}\}$ is an $\mathbb{F}_q$-basis of $\mathbb{F}_{q^n}$, and let

$$f(x) = \mathrm{Tr}_{q^n/q}(c_{n-2}x) + \lambda \mathrm{Tr}_{q^n/q}(c_{n-1}x) \tag{8}$$

where $\omega \in \mathbb{F}_{q^n}$ is such that $(1, \lambda, \ldots, \lambda^{n-3}, \lambda^{n-2} + \omega, \omega\lambda)$ is an ordered $\mathbb{F}_q$-basis of $\mathbb{F}_{q^n}$ and $(c_0, \ldots, c_{n-1})$ is its dual basis. Then $f(x)$ is a 1-fat polynomial with maximum weight $n-2$. In particular, if q is odd then we may choose $\omega = \lambda^{n-2}$ and in this case

$$c_i = \frac{1}{\delta} \sum_{j=0}^{n-i-1} \lambda^j a_{i+j+1},$$

for $i \in \{n-2, n-1\}$, where $p(x) = a_0 + a_1 x + \ldots + a_{n-1} x^{n-1} + x^n$ is the minimal polynomial of λ over $\mathbb{F}_q$ and $\delta = p'(\lambda)$.

The interest for these polynomials relies on the fact that when $q = 2$, they can be used to construct special type of arcs; see [11]. Another family of 1-fat polynomials, which extends (7), is the following.

Examples of 1-fat polynomials in $\mathbb{F}_{q^n}$ with $M \in \{n-t, n-t+1\}$ where $t \mid n$

Let $n = tr$, $t, r > 1$, $s(x) \in \tilde{\mathcal{L}}_{t,q}[x]$ a scattered polynomial. In [11, Lemma 3.6], the authors provided examples of i-clubs in $\tilde{\mathcal{L}}_{n,q}[x]$. More precisely, with the notation of [11, Lemma 3.6], they considered

$$U_{a,b} = \left\{ \left(s(x_0) - ax_0, bx_0 + \sum_{i=1}^{r-1} x_i \omega^i\right) : x_i \in \mathbb{F}_{q^t} \right\}, \tag{9}$$

for some fixed $a, b \in \mathbb{F}_{q^t}$ with $b \neq 0$, where $\{1, \omega, \ldots, \omega^{r-1}\}$ is an $\mathbb{F}_{q^t}$-basis of $\mathbb{F}_{q^n}$. So, they proved that $L_{U_{a,b}}$ is an i-club, with

$$i = \begin{cases} t(r-1), & \text{if } s(x) - ax \text{ is invertible over } \mathbb{F}_{q^t}, \\ t(r-1) + 1, & \text{otherwise}. \end{cases} \tag{10}$$

In [17, Theorem 5.3], the authors proved that $U_{a,b}$ is $\mathrm{GL}(2, q^n)$-equivalent to the graph of $f(x)$ where

$$f(x) = \mathrm{Tr}_{q^n/q^t}(s(x) - x). \tag{11}$$

Therefore, L_f defines an i-club and so f is a 1-fat polynomial with maximum weight $M \in \{n - t, n - t + 1\}$.

For sake of completeness, we will show that polynomials of the form (11) are 1-fat polynomials with maximum weight $M \in \{n - t, n - t + 1\}$, by using a slightly different approach. Here we will use the fact that the trace function $\mathrm{Tr}_{q^n/q}$ defines a symmetric non-degenerate and reflexive bilinear form over $\mathbb{F}_{q^n}$

$$(a, b) \in \mathbb{F}_{q^n} \times \mathbb{F}_{q^n} \mapsto \mathrm{Tr}_{q^n/q}(ab) \in \mathbb{F}_q.$$

This allows us to define the orthogonal complement map which is denoted by $\perp$.

We first prove the following lemma.

Lemma 2. *Let $(b_0 = 1, \ldots, b_{r-1})$ be an $\mathbb{F}_{q^t}$-basis of $\mathbb{F}_{q^{rt}}$ and let $(c_0, \ldots, c_{r-1})$ be its dual basis with respect to the non-degenerate symmetric bilinear form defined by the trace function. Let $s(x) \in \tilde{\mathcal{L}}_{t,q}[x]$, then for any $x = c_0 x_0 + c_1 x_1 + \ldots + c_{r-1} x_{r-1}$, with $x_0, \ldots, x_{r-1} \in \mathbb{F}_{q^t}$, then*

$$\mathrm{Tr}_{q^n/q^t}(s(x)) = s(x_0).$$

Proof. We have that

$$\mathrm{Tr}_{q^n/q^t}(s(x)) = s(\mathrm{Tr}_{q^n/q^t}(c_0 x_0 + c_1 x_1 + \ldots + c_{r-1} x_{r-1})) =$$

$$s\left(\sum_{i=0}^{r-1} x_i \mathrm{Tr}_{q^n/q^t}(c_i)\right) = s(x_0),$$

as $\mathrm{Tr}_{q^n/q^t}(c_i) = 0$ if $i \neq 0$ and $\mathrm{Tr}_{q^n/q^t}(c_0) = 1$.

We are now ready to prove that polynomials of the form (11) are 1-fat polynomials.

Proposition 3. *Let $n = tr$, $t, r > 1$ and let $s(x) \in \tilde{\mathcal{L}}_{t,q}[x]$ be a scattered polynomial. Then $\mathrm{Tr}_{q^n/q^t}(s(x) - x)$ is a 1-fat polynomial with maximum weight*

$$i = \begin{cases} t(r-1), & \text{if } s(x) - x \text{ is invertible over } \mathbb{F}_{q^t}, \\ t(r-1) + 1, & \text{otherwise.} \end{cases} \quad (12)$$

Proof. It is easy to see that $\dim_{\mathbb{F}_q}(\ker(\mathrm{Tr}_{q^n/q^t}(s(x) - x)))$ is i, so we just need to prove that for any $m \in \mathbb{F}_{q^n}^*$ we have

$$\dim_{\mathbb{F}_q}(\ker(\mathrm{Tr}_{q^n/q^t}(s(x) - x)) - mx) \leq 1.$$

Let $m \in \mathbb{F}_{q^n}^*$ and let $y, z \in \ker(\mathrm{Tr}_{q^n/q^t}(s(x) - x)) - mx)$. Let $(b_0 = 1, \ldots, b_{r-1})$ be an $\mathbb{F}_{q^t}$-basis of $\mathbb{F}_{q^{rt}}$ and let $(c_0, \ldots, c_{r-1})$ be its dual basis with respect to the non-degenerate symmetric bilinear form defined by the trace function. Let $y = c_0 y_0 + c_1 y_1 + \ldots + c_{r-1} y_{r-1}$ and $z = c_0 z_0 + c_1 z_1 + \ldots + c_{r-1} z_{r-1}$, with $y_0, \ldots, y_{r-1}, z_0, \ldots, z_{r-1} \in \mathbb{F}_{q^t}$, then by Lemma 2 we have

$$\begin{aligned} \mathrm{Tr}_{q^n/q^t}(s(y) - y) - my = s(y_0) - y_0 + my = 0, \\ \mathrm{Tr}_{q^n/q^t}(s(z) - z) - mz = s(z_0) - z_0 + mz = 0. \end{aligned} \quad (13)$$

Let $1/m = \sum_{i=0}^{r-1} \overline{m}_i c_i$ where $\overline{m}_i \in \mathbb{F}_{q^t}$. Using (13) we obtain that

$$s(y_0) - y_0 = y_0/m_0, \qquad (14)$$
$$s(z_0) - z_0 = z_0/m_0.$$

Using (14) and the fact that s is scattered we obtain that $\alpha = z_0/y_0 \in \mathbb{F}_q$. Therefore, (13) implies $z_i/y_i = \alpha \in \mathbb{F}_q$ for any i. As a consequence $z/y = \alpha$, which proves the assertion.

We can give the following characterization of 1-fat polynomials in $\tilde{\mathcal{L}}_{n,q}[x]$ with maximum weight $n-2$.

Theorem 2. *Let $f(x)$ be a 1-fat polynomial in $\tilde{\mathcal{L}}_{n,q}[x]$ with maximum weight $n-2$. There exist $b \in \mathbb{F}_{q^n} \setminus \mathbb{F}_q$ and $\xi, \eta \in \mathbb{F}_{q^n}^*$ such that $f(x)$ is equivalent to*

$$g(x) = \mathrm{Tr}_{q^n/q}(\xi x) + b \mathrm{Tr}_{q^n/q}(\eta x), \qquad (15)$$

with $\langle \xi, \eta \rangle_{\mathbb{F}_q}^\perp = Z \oplus c \langle 1, b, \ldots, b^{t-3} \rangle_{\mathbb{F}_q}$, where $t = [\mathbb{F}_q(b) : \mathbb{F}_q]$, $c \in \mathbb{F}_{q^n}^*$ and Z is an $\mathbb{F}_{q^t}$-subspace of $\mathbb{F}_{q^n}$ of dimension $n/t - 1$.

Proof. The proof follows from the fact that f is a 1-fat polynomial with maximum weight $n-2$ if and only if L_f is an $(n-2)$-club. By [17, Theorem 5.4] this can happen if and only if the graph of f is $\mathrm{GL}(2, q^n)$-equivalent to the graph of g, where g is one of the polynomials described in (15).

Note that when $t = n$, the examples described in (15) are of type described in (8).

We investigate the equivalence problem of the polynomials as in (11).

Theorem 3. *Let $n = tr$, $t, r > 1$, $s_1(x), s_2(x) \in \tilde{\mathcal{L}}_{t,q}[x]$ scattered polynomials. Then*

$$f_1(x) = \mathrm{Tr}_{q^n/q^t}(s_1(x) - x)$$

and

$$f_2(x) = \mathrm{Tr}_{q^n/q^t}(s_2(x) - x)$$

are equivalent if and only if $s_1(x) - x$ and $s_2(x) - x$ are equivalent in $\tilde{\mathcal{L}}_{t,q}$ via an element $\phi \in \Gamma\mathrm{L}(2, q^t)$ that fixes the subspace $\langle (1, 0) \rangle_{\mathbb{F}_{q^t}}$.

Proof. Let $(1, \omega, \ldots, \omega^{r-1})$ be an $\mathbb{F}_{q^t}$-basis of $\mathbb{F}_{q^{rt}}$ and let $(c_0, \ldots, c_{r-1})$ be its dual basis and let $S = \langle \omega, \ldots, \omega^{r-1} \rangle_{\mathbb{F}_{q^t}}$. First, suppose that $f_1(x)$ and $f_2(x)$ are equivalent. Clearly, $f_1(c_0 x)$ and $f_2(c_0 x)$ are also equivalent. By definition, there exists a matrix $\begin{pmatrix} A & B \\ C & D \end{pmatrix} \in \mathrm{GL}(2, q^n)$ and $\rho \in \mathrm{Aut}(\mathbb{F}_{q^n})$ such that

$$\begin{pmatrix} A & B \\ C & D \end{pmatrix} (\mathcal{G}_{f_1(c_0 x)}^\rho)^T = (\mathcal{G}_{f_2(c_0 x)})^T, \qquad (16)$$

where $\mathcal{G}_{f_i(c_0x)} = \{(x, \mathrm{Tr}_{q^n/q^t}(s_i(c_0x) - c_0x)) \colon x \in \mathbb{F}_{q^n}\}$, for $i \in \{1, 2\}$. Since $s_1(x), s_2(x) \in \tilde{\mathcal{L}}_{t,q}[x]$, we have that

$$\mathrm{Tr}_{q^n/q^t} \circ (s_i - \mathrm{id}) = (s_i - \mathrm{id}) \circ \mathrm{Tr}_{q^n/q^t},$$

for $i \in \{1, 2\}$. So, writing each $x \in \mathbb{F}_{q^n}$, with respect to the basis $(1, \omega, \ldots, \omega^{r-1})$ and using Lemma 2, we get that

$$f_i(c_0x) = s_i(x_0) - x_0, \text{ for each } x = \sum_{i=0}^{r-1} x_i\omega^i \in \mathbb{F}_{q^n},$$

and

$$\mathcal{G}_{f_i(c_0x)} = \left\{ \left(x_0 + \sum_{i=1}^{r-1} x_i\omega^i, s_i(x_0) - x_0\right) \colon x = \sum_{i=0}^{r-1} x_i\omega^i \in \mathbb{F}_{q^n} \right\},$$

for $i \in \{1, 2\}$. Therefore, $S = \langle \omega, \ldots, \omega^{r-1} \rangle_{\mathbb{F}_{q^t}} \subseteq \ker(f_i(c_0x))$, implying that

$$\dim_{\mathbb{F}_q}(\ker(f_i(c_0x))) \geq (r-1)t.$$

Moreover, by (8), we know that $f_1(x)$ and $f_2(x)$ are 1-fat polynomials and so $f_1(c_0x)$ and $f_2(c_0x)$ are 1-fat polynomials, as well. This implies that $\dim_{\mathbb{F}_q}(\ker(f_i(c_0x))) = M$, where M is as in (12) and $\dim_{\mathbb{F}_q}(\ker(f_i(c_0x)) - mx) \in \{0, 1\}$, for every $m \in \mathbb{F}_{q^n}^*$. Then, by (5), we get that $\langle (1, 0) \rangle_{\mathbb{F}_{q^n}}$ is the only point in $L_{f_1(c_0x)}$ and $L_{f_2(c_0x)}$ having weight greater than 1, then $C = 0, A \neq 0$ and

$$AZ_1^\rho = Z_2, \tag{17}$$

where $Z_i = \ker(f_i(c_0x))$. Note that $\dim_{\mathbb{F}_q}(Z_i) = M \in \{(r-1)t, (r-1)t+1\}$, and so $Z_i = S \oplus \alpha_i\mathbb{F}_q$, where $\alpha_i \notin S$ or $\alpha_i = 0$ according whether $M = (r-1)t+1$ or $M = (r-1)t$, respectively. Note also that since AS^ρ is an $\mathbb{F}_{q^t}$-subspace of Z_2 and S is the maximum $\mathbb{F}_{q^t}$-subspace contained in Z_i, by (17), we also have

$$AS^\rho = S. \tag{18}$$

We have that (16) is satisfied if and only if for every $x \in \mathbb{F}_{q^n}$ there exist $y \in \mathbb{F}_{q^n}$ such that

$$Ax^\rho + B(f_1(c_0x))^\rho = y \tag{19}$$

and

$$D(f_1(c_0x))^\rho = f_2(c_0y). \tag{20}$$

Write $x = \sum_{i=0}^{r-1} x_i\omega^i$, $y = \sum_{i=0}^{r-1} y_i\omega^i$, $A = \sum_{i=0}^{r-1} A_i\omega^i$ and $B = \sum_{i=0}^{r-1} B_i\omega^i$, with x_i, y_i, A_i, B_i's in $\mathbb{F}_{q^t}$. Now, (19) reads as follows

$$\sum_{i=0}^{r-1} A_i\omega^i x^\rho + \sum_{i=0}^{r-1} B_i\omega^i (f_1(c_0x))^\rho = \sum_{i=0}^{r-1} y_0\omega^i.$$

Moreover, (18) implies that $A(\sum_{i=1}^{r} x_i \omega^i)^\rho \in S$, and so, the above equality implies
$$A_0 x_0^\rho + B_0 (f_1(c_0 x))^\rho = y_0. \tag{21}$$

Note also that $D \in \mathbb{F}_{q^t}$, because of Equation (20). Suppose that $A_0 = 0$, then (21) together with (20) implies that $s_2(y_0) - y_0 = f_2(c_0 y) = \frac{D}{B_0} y_0$, for every $y_0 \in \mathbb{F}_{q^t}$. So, $s_2(x)$ is not a scattered polynomial, a contradiction. So $A_0 \neq 0$. This means that the spaces

$$\mathcal{G}_{s_i} = \{(y_0, s_i(y_0) - y_0) : y_0 \in \mathbb{F}_{q^t}\}$$

are $\Gamma L(2, q^t)$-equivalent via the map of $\Gamma L(2, q^t)$ defined by the matrix $\begin{pmatrix} A_0 & B_0 \\ 0 & D \end{pmatrix}$ and the restriction of the automorphism ρ to $\mathbb{F}_{q^t}$.

Conversely, suppose that $s_1(x) - x$ and $s_2(x) - x$ are equivalent over $\tilde{\mathcal{L}}_{t,q}[x]$ via an element $\phi \in \Gamma L(2, q^t)$ such that $\phi(\langle(1,0)\rangle_{\mathbb{F}_{q^t}}) = \langle(1,0)\rangle_{\mathbb{F}_{q^t}}$. Up to replacing $s_1(x) = \sum_{i=0}^{t-1} a_i x^{q^i}$ (or s_2) with $s_1^\rho(x) = \sum_{i=0}^{t-1} \rho(a_i) x^{q^i}$, we can assume that $\phi \in \mathrm{GL}(2, q^n)$. Then ϕ can be represented by a matrix $\begin{pmatrix} A_0 & B_0 \\ 0 & D_0 \end{pmatrix} \in \mathrm{GL}(2, q^t)$ such that

$$\begin{pmatrix} A_0 & B_0 \\ 0 & D_0 \end{pmatrix} \mathcal{G}_{s_1(x)-x}^T = \mathcal{G}_{s_2(x)-x}^T, \tag{22}$$

where $\mathcal{G}_{s_i(x)-x} = \{(x, s_i(x) - x) : x \in \mathbb{F}_{q^t}\}$, for $i \in \{1, 2\}$. It is easy to check that $\phi(U_1) = U_2$, where

$$U_1 = \left\{ \left(x_0 + \sum_{i=1}^{r-1} x_i \omega^i, s_1(x_0) - x_0 \right) : x_i \in \mathbb{F}_{q^t} \right\}$$

and

$$U_2 = \left\{ \left(x_0 + \sum_{i=1}^{r-1} x_i \omega^i, s_2(x_0) - x_0 \right) : x_i \in \mathbb{F}_{q^t} \right\}.$$

Arguing as in the previous implication, U_i is $\mathrm{GL}(2, q^n)$-equivalent to

$$\{(x, \mathrm{Tr}_{q^n/q^t}(s_1(c_0 x) - c_0 x)) : x \in \mathbb{F}_{q^n}\} = \mathcal{G}_{f_i(x)}$$

and hence $f_1(x)$ and $f_1(x)$ are equivalent.

Therefore, the number of inequivalent 1-fat polynomials is at least as large as the set of inequivalent scattered polynomials.

Corollary 1. *The number of inequivalent 1-fat polynomials over $\mathbb{F}_{q^{rt}}$ is at least the number of inequivalent scattered polynomials over $\mathbb{F}_{q^t}$ under the action of the subgroup of $\Gamma L(2, q^t)$ fixing $\langle(1,0)\rangle_{\mathbb{F}_{q^t}}$. In particular, the number of inequivalent 1-fat polynomials over $\mathbb{F}_{q^{rt}}$ is at least the number of inequivalent scattered polynomials over $\mathbb{F}_{q^t}$.*

5 2-fat Polynomials and Complementary Weights

In this section we provide some necessary conditions to guarantee the existence of certain 2-fat polynomials and we will show some examples.

For $n = 2$ there are no 2-fat polynomials and for $n = 3$ the only 2-fat polynomials are of the form $\alpha \mathrm{Tr}_{q^3/q}(\beta x)$ for some $\alpha, \beta \in \mathbb{F}_{q^3}^*$. For $n = 4$, we have a complete classification of all the r-fat polynomials and they all satisfy the conditions of Theorem 5 (see [2, Section 4]) and there are two inequivalent classes.

For $n = 5$, the following holds.

Theorem 4. *If $q \geq 5$ then there do not exist 2-fat polynomials in $\tilde{\mathcal{L}}_{5,q}[x]$.*

Proof. By [12, Main Theorem], the possible values for r are

$$r \in \{0, 1\} \cup [q - 2\sqrt{q} + 1, q + 2\sqrt{q} + 1] \cup \{2q, 2q + 1, 2q + 2, 3q, 3q + 1, q^2 + 1\}.$$

It is easy to check that when $q \geq 5$ then

$$2 \notin \{0, 1\} \cup [q - 2\sqrt{q} + 1, q + 2\sqrt{q} + 1] \cup \{q+1, 2q, 2q+1, 2q+2, 3q, 3q+1, q^2+1\}.$$

When $q \in \{2, 3, 4\}$, by [12, Theorems 1.2, 1.3 and 1.4].

With the aid of MAGMA computations, we show some examples of 2-fat polynomials in $\tilde{\mathcal{L}}_{5,q}[x]$ when $q \in \{2, 3, 4\}$.

For $q = 2$, let $w \in \mathbb{F}_{q^5}$ be a root of $x^5 + x^2 + 1$. Then $f(x) = x^q + w^{11}x^{q^2} + wx^{q^3} + w^2x^{q^4}$ is a 2-fat polynomial. For $q = 3$, let $w \in \mathbb{F}_{q^5}$ be a root of $x^5 + 2x + 1$. Then $f(x) = x^q + w^2 x^{q^2} + w^{127} x^{q^3} + w^{214} x^{q^4}$ is a 2-fat polynomial. For $q = 4$, let $z \in \mathbb{F}_q$ such that $z^2 + z + 1 = 0$ and let $w \in \mathbb{F}_{q^5}$ be such that $w^5 + w^3 + w + z = 0$. Then $f(x) = x^q + w^{68} x^{q^2} + w^{460} x^{q^3} + w^{837} x^{q^4}$ is a 2-fat polynomial.

So we got the following result.

Corollary 2. *There exists a 2-fat polynomial in $\tilde{\mathcal{L}}_{5,q}[x]$ if and only if $q \in \{2, 3, 4\}$.*

When n is larger very few is known and the constructions of 2-fat polynomials seem to be very hard to describe. We start by proving a necessary condition on the existence of 2-fat polynomials involving the only two values w_1 and w_2, where the w_i's are as in (1). In terms of linear set, this corresponds to talk about linear sets with complementary weights.

Theorem 5. *Let $f(x) \in \tilde{\mathcal{L}}_{n,q}[x]$ be a 2-fat polynomial. Let $m_1, m_2 \in \mathbb{F}_{q^n}$ with $m_1 \neq m_2$, such that $w_1 = \dim_{\mathbb{F}_q}(\ker(f(x) - m_1 x))) > 1$, $w_2 = \dim_{\mathbb{F}_q}(\ker(f(x) - m_2 x))) > 1$. Assume that $w_1 + w_2 = n$. Then n is even and $w_1 = w_2 = n/2$.*

Proof. The polynomial $f(x)$ defines the $\mathbb{F}_q$-linear set

$$L_f = \{\langle (x, f(x)) \rangle_{\mathbb{F}_{q^n}} : x \in \mathbb{F}_{q^n}^*\}.$$

Since $f(x)$ is a 2-fat polynomial, we have that L_f has exactly two points of weight greater than one. By [18, Corollary 4.3] the assertion follows.

We will show some examples of 2-fat polynomials as in the above theorem. Therefore, we will assume that $n = 2t$.

In the next lines, we will provide an easy way to describe r-fat polynomials with $r \geq 2$, relying on the projection of certain subspaces.

Let S and T be $\mathbb{F}_q$-subspaces of $\mathbb{F}_{q^n}$ such that $\mathbb{F}_{q^n} = S \oplus T$, $\dim_{\mathbb{F}_q}(T) = t$ and let $p_{T,S}$ be the $\mathbb{F}_q$-linear map of $\mathbb{F}_{q^n}$ projecting an element of $\mathbb{F}_{q^n}$ from T onto S. The map $p_{T,S}$ can be also written as a linearized polynomial as follows

$$p_{T,S}(x) = \sum_{i=t}^{n-1} \xi_i \mathrm{Tr}_{q^n/q}(\xi_i^* x) = \sum_{j=0}^{n-1} \left(\sum_{i=t}^{n-1} \xi_i \cdot \xi_i^{*q^j} \right) x^{q^j},$$

where $\mathcal{B}_T = \{\xi_0, \ldots, \xi_{t-1}\}$ is an $\mathbb{F}_q$-basis of T, $\mathcal{B}_S = \{\xi_t, \ldots, \xi_{n-1}\}$ is an $\mathbb{F}_q$-basis of S and $\mathcal{B}^* = (\xi_0^*, \ldots, \xi_{n-1}^*)$ is the dual basis of the ordered basis $\mathcal{B} = (\xi_0, \ldots, \xi_{n-1})$. Since $\ker(p_{T,S}) = T$ and $p_{T,S}(y) = y$ if and only if $y \in S$ then

$$\dim_{\mathbb{F}_q}(\ker(p_{T,S}(x))) = \dim_{\mathbb{F}_q}(T) \text{ and } \dim_{\mathbb{F}_q}(\ker(p_{T,S}(x) - x)) = \dim_{\mathbb{F}_q}(S),$$

and so $p_{T,S}(x)$ is an r-fat polynomials with $r \geq 2$. In the next we will describe possible choices for the subspace T and S for which the polynomial $p_{T,S}$ gives a 2-fat polynomial.

By [18, Proposition 4.18] we have a first example of 2-fat polynomial with maximum weight t.

Proposition 4. *Let $n = 2t$ be an even positive integer, let $f(z) = \sum_{i=0}^{t-1} A_i z^{q^i} \in \tilde{\mathcal{L}}_{t,q}[x]$ be a scattered polynomial and let $\epsilon \in \mathbb{F}_{q^n} \setminus \mathbb{F}_{q^t}$. Then*

$$p(x) = \mathrm{Tr}_{q^n/q^t}\left(\frac{A_0 + \epsilon^{q^t}}{\epsilon^{q^t} - \epsilon} x + \sum_{i=1}^{t-1} \frac{A_i}{\epsilon^{q^{i+t}} - \epsilon^{q^i}} x^{q^i} \right) =$$

$$\mathrm{Tr}_{q^n/q^t}\left(f\left(\frac{x}{\epsilon^{q^t} - \epsilon} \right) \right) + \mathrm{Tr}_{q^n/q^t}\left(\frac{\epsilon^{q^t} x}{\epsilon^{q^t} - \epsilon} \right),$$

is a 2-fat polynomial with maximum weight t.

Note that p can be described as projection from $\epsilon S_f = \{\epsilon z + f(z) \colon z \in \mathbb{F}_{q^t}\}$ onto $\mathbb{F}_{q^t}$.

Choosing $f(z) = z^{q^s}$ with $\gcd(s,t) = 1$ we obtain the following result.

Corollary 3. *Let $n = 2t$ be an even positive integer and let s be a positive integer such that $\gcd(s,t) = 1$. Let $\epsilon \in \mathbb{F}_{q^n} \setminus \mathbb{F}_{q^t}$, then*

$$p(x) = \mathrm{Tr}_{q^n/q^t}\left(\frac{\epsilon^{q^t}}{\epsilon^{q^t} - \epsilon} x + \frac{1}{\epsilon^{q^{s+t}} - \epsilon^{q^s}} x^{q^s} \right),$$

is a 2-fat polynomial in $\mathbb{F}_{q^n}$ with maximum weight t.

Another possible choice for $f(z)$ is given by

$$f(z) = z^{q^s} + \delta z^{q^{s(t-1)}},$$

for some s such that $\gcd(s,t) = 1$ and $\delta \in \mathbb{F}_{q^t}$ such that $N_{q^t/q}(\delta) \neq 1$.
From [18, Theorem 4.20] we can derive the following example.

Proposition 5. *Let s be a positive integer coprime with t and $\xi \in \mathbb{F}_{q^{2t}}$ such that $\{1,\xi\}$ is an $\mathbb{F}_{q^t}$-basis of $\mathbb{F}_{q^{2t}}$. Let $f(x) = x^{q^s}$ and $g(x) = \mu x^{q^s}$ with $\mu \in \mathbb{F}_{q^t}$ such that $N_{q^t/q}(\mu) \neq 1$ and $N_{q^t/q}(-\xi^{q^t+1}\mu) \neq (-1)^t$. Let*

$$p(x) = \sum_{k=0}^{n-1} \left(\sum_{\ell=0}^{t-1} (u_\ell + u_\ell^{q^s} \xi) \lambda_\ell^{*q^k} \right) x^{q^k},$$

$\{u_0, \ldots, u_{t-1}\}$ is an $\mathbb{F}_q$-basis of $\mathbb{F}_{q^t}$ and $(\lambda_0^*, \ldots, \lambda_{n-1}^*)$ is the dual basis of $(u_0 + \mu u_0^{q^s}\xi, \ldots, u_{t-1} + \mu u_{t-1}^{q^s}\xi, u_0 + u_0^{q^s}\xi, \ldots, u_{t-1} + u_{t-1}^{q^s}\xi)$. Then $p(x)$ is a 2-fat polynomial with maximum weight t.

6 r-fat Polynomials and Rank-Metric Codes

Rank-metric codes have obtained significant attention, because of their applications and intriguing connections with various mathematical objects, such as in linear random network coding, see [6]. Nevertheless, the origin of rank-metric codes traces its roots to Delsarte [13] in 1978, and later to Gabidulin in [15] and Roth in [20]. A **rank-metric code** is a subset of matrices of $\mathbb{F}_q^{m \times n}$ equipped with the rank distance. When $m = n$, a rank-metric code can be also seen as a subset of linearized polynomials, as $\mathbb{F}_q^{n \times n}$ is an $\mathbb{F}_q$-algebra isomorphic to $\tilde{\mathcal{L}}_{n,q}[x]$. So, given a polynomial $f(x) \in \tilde{\mathcal{L}}_{n,q}[x]$ we can define the following subspace

$$\mathcal{C}_f = \langle x, f(x) \rangle_{\mathbb{F}_{q^n}} \subseteq \tilde{\mathcal{L}}_{n,q}[x].$$

The matrices associated with the polynomials in $\mathcal{C}_f$ form a rank-metric code with minimum distance equals to

$$\min\{\dim_{\mathbb{F}_q}(\mathrm{Im}(\alpha x + \beta f(x))) \colon \alpha, \beta \in \mathbb{F}_{q^n}, (\alpha, \beta) \neq (0,0)\}.$$

Denote by A_i the number of polynomials in $\mathcal{C}_f$ having rank i.

Proposition 6. *Let $f(x) \in \tilde{\mathcal{L}}_{n,q}[x]$. Then $f(x)$ is an r-fat polynomial with maximum weight M if and only if $\mathcal{C}_f$ has minimum distance $n - M$ and*

$$\sum_{i=n-M}^{n-2} A_i = (q^n - 1)r.$$

Proof. The assertion follows from the fact that

$$\dim_{\mathbb{F}_q}(\mathrm{Im}(\alpha x + \beta f(x))) = n - \dim_{\mathbb{F}_q}(\ker(\alpha x + \beta f(x))),$$

for any $\alpha, \beta \in \mathbb{F}_{q^n}$.

Therefore, by the above proposition, the previous results immediately can be translated in terms of rank-metric codes.

Corollary 4. *Let f be a 1-fat polynomial in $\mathbb{F}_{q^n}$ with maximum weight M. Then $\mathcal{C}_f$ has*

- *$q^n - 1$ codewords of weight $n - M$;*
- *$(q^n - 1)(q^{n-1} + \ldots + q^M)$ of weight $n - 1$;*
- *$(q^n - 1)(q^n - q^{n-1} - \ldots - q^M)$ of weight n.*

For the codes associated with 2-fat polynomials the following holds.

Corollary 5. *Let f be a 2-fat polynomial in $\mathbb{F}_{q^n}$. Let $m_1, m_2 \in \mathbb{F}_{q^n}$ such that $w_1 = \dim_{\mathbb{F}_q}(\ker(f(x) - m_1 x))) > 1$, $w_2 = \dim_{\mathbb{F}_q}(\ker(f(x) - m_2 x))) > 1$. Then, if $w_1 \neq w_2$, $\mathcal{C}_f$ has*

- *$q^n - 1$ codewords of weight $n - w_1$;*
- *$q^n - 1$ codewords of weight $n - w_2$;*
- *$(q^n - 1)(q^{n-1} + \ldots + q^{\max\{w_1,w_2\}} - q^{\min\{w_1,w_2\}-1} - \ldots - q - 1)$ of weight $n - 1$;*
- *$(q^n - 1)(q^n - (q^{n-1} + \ldots + q^{\max\{w_1,w_2\}} - q^{\min\{w_1,w_2\}-1} - \ldots - q))$ of weight n.*

While, if $M = w_1 = w_2$, $\mathcal{C}_f$ has

- *$2(q^n - 1)$ codewords of weight $n - M$;*
- *$(q^n - 1)(q^{n-1} + \ldots + q^M - q^{M-1} - \ldots - q - 1)$ of weight $n - 1$;*
- *$(q^n - 1)(q^n - (q^{n-1} + \ldots + q^M - q^{M-1} - \ldots - q))$ of weight n.*

Moreover, if $w_1 + w_2 = n$. Then n is even and $w_1 = w_2 = n/2$.

Corollary 2 has natural implications on rank-metric codes with the parameters and weight distribution as in the above corollary.

Corollary 6. *Let $n = 5$. A rank-metric code $\mathcal{C}$ with the parameters and the weight distribution as in Corollary 5 exists if and only if $q \leq 4$.*

Proof. Every such a code, up to equivalence, can be always written as a $\mathcal{C}_f$ for some $f \in \tilde{\mathcal{L}}_{5,q}[x]$. The weight distribution of $\mathcal{C}$, by Proposition 6, immediately implies that f is a 2-fat polynomial in $\tilde{\mathcal{L}}_{5,q}[x]$. The assertion follows by Corollary 2.

Acknowledgements. The authors are thankful to Lukas Koelsch for fruitful discussions on the CCZ equivalence, and also to the reviewers which sensibly improve the paper. The research was supported by the project "COMBINE" of the University of Campania "Luigi Vanvitelli" and was partially supported by the Italian National Group for Algebraic and Geometric Structures and their Applications (GNSAGA - INdAM).

References

1. Bartoli, D., Giulietti, M., Zini, G.: Towards the classification of exceptional scattered polynomials. arXiv preprint arXiv:2206.13795, 2022
2. Bartoli, D., Micheli, G., Zini, G., Zullo, F.: r-fat linearized polynomials over finite fields. J. Comb. Theory Ser. A **189**, 105609 (2022)
3. Bartoli, D., Montanucci, M.: On the classification of exceptional scattered polynomials. J. Comb. Theory Ser. A **179**, 105386 (2021)
4. Bartoli, D., Zhou, Y.: Exceptional scattered polynomials. J. Algebra **509**, 507–534 (2018)
5. Bartoli, D., Zini, G., Zullo, F.: Investigating the exceptionality of scattered polynomials. Finite Fields Appl. **77**, 101956 (2022)
6. Bartz, H., Holzbaur, L., Liu, H., Puchinger, S., Renner, J., Wachter-Zeh, A.: Rank-metric codes and their applications. Found. Trends® Commun. Inf. Theory **19**(3), 390–546 (2022)
7. Blokhuis, A., Lavrauw, M.: Scattered spaces with respect to a spread in $\mathrm{PG}(n,q)$. Geom. Dedicata. **81**(1), 231–243 (2000)
8. Carlet, C., Charpin, P., Zinoviev, V.: Codes, bent functions and permutations suitable for DES-like cryptosystems. Des. Codes Crypt. **15**, 125–156 (1998)
9. Csajbók, B., Marino, G., Polverino, O.: A Carlitz type result for linearized polynomials. Ars Math. Contemp. **16**, 585–608 (2019)
10. Csajbók, B., Marino, G., Polverino, O.: Classes and equivalence of linear sets in $\mathrm{PG}(1,q^n)$. J. Comb. Theory Ser. A **157**, 402–426 (2018)
11. De Boeck, M., Van de Voorde, G.: A linear set view on KM-arcs. J. Algebraic Comb. **44**(1), 131–164 (2016)
12. De Boeck, M., Van de Voorde, G.: The weight distributions of linear sets in $\mathrm{PG}(1,q^5)$. Finite Fields Appl. **82**, 102034 (2022)
13. Delsarte, P.: Bilinear forms over a finite field, with applications to coding theory. J. Comb. Theory Ser. A **25**(3), 226–241 (1978)
14. Ferraguti, A., Micheli, G.: Exceptional scatteredness in prime degree. J. Algebra **565**, 691–701 (2021)
15. Gabidulin, E.M.: Theory of codes with maximum rank distance. Problemy Peredachi Informatsii **21**(1), 3–16 (1985)
16. Longobardi, G., Zanella, C.: Partially scattered linearized polynomials and rank metric codes. Finite Fields Appl. **76**, 101914 (2021)
17. Napolitano, V., Polverino, O., Santonastaso, P., Zullo, F.: Clubs and their applications. SIAM J. Appl. Algebra Geom. **8**(3), 493–518 (2024)
18. Napolitano, V., Polverino, O., Santonastaso, P., Zullo, F.: Linear sets on the projective line with complementary weights. Discret. Math. **345**(7), 112890 (2022)
19. Polverino, O.: Linear sets in finite projective spaces. Discret. Math. **310**(22), 3096–3107 (2010)
20. Roth, R.M.: Maximum-rank array codes and their application to crisscross error correction. IEEE Trans. Inf. Theory **37**(2), 328–336 (1991)
21. Sheekey, J.: A new family of linear maximum rank distance codes. Adv. Math. Commun. **10**(3), 475 (2016)

Counting Polynomials with Distinct Roots Using Subset Sum

Simon Kuttner(✉), Zhicheng Gao(✉), and Qiang Wang(✉)

School of Mathematics and Statistics, Carleton University, Ottawa, Canada
simonkuttner@cmail.carleton.ca, {zgao,wang}@math.carleton.ca

Abstract. Given a finite abelian group G, a finite set D, and a mapping $f : D \to G$, we find the number of r-subsets $S \subseteq D$ where for $b \in G$,

$$\sum_{x \in S} f(x) = b.$$

We count degree n monic polynomials over $\mathbb{F}_q$ with r distinct roots in a set $D \subseteq \mathbb{F}_q$ when the leading terms of degree at least $n - \ell$ are fixed. We obtain new formulas for $\ell = 2$ when D is an arbitrary subfield of $\mathbb{F}_q$ with q odd.

Keywords: subset sum · prescribed coefficients · distinct roots

Let G be a finite abelian group under addition with identity 0. Suppose D is a finite set and $f : D \to G$ is any mapping. For $b \in G$, we define

$$S_G\left(D^{r,f}, b\right) = \left|\left\{S \subseteq D : |S| = r, \sum_{x \in S} f(x) = b\right\}\right|. \quad (1)$$

For convenience, the mapping f will be omitted when $D \subseteq G$ and $f(x) = x$. Thus $S_G(D^r, b)$ is the number of r-subsets of D with sum b. By convention the sum over the empty set is equal to 0. In the rest of the paper, we shall use p_n to denote the power mapping $p_n(x) = x^n$.

This setup generalizes on the usual subset sum problem, which is finding $S_G(G^r, b)$. For a general finite abelian group G, Li and Wan [6] found $S_G(G^r, b)$ using a "coordinate sieve" formula. Later, Kosters [4] found $S_G(G^r, b)$ and $S_G(G\backslash\{0\})^r, b)$ using character sums. In the finite field case, Li and Wan [5] used recursion and found explicit formulas for $S_{\mathbb{F}_q}(D^r, b)$ when $D \subseteq \mathbb{F}_q$ has order $q - 2$, $q - 1$, or q. When q is odd, Zhou, Wang, and Wang [10] found a formula for $S_{\mathbb{F}_q}(H^r, b)$ using properties of Gauss sums when $H = \{\alpha^2 : \alpha \in \mathbb{F}_q^*\}$.

The subset sum problem has applications in Reed-Solomon codes. Given a finite set $D = \{x_1, x_2, ..., x_t\} \subseteq \mathbb{F}_q$, a Reed-Solomon code of dimension k with evaluation set D has codewords

$$RS = \{(g(x_1), g(x_2), ..., g(x_t)) \in \mathbb{F}_q^t : g \in \mathbb{F}_q[x], \deg(g) \le k - 1\}.$$

Using the Lagrange interpolation formula, given $F \in \mathbb{F}_q^t$, there is a unique polynomial $f \in \mathbb{F}_q[x]$ of degree $n \leq t-1$ where

$$F = (f(x_1), f(x_2), ..., f(x_t)).$$

The distance between F and a codeword $G \in RS$ is the number of nonzero components of the vector $F - G$. If $\deg(f) = n = k + \ell$ where $\ell \geq 0$ the number of codewords at distance $t - r$ from F equals the number of polynomials $g \in \mathbb{F}_q[x]$ of degree at most $n - \ell - 1$ where $f - g$ has r distinct roots in D.

Definition 1. *Let $D \subseteq \mathbb{F}_q$. Let n, r, ℓ be non-negative integers where $n \geq \ell$. We define $N_{n,r,D}(\gamma_1, \gamma_2, \ldots, \gamma_\ell)$ as the number of degree n polynomials over $\mathbb{F}_q$ with r distinct roots in D of the form*

$$f(x) = x^n - \gamma_1 x^{n-1} + \ldots + (-1)^\ell \gamma_\ell x^{n-\ell} + g(x)$$

where g has degree at most $n - \ell - 1$.

In [2], Gao related $N_{n,r,D}(\gamma_1, \gamma_2, \ldots, \gamma_\ell)$ to the subset sum problem. When $\ell = 1$, Gao proved that for $b \in \mathbb{F}_q$,

$$N_{n,r,D}(b) = q^{n-1-r} \binom{|D|}{r} \sum_{j=0}^{n-r-1} (-q)^{-j} \binom{|D|-r}{j}$$
$$+ (-1)^{n-r} \binom{n}{r} S_{\mathbb{F}_q}(D^n, b).$$

When $\ell = 2$ and q is odd, Gao [2,3] expressed $N_{n,r,\mathbb{F}_m}(\gamma_1, \gamma_2)$ when $\mathbb{F}_m$ is a subfield of $\mathbb{F}_q$ in terms of

$$V_k(a,b) = \left| \left\{ S \subseteq \mathbb{F}_m : |S| = k, \sum_{x \in S} x = a, \sum_{x \in S} x^2 = b \right\} \right|. \tag{2}$$

An explicit formula for $N_{n,r,\mathbb{F}_m}(\gamma_1, \gamma_2)$ when $\gamma_1, \gamma_2 \in \mathbb{F}_m$ and m is an even power of the characteristic p was also found. Formulas for $V_k(0,0)$ were obtained when m is an odd power of p in [8].

In this paper, we find a general formula for $S_G(D^{r,f}, b)$ in terms of generating functions. When q is odd and $\mathbb{F}_m$ is a subfield of $\mathbb{F}_q$, we compute $S_{\mathbb{F}_m}(\mathbb{F}_m^{r,p_2}, b)$, which is the number of r-subsets $\{x_1, x_2, ..., x_r\} \subseteq \mathbb{F}_m$ where $x_1^2 + x_2^2 + \cdots + x_r^2 = b$. Using this result and the relation between $V_k(a,b)$ and $N_{n,r,\mathbb{F}_m}(\gamma_1, \gamma_2)$, we find explicit formulas for $N_{n,r,\mathbb{F}_m}(\gamma_1, \gamma_2)$. These formulas are valid for any $\gamma_1, \gamma_2 \in \mathbb{F}_q$ and any subfield $\mathbb{F}_m$ of $\mathbb{F}_q$, which generalizes on [3]. When m is an even power of an odd prime p and $\gamma_1, \gamma_2 \in \mathbb{F}_m$ we recover Theorem 4 in [3] for $N_{n,r,\mathbb{F}_m}(\gamma_1, \gamma_2)$.

This paper is organized as follows. In Sect. 2, we derive formulas for $S_G(D^{r,f}, b)$ in terms of generating functions. In Sect. 3, we obtain formulas for $S_{\mathbb{F}_m}(\mathbb{F}_m^{r,p_2}, b)$. In Sect. 4, we find formulas for $N_{n,r,\mathbb{F}_m}(\gamma_1, \gamma_2)$.

1 Formulas for $S_G\left(D^{r,f}, b\right)$

In this section, we derive a formula for the number of r-subsets $\{x_1, x_2, ..., x_r\}$ of a finite set D where $f(x_1) + f(x_2) + \cdots + f(x_r) = b$ when G is a finite abelian group and $f : D \to G$ is a fixed mapping. We use properties of characters over the finite abelian group G as in [4].

Definition 2. *Let G be a finite abelian group under addition. We define $\widehat{G} = \{\chi : G \to \mathbb{C}\backslash\{0\} : \chi(a+b) = \chi(a)\chi(b)\}$ to be the character group of G.*

For $\chi \in \widehat{G}$, we call χ the trivial character of G and write $\chi = 1$ if $\chi(g) = 1$ for all $g \in G$. For $\chi \in \widehat{G}$, we have $\chi^{-1} \in \widehat{G}$ where for $g \in G$, $\chi^{-1}(g) = \overline{\chi(g)}$. We define the order of χ, denoted by $\mathrm{ord}(\chi)$ to be the smallest positive integer k where $\chi^k = 1$.

We note that $\widehat{G}$ is a group under the multiplication of functions. For convenience, we use the Iverson bracket $[\![P]\!]$ which equals 1 if P is true and 0 otherwise. The following proposition is well known.

Proposition 1 ([9] Chap. 6.1). *Let G be a finite abelian group. We have $\widehat{G} \cong G$. For $g_1, g_2 \in G$,*

$$\frac{1}{|G|} \sum_{\chi \in \widehat{G}} \chi(g_1)\overline{\chi(g_2)} = [\![g_1 = g_2]\!].$$

We have the following general formula for $S_G\left(D^{r,f}, b\right)$.

Lemma 1. *Let G be a finite abelian group and D be a finite set. Then for each non-negative integer r and $b \in G$,*

$$S_G\left(D^{r,f}, b\right) = \frac{1}{|G|}\binom{|D|}{r} + \frac{1}{|G|} \sum_{\chi \in \widehat{G}\backslash\{1\}} \overline{\chi(b)}(-1)^r [z^r] \prod_{\alpha \in D}(1 - \chi(f(\alpha))z).$$

Proof. Assume without loss of generality that the operation on G is addition. Then from Lemma 1,

$$\begin{aligned}
S_G\left(D^{r,f}, b\right) &= \sum_{S \subseteq D : |S| = r} \left[\!\!\left[\sum_{\alpha \in S} f(\alpha) = b\right]\!\!\right] \\
&= \sum_{S \subseteq D : |S| = r} \frac{1}{|G|} \sum_{\chi \in \widehat{G}} \chi\left(-b + \sum_{\alpha \in S} f(\alpha)\right) \\
&= \frac{1}{|G|} \sum_{\chi \in \widehat{G}} \overline{\chi(b)} \sum_{S \subseteq D : |S| = r} \prod_{\alpha \in S} \chi(f(\alpha)) \\
&= \frac{1}{|G|} \sum_{\chi \in \widehat{G}} \overline{\chi(b)} (-1)^r [z^r] \prod_{\alpha \in D}(1 - \chi(f(\alpha))z).
\end{aligned}$$

Noting that $[z^r] \prod_{\alpha \in D}(1-z) = (-1)^r \binom{|D|}{r}$, we obtain the result. $\square$

The following result is useful for simplifying the products in Lemma 1.

Lemma 2. *Let G be a finite abelian group and D be a finite set. For any mapping $f: D \to G$ and any character $\chi \in \widehat{G}$ with $\text{ord}(\chi) = m$,*

$$\prod_{\alpha \in D}(1 - \chi(f(\alpha))z) = (1 - z^m)^{\frac{|D|}{m}} \exp\left(-\sum_{k \geq 1, m \nmid k} \sum_{\alpha \in D} \chi^k(f(\alpha)) \frac{z^k}{k}\right).$$

Proof. Using $\ln(1-z) = -\sum_{k \geq 1} \frac{z^k}{k}$ and $\chi^m = 1$, we have

$$\prod_{\alpha \in D}(1 - \chi(f(\alpha))z) = \exp\left(\sum_{\alpha \in D} \ln(1 - \chi(f(\alpha))z)\right)$$

$$= \exp\left(-\sum_{k \geq 1} \sum_{\alpha \in D} \chi^k(f(\alpha)) \frac{z^k}{k}\right)$$

$$= \exp\left(-\sum_{k \geq 1, m \nmid k} \sum_{\alpha \in D} \chi^k(f(\alpha)) \frac{z^k}{k} - |D| \sum_{k \geq 1} \frac{z^{mk}}{mk}\right)$$

$$= (1 - z^m)^{\frac{|D|}{m}} \exp\left(-\sum_{k \geq 1, m \nmid k} \sum_{\alpha \in D} \chi^k(f(\alpha)) \frac{z^k}{k}\right).$$

Thus, we obtain the result. □

2 Computing $S_{\mathbb{F}_m}\left(\mathbb{F}_m^{r,p_2}, b\right)$

In this section we assume $m = p^e$ for some odd prime p and positive integer e. We use the quadratic character to find $S_{\mathbb{F}_m}(\mathbb{F}_m^{r,p_2}, b)$, which is the number of r-subsets $\{x_1, x_2, ..., x_r\} \subseteq \mathbb{F}_m$ where $x_1^2 + x_2^2 + \cdots + x_r^2 = b$. This is important for computing $N_{n,r,\mathbb{F}_m}(\gamma_1, \gamma_2)$, which is the number of degree n monic polynomials over $\mathbb{F}_q$ with trace γ_1 and subtrace γ_2, and r distinct roots in the subfield $\mathbb{F}_m$ of $\mathbb{F}_q$. We note that formulas for the number of r-subsets $\{x_1, x_2, ..., x_r\} \subseteq \mathbb{F}_m^*$ where $x_1^2 + x_2^2 + \cdots + x_r^2 = b$ have been found in [10].

Definition 3. *Let $m = p^e$ for an odd prime p and positive integer e. We define η_m to be the quadratic character of $\mathbb{F}_m$ where for $b \in \mathbb{F}_m^*$, $\eta_m(b) = 1$ if b is a quadratic residue of $\mathbb{F}_m$, $\eta_m(b) = -1$ if b is a quadratic nonresidue of $\mathbb{F}_m$, and $\eta_m(0) = 0$.*

It is convenient to introduce the canonical additive character.

Definition 4. *([7] Chap. 5) Suppose $m = p^e$ for a prime p and positive integer e. The Trace function $Tr : \mathbb{F}_m \to \mathbb{F}_p$ is defined by*

$$Tr(\alpha) = \sum_{j=0}^{e-1} \alpha^{p^j}.$$

158   S. Kuttner et al.

In the rest of the paper, we shall use χ_1 to denote the canonical additive character of $\mathbb{F}_m$, which is defined by $\chi_1(\alpha) = \exp\left(\frac{2\pi i\, \mathrm{Tr}(\alpha)}{p}\right)$.

We note the following result.

Proposition 2. *([1] Lemma 1) Suppose $m = p^e$ for an odd prime p and positive integer e. Then for $\beta \in \mathbb{F}_m^*$*

$$\sum_{\alpha \in \mathbb{F}_m} \chi_1(\beta \alpha^2) = \eta_m(\beta)(-1)^{e-1} i^{[\![4|p+1]\!]} \sqrt{m}.$$

Define

$$F_r(m,w) := [z^r](1-z^p)^{\frac{m}{p}} \exp\left(-w \sum_{k \geq 1} \eta_m(k) \frac{z^k}{k}\right). \qquad (3)$$

When m is an even power of an odd prime p, it is known that $\eta_m(k) = [\![p \nmid k]\!]$ for any integer k (see [7] Chap. 5). Since $(-1)^{p-1} = 1$, we have

$$F_r(m,w) = [z^r](1-z)^w(1-z^p)^{(m-w)/p} \qquad (4)$$

$$= (-1)^r \sum_{j=0}^{\lfloor r/p \rfloor} \binom{(m-w)/p}{j}\binom{w}{r-pj}. \qquad (5)$$

Lemma 3. *Suppose $m = p^e$ for an odd prime p and positive integer e. For $b \in \mathbb{F}_m$,*

$$S_{\mathbb{F}_m}(\mathbb{F}_m^{r,p_2}, b) = \frac{1}{m}\binom{m}{r} + \frac{m[\![b=0]\!] + \eta_m(-b)\epsilon\sqrt{m} - 1}{2m}(-1)^r F_r(m, \epsilon\sqrt{m})$$
$$+ \frac{m[\![b=0]\!] - \eta_m(-b)\epsilon\sqrt{m} - 1}{2m}(-1)^r F_r(m, -\epsilon\sqrt{m}),$$

where $\epsilon = i^{[\![4|m+1]\!]}$.

Proof. Suppose first e is even. Then $m \equiv 1 \pmod{4}$, so $\epsilon = 1$. If e is odd, then $m \equiv p \pmod{4}$, so $\epsilon = i^{[\![4|p+1]\!]}$. Regardless of whether e is even or odd, $\epsilon = i^{[\![2\nmid e, 4|p+1]\!]}$. Thus,

$$i^{[\![4|p+1]\!]} = \epsilon(-1)^{\lfloor e/2 \rfloor [\![4|p+1]\!]}$$

Let $\sigma \in \mathbb{F}_m^*$ where $\eta_m(\sigma) = (-1)^{\lfloor e/2 \rfloor [\![4|p+1]\!] + e - 1}$. For $g \in \mathbb{F}_m$, we have $\eta_m(\sigma g) = \eta_m(\sigma)\eta_m(g)$. By Proposition 2, using the fact that $\chi_1(0) = 1$,

$$\sum_{\alpha \in \mathbb{F}_m} \chi_1(\sigma g \alpha^2) = m[\![g=0]\!] + \epsilon\sqrt{m}\,\eta_m(g).$$

Let $H = \{\alpha^2 : \alpha \in \mathbb{F}_m^*\}$. Then H is a subgroup of $\mathbb{F}_m^*$ of index 2, so

$$\sum_{h \in H} \chi_1(\sigma g h) = \frac{m[\![g=0]\!] + \eta_m(g)\epsilon\sqrt{m} - 1}{2}.$$

For $b \in \mathbb{F}_m$,
$$\sum_{\alpha \in \mathbb{F}_m^*} \chi_1(b\alpha) = m[\![b = 0]\!] - 1.$$

Thus,
$$\sum_{g \in H} \chi_1(-b\sigma g) = \frac{m[\![b = 0]\!] + \eta_m(-b)\epsilon\sqrt{m} - 1}{2},$$
$$\sum_{g \in \mathbb{F}_m^* \setminus H} \chi_1(-b\sigma g) = \frac{m[\![b = 0]\!] - \eta_m(-b)\epsilon\sqrt{m} - 1}{2}.$$

For each integer k with $p \nmid k$ and $g \in \mathbb{F}_m^*$, we have from the multiplicative property of the quadratic character that
$$\sum_{\alpha \in \mathbb{F}_m} \chi_1(k\sigma g \alpha^2) = \epsilon\sqrt{m}\eta_m(g)\eta_m(k).$$

By Lemma 2, since the canonical additive character χ_1 has order p,
$$\prod_{\alpha \in \mathbb{F}_m}(1 - \chi_1(\sigma g \alpha^2)z) = (1 - z^p)^{\frac{m}{p}}\exp\left(-\sum_{k \geq 1, p\nmid k}\sum_{\alpha \in \mathbb{F}_m}\chi_1(k\sigma g\alpha^2)\frac{z^k}{k}\right)$$
$$= (1 - z^p)^{\frac{m}{p}}\exp\left(-\epsilon\sqrt{m}\eta_m(g)\sum_{k \geq 1, p\nmid k}\eta_m(k)\frac{z^k}{k}\right).$$

Each $\eta_m(k) = 0$ for $p \mid k$. We note that each $\beta \in \mathbb{F}_m^*$ can be written uniquely as $\beta = \sigma g$ for some $g \in \mathbb{F}_m^*$. It is well known (see, e.g., [7, Chap. 5]) that $\chi_\beta : \alpha \mapsto \chi_1(\beta\alpha)$ for $\beta \in \mathbb{F}_m$ gives all the additive characters of $\mathbb{F}_m$. Using Lemma 1, we obtain the result. □

The following result can be viewed as a generalization of Lemma 3 (where $k = 0$).

Lemma 4. Let $E_r(m, b) = S_{\mathbb{F}_m}(\mathbb{F}_m^{r,p^2}, b) - \frac{1}{m}\binom{m}{r}$. For $n_1, ..., n_k \in \mathbb{F}_m^*$,
$$\sum_{\sigma_1,...,\sigma_k \in \mathbb{F}_m} E_r\left(m, b - \sum_{t=1}^k n_t\sigma_t^2\right)$$
$$= \frac{m[\![b = 0]\!] + \eta_m(-b)\epsilon\sqrt{m} - 1}{2m}(-1)^r\eta_m(w)(\epsilon\sqrt{m})^k F_r(m, \epsilon\sqrt{m})$$
$$+ \frac{m[\![b = 0]\!] - \eta_m(-b)\epsilon\sqrt{m} - 1}{2m}(-1)^r\eta_m(w)(-\epsilon\sqrt{m})^k F_r(m, -\epsilon\sqrt{m}),$$
where $w = n_1 \cdots n_k$, In particular, when $k = 0$, we have
$$E_r(m, b) = \frac{m[\![b = 0]\!] + \eta_m(-b)\epsilon\sqrt{m} - 1}{2m}(-1)^r F_r(m, \epsilon\sqrt{m})$$
$$+ \frac{m[\![b = 0]\!] - \eta_m(-b)\epsilon\sqrt{m} - 1}{2m}(-1)^r F_r(m, -\epsilon\sqrt{m}).$$

Here, $\epsilon = 1$ if $m \equiv 1 \pmod 4$ and $\epsilon = i$ if $m \equiv 3 \pmod 4$.

Proof. Let χ be the canonical additive character of $\mathbb{F}_q$. By Proposition 2, there exists $\beta_1, \beta_2 \in \mathbb{F}_m^*$ representatives of the cosets of $H = \{\alpha^2 : \alpha \in \mathbb{F}_m^*\}$ where for $\gamma \in \mathbb{F}_m^*$

$$\sum_{\sigma \in \mathbb{F}_m} \chi(\gamma \beta_j \sigma^2) = \eta_m(\gamma)(-1)^{j-1}\epsilon\sqrt{m}. \tag{6}$$

Since $H = \{\alpha^2 : \alpha \in \mathbb{F}_m^*\}$ has index 2 in $\mathbb{F}_m^*$, by Eq. (6),

$$\sum_{\alpha \in H} \chi(-b\beta_j \alpha) = M_{m,j}(b)$$

where

$$M_{m,j}(b) = \frac{m[\![b=0]\!] - (-1)^j \eta_m(-b)\epsilon\sqrt{m} - 1}{2}.$$

Let $n_1, ..., n_k \in \mathbb{F}_m^*$. By the multiplicative property of the quadratic character and Eq. (6), since $\eta_m(\alpha) = 1$ for $\alpha \in H$,

$$\sum_{\sigma_1,...,\sigma_k \in \mathbb{F}_m} M_{m,j}\left(b - \sum_{t=1}^{k} n_t \sigma_t^2\right)$$

$$= \sum_{\alpha \in H} \chi(-b\beta_j \alpha) \prod_{t=1}^{k} \sum_{\sigma_t \in \mathbb{F}_m} \chi(n_t \sigma_t^2 \beta_j \alpha)$$

$$= \eta_m(n_1 \cdots n_k)(-1)^{k(j-1)}(\epsilon\sqrt{m})^k M_{m,j}(b). \tag{7}$$

By Lemma 3,

$$E_r(m,b) = \frac{M_{m,1}(b)}{m}(-1)^r F_r(m, \epsilon\sqrt{m}) + \frac{M_{m,2}(b)}{m}(-1)^r F_r(m, -\epsilon\sqrt{m}).$$

The result follows from Eq. (7). □

Next, we simplify the formula for $F_r(m,w)$ when m is an odd power of an odd prime p.

We set $\omega_p = \exp(2\pi i/p)$. For convenience, we define

$$J(z;p) = \prod_{j=1}^{\frac{p-1}{2}}\left(1 + \omega_p^{t^2} z\right), K(z;p) = \prod_{j=1}^{\frac{p-1}{2}}\left(1 + \omega_p^{\xi t^2} z\right), \tag{8}$$

where ξ is a fixed element of $\mathbb{F}_p$ with $\eta_p(\xi) = -1$. We also define

$$A_r = \frac{p-1}{2}[z^r](1+z)^{\sqrt{\frac{m}{p}}}(1+z^p)^{\frac{m}{p}-\sqrt{\frac{m}{p}}}(J^{2\sqrt{\frac{m}{p}}}(z;p) + K^{2\sqrt{\frac{m}{p}}}(z;p)), \tag{9}$$

$$B_r = \frac{p-1}{2\epsilon\sqrt{p}}[z^r](1+z)^{\sqrt{\frac{m}{p}}}(1+z^p)^{\frac{m}{p}-\sqrt{\frac{m}{p}}}(J^{2\sqrt{\frac{m}{p}}}(z;p) - K^{2\sqrt{\frac{m}{p}}}(z;p)), \tag{10}$$

where $\epsilon = i^{[\![4|p+1]\!]}$. We have the following result.

Theorem 1. *Let m be an odd power of an odd prime p and*
$$\epsilon = i^{\lfloor 4 \mid p+1 \rfloor}.$$
Then
$$F_r(m, \epsilon\sqrt{m}) = (-1)^r \frac{A_r + \epsilon\sqrt{p}B_r}{p-1}, \tag{11}$$
$$F_r(m, -\epsilon\sqrt{m}) = (-1)^r \frac{A_r - \epsilon\sqrt{p}B_r}{p-1}. \tag{12}$$

Proof. Since m is an odd power of an odd prime p, we have from Chap. 5 in [7] that $\eta_m(k) = \eta_p(k)$ for any integer k. Hence
$$F_r(m, w) = [z^r](1 - z^p)^{\frac{m}{p}} \exp\left(-w \sum_{k \geq 1} \eta_p(k) \frac{z^k}{k}\right).$$

From Proposition 2, when $p \nmid k$,
$$\sum_{t=0}^{p-1} \omega_p^{k\beta t^2} = \eta_p(k\beta)\epsilon\sqrt{p},$$

By Lemma 2, for $\beta \in \mathbb{F}_p^*$,
$$\prod_{t=0}^{p-1}(1 - \omega_p^{\beta t^2} z) = (1 - z^p) \exp\left(-\sum_{k \geq 1, p \nmid k} \sum_{t=0}^{p-1} \omega_p^{k\beta t^2} \frac{z^k}{k}\right)$$
$$= (1 - z^p) \exp\left(-\eta_p(\beta) \sum_{k \geq 1, p \nmid k} \eta_p(k)\epsilon\sqrt{p} \frac{z^k}{k}\right).$$

Thus,
$$[z^r](1 - z^p)^{\frac{m}{p} - \sqrt{\frac{m}{p}}} \prod_{t=0}^{p-1}\left(1 - \omega_p^{\beta t^2} z\right)^{\sqrt{\frac{m}{p}}} = F_r(m, \eta_p(\beta)\epsilon\sqrt{m}).$$

The result follows from Equations (9) and (10) since $(-z)^p = -z^p$. □

We remark that the result for m is an even power of an odd prime p is given in Equation (4). Now we have the following result for A_r and B_r.

Theorem 2. *Let m be an odd power of an odd prime p and ξ be a primitive element of $\mathbb{F}_p$. In Eqs. (9) and (10),*

$$A_r = \sum_{pt+u+a_1+\cdots+a_{\frac{p-1}{2}}=r} \upsilon_p\left(\sum_{\sigma=1}^{\frac{p-1}{2}} a_\sigma \xi^{2\sigma}\right)\left(\frac{m}{p} - \sqrt{\frac{m}{p}}\right)_t \left(\sqrt{\frac{m}{p}}\right)_u \prod_{\sigma=1}^{\frac{p-1}{2}}\binom{2\sqrt{\frac{m}{p}}}{a_\sigma},$$

$$B_r = \sum_{pt+u+a_1+\cdots+a_{\frac{p-1}{2}}=r} \eta_p\left(\sum_{\sigma=1}^{\frac{p-1}{2}} a_\sigma \xi^{2\sigma}\right)\left(\frac{m}{p} - \sqrt{\frac{m}{p}}\right)_t \left(\sqrt{\frac{m}{p}}\right)_u \prod_{\sigma=1}^{\frac{p-1}{2}}\binom{2\sqrt{\frac{m}{p}}}{a_\sigma},$$

where the sums run over all tuples $(t, u, a_1, \cdots, a_{\frac{p-1}{2}})$ of non-negative integers, $v_p(k) = p[\![p \mid k]\!] - 1$, and η_p is the quadratic character of $\mathbb{F}_p$.

Proof. Since ξ is a primitive element of $\mathbb{F}_p$, by Eq. (8), for any integer c,

$$J(z;p) = \prod_{\sigma=1}^{\frac{p-1}{2}} \left(1 + \omega_p^{\xi^{2(\sigma+c)}} z\right), K(z;p) = \prod_{\sigma=1}^{\frac{p-1}{2}} \left(1 + \omega_p^{\xi^{2(\sigma+c)-1}} z\right).$$

Set $\tau = \frac{p-1}{2}$. Then

$$[z^j](J^n(z;p) + K^n(z;p))$$

$$= \frac{1}{\tau} \sum_{c=1}^{\tau} \sum_{a_1+\cdots+a_\tau=j} \left(\prod_{\sigma=1}^{\tau} \binom{n}{a_\sigma} \omega_p^{a_\sigma \xi^{2(\sigma+c)}} + \prod_{\sigma=1}^{\tau} \binom{n}{a_\sigma} \omega_p^{a_\sigma \xi^{2(\sigma+c)-1}} \right)$$

$$= \frac{1}{\tau} \sum_{a_1+\cdots+a_\tau=j} \binom{n}{a_1} \cdots \binom{n}{a_\tau} \sum_{c=1}^{p-1} \prod_{\sigma=1}^{\tau} \omega_p^{a_\sigma \xi^{2\sigma+c}}$$

$$= \frac{1}{\tau} \sum_{a_1+\cdots+a_\tau=j} \binom{n}{a_1} \cdots \binom{n}{a_\tau} \sum_{c=1}^{p-1} \omega_p^{c \sum_{\sigma=1}^{\tau} a_\sigma \xi^{2\sigma}}$$

$$= \frac{1}{\tau} \sum_{a_1+\cdots+a_\tau=j} \binom{n}{a_1} \cdots \binom{n}{a_\tau} \left(p \left[\!\left[p \mid \sum_{\sigma=1}^{\tau} a_\sigma \xi^{2\sigma} \right]\!\right] - 1 \right).$$

It is well known (see, e.g., [7, Chap. 5]) that for $a \in \mathbb{F}_p$,

$$\sum_{c=1}^{p-1} \omega_p^{ac} \eta_p(c) = \epsilon \sqrt{p} \eta_p(a), \tag{13}$$

where $\epsilon = i^{[\![4 \mid p+1]\!]}$. Moreover, $\eta_p(\xi^c) = (-1)^c$. Hence,

$$[z^j](J^n(z;p) - K^n(z;p))$$

$$= \frac{1}{\tau} \sum_{c=1}^{\tau} \sum_{a_1+a_2+\cdots+a_\tau=j} \left(\prod_{\sigma=1}^{\tau} \binom{n}{a_\sigma} \omega_p^{a_\sigma \xi^{2(\sigma+c)}} - \prod_{\sigma=1}^{\tau} \binom{n}{a_\sigma} \omega_p^{a_\sigma \xi^{2(\sigma+c)-1}} \right)$$

$$= \frac{1}{\tau} \sum_{a_1+a_2+\cdots+a_\tau=j} \binom{n}{a_1} \cdots \binom{n}{a_\tau} \sum_{c=1}^{p-1} (-1)^c \prod_{\sigma=1}^{\tau} \omega_p^{a_\sigma \xi^{2\sigma+c}}$$

$$= \frac{1}{\tau} \sum_{a_1+a_2+\cdots+a_\tau=j} \binom{n}{a_1} \cdots \binom{n}{a_\tau} \sum_{c=1}^{p-1} \omega_p^{c \sum_{\sigma=1}^{\tau} a_\sigma \xi^{2\sigma}} \eta_p(c)$$

$$= \frac{\epsilon \sqrt{p}}{\tau} \sum_{a_1+a_2+\cdots+a_\tau=j} \binom{n}{a_1} \cdots \binom{n}{a_\tau} \eta_p \left(\sum_{\sigma=1}^{\tau} a_\sigma \xi^{2\sigma} \right).$$

Thus, we obtain the result. □

We have the following for $p = 3$.

Example 1. Let m be an odd power of 3. In Eqs. (9) and (10),

$$A_r = \sum_{3t+u+j=r} \binom{\frac{m}{3} - \sqrt{\frac{m}{3}}}{t}\binom{\sqrt{\frac{m}{3}}}{u}\binom{2\sqrt{\frac{m}{3}}}{j}(3[\![3 \mid j]\!] - 1),$$

$$B_r = \sum_{3t+u+j=r} \binom{\frac{m}{3} - \sqrt{\frac{m}{3}}}{t}\binom{\sqrt{\frac{m}{3}}}{u}\binom{2\sqrt{\frac{m}{3}}}{j}([\![3 \mid (j-1)]\!] - [\![3 \mid (j-2)]\!]),$$

where the sums run over all 3-tuples (t, u, j) of non-negative integers.

3 Polynomials with First Two Coefficients and Number of Distinct Roots in $\mathbb{F}_m$ Prescribed

In this section, we work out an explicit formula for $N_{n,r,\mathbb{F}_m}(\gamma_1, \gamma_2)$ when $\mathbb{F}_m$ is a subfield of $\mathbb{F}_q$ and q is odd. This generalizes Gao's results [3] for $N_{n,r,\mathbb{F}_m}(\gamma_1, \gamma_2)$ when m is an even power of an odd prime p and $\gamma_1, \gamma_2 \in \mathbb{F}_m$. The following result from [2] converts the problem of finding $N_{n,r,\mathbb{F}_m}(\gamma_1, \gamma_2)$ in terms of the number of solutions to some systems of equations.

Lemma 5. *Let $q = p^e$ for an odd prime p and $\gamma_1, \gamma_2 \in \mathbb{F}_q$. For $n \geq 2$,*

$$N_{n,r,\mathbb{F}_m}(\gamma_1, \gamma_2) = q^{n-2-r}\binom{m}{r}\sum_{j=0}^{n-2-r}(-q)^{-j}\binom{m-r}{j}$$
$$+ (-1)^{n-1-r}\binom{n-1}{r}W_n(\gamma_1, \gamma_1^2 - 2\gamma_2)$$
$$+ (-1)^{n-r}\binom{n}{r}V_n(\gamma_1, \gamma_1^2 - 2\gamma_2),$$

where

$$V_r(\gamma_1, \gamma_2) = \left|\left\{S \subseteq \mathbb{F}_m : |S| = r, \sum_{x \in S} x = \gamma_1, \sum_{x \in S} x^2 = \gamma_2\right\}\right|, \quad (14)$$

$$W_r(\gamma_1, \gamma_2) = \sum_{y \in \mathbb{F}_q} V_{r-1}(\gamma_1 - y, \gamma_2 - y^2). \quad (15)$$

Formulas for $V_r(\gamma_1, \gamma_2)$ and $W_r(\gamma_1, \gamma_2)$ have been worked out explicitly in [2] when $\gamma_1, \gamma_2 \in \mathbb{F}_m$ using the Li-Wan sieve formula. In this paper, we work out these numbers in terms of $S_{\mathbb{F}_m}(\mathbb{F}_m^{r,p^2}, b)$ and Lemma 3. We need the following result.

Lemma 6 *(Corollary 1.4 [6]). Suppose $m = p^e$ for some prime p and positive integer e. Then for $b \in \mathbb{F}_m$, the number of r-subsets of $\mathbb{F}_m$ whose elements sum to b is*

$$S_{\mathbb{F}_m}(\mathbb{F}_m^r, b) = \frac{1}{m}\binom{m}{r} + \frac{m[\![b = 0]\!] - 1}{m}(-1)^{r+r/p}\binom{m/p}{r/p}[\![p \mid r]\!].$$

In the following lemma, we give $V_r(0,\gamma)$ and $W_r(0,\gamma)$ in terms of $S_{\mathbb{F}_m}(\mathbb{F}_m^{r,p_2}, b)$ when $\gamma \in \mathbb{F}_m$.

Lemma 7. *Suppose $q = p^e$ for an odd prime p and positive integer e. Let $\mathbb{F}_m$ be a subfield of $\mathbb{F}_q$ and $\gamma \in \mathbb{F}_m$.*

$$V_r(0,\gamma) = \frac{1}{m^2}\binom{m}{r} + \frac{m-1}{m^2}\binom{m/p}{r/p}[\![p\mid r]\!] + \frac{1}{m}\sum_{c\in\mathbb{F}_m} E_r\left(m,\gamma+rc^2\right),$$

$$W_r(0,\gamma) = \frac{1}{m}\binom{m}{r-1} + \frac{1}{m}\sum_{c,y\in\mathbb{F}_m} E_{r-1}\left(m,\gamma+rc^2-y^2\right),$$

where $E_r(m,b) = S_{\mathbb{F}_m}(\mathbb{F}_m^{r,p_2}, b) - \frac{1}{m}\binom{m}{r}$.

Proof. For any function $\phi : \mathbb{F}_m \times \mathbb{F}_m \to \mathbb{C}$, we have

$$\phi(0,\gamma) = \frac{1}{m}\sum_{a\in\mathbb{F}_m}\sum_{c\in\mathbb{F}_m} \phi(a,\gamma+2ca) - \frac{1}{m}\sum_{a\in\mathbb{F}_m^*}\sum_{c\in\mathbb{F}_m} \phi(a,\gamma+2ca)$$

$$= \frac{1}{m}\sum_{c\in\mathbb{F}_m}\sum_{a\in\mathbb{F}_m} \phi(a,\gamma+2ca) - \frac{1}{m}\sum_{a\in\mathbb{F}_m^*}\sum_{b\in\mathbb{F}_m} \phi(a,b). \tag{16}$$

Since p is odd, $(-1)^{r+r/p} = 1$, by Lemma 6,

$$\sum_{a\in\mathbb{F}_m^*}\sum_{b\in\mathbb{F}_m} V_r(a,b) = \sum_{a\in\mathbb{F}_m^*} S_{\mathbb{F}_m}(\mathbb{F}_m^r, a) = \frac{m-1}{m}\left(\binom{m}{r} - \binom{m/p}{r/p}[\![p\mid r]\!]\right).$$

We have

$$\sum_{a\in\mathbb{F}_m} V_r(a,\gamma+2ca) = \left|\left\{S\subseteq\mathbb{F}_m : |S|=r, \sum_{x\in S} x^2 = \gamma + 2c\sum_{x\in S} x\right\}\right|$$

$$= \left|\left\{S\subseteq\mathbb{F}_m : |S|=r, \sum_{x\in S}(x^2-2cx) = \gamma\right\}\right|$$

$$= \left|\left\{S\subseteq\mathbb{F}_m : |S|=r, \sum_{x\in S}(x-c)^2 = \gamma + rc^2\right\}\right|$$

$$= S_{\mathbb{F}_m}\left(\mathbb{F}_m^{r,p_2}, \gamma+rc^2\right). \tag{17}$$

Since $\mathbb{F}_m$ is closed under addition and multiplication,

$$\sum_{a\in\mathbb{F}_m^*}\sum_{b\in\mathbb{F}_m} W_r(a,b) = \sum_{a\in\mathbb{F}_m^*}\sum_{b\in\mathbb{F}_m}\sum_{y\in\mathbb{F}_m} V_{r-1}(a-y,b-y^2)$$

$$= \sum_{a\in\mathbb{F}_m^*}\sum_{y\in\mathbb{F}_m}\sum_{b\in\mathbb{F}_m} V_{r-1}(a-y,b-y^2).$$

Using the substitutions $b \to b + y^2$, then $y \to a - y$, it follows that

$$\sum_{a \in \mathbb{F}_m^*} \sum_{b \in \mathbb{F}_m} W_r(a,b) = \sum_{a \in \mathbb{F}_m^*} \sum_{y \in \mathbb{F}_m} \sum_{b \in \mathbb{F}_m} V_{r-1}(y,b) = (m-1)\binom{m}{r-1}.$$

We have

$$\sum_{a \in \mathbb{F}_m} W_r(a, \gamma + 2ca) = \sum_{y \in \mathbb{F}_m} \sum_{a \in \mathbb{F}_m} V_{r-1}(a - y, \gamma + 2ca - y^2)$$

$$= \sum_{y \in \mathbb{F}_m} \sum_{a \in \mathbb{F}_m} V_{r-1}(a, \gamma + 2c(a+y) - y^2)$$

$$= \sum_{y \in \mathbb{F}_m} \sum_{a \in \mathbb{F}_m} V_{r-1}(a, \gamma - (y-c)^2 + c^2 + 2ca)$$

$$= \sum_{y \in \mathbb{F}_m} \sum_{a \in \mathbb{F}_m} V_{r-1}(a, \gamma - y^2 + c^2 + 2ca).$$

By Eq. (17),

$$\sum_{a \in \mathbb{F}_m} W_r(a, \gamma + 2ca) = \sum_{y \in \mathbb{F}_m} S_{\mathbb{F}_m}\left(\mathbb{F}_m^{r-1, p_2}, \gamma - y^2 + c^2 + (r-1)c^2\right)$$

$$= \sum_{y \in \mathbb{F}_m} S_{\mathbb{F}_m}\left(\mathbb{F}_m^{r-1, p_2}, \gamma + rc^2 - y^2\right).$$

The result follows from Eq. (16). □

More generally, we have the following result.

Lemma 8. *Suppose $q = p^e$ for an odd prime p and positive integer e. Let $\mathbb{F}_m$ be a subfield of $\mathbb{F}_q$, $\gamma_1, \gamma_2 \in \mathbb{F}_m$ and*

$$E_r(m, b) = S_{\mathbb{F}_m}(\mathbb{F}_m^{r, p_2}, b) - \frac{1}{m}\binom{m}{r}.$$

If $p \mid r$, then

$$V_r(\gamma_1, \gamma_2) = \frac{1}{m^2}\binom{m}{r} + \frac{m[\![\gamma_1 = 0]\!] - 1}{m^2}\binom{m/p}{r/p} + [\![\gamma_1 = 0]\!] E_r(m, \gamma_2),$$

$$W_r(\gamma_1, \gamma_2) = \frac{1}{m}\binom{m}{r-1} + [\![\gamma_1 = 0]\!] \sum_{y \in \mathbb{F}_m} E_{r-1}(m, \gamma_2 - y^2).$$

If $p \nmid r$, then

$$V_r(\gamma_1, \gamma_2) = \frac{1}{m^2}\binom{m}{r} + \frac{1}{m}\sum_{c \in \mathbb{F}_m} E_r(m, \gamma + rc^2),$$

$$W_r(\gamma_1, \gamma_2) = \frac{1}{m}\binom{m}{r-1} + \frac{1}{m}\sum_{c, y \in \mathbb{F}_m} E_{r-1}(m, \gamma + rc^2 - y^2),$$

where $\gamma = \gamma_2 - r^{-1}\gamma_1^2$.

Proof. We note that for $c \in \mathbb{F}_m$ and $x_1, x_2, \ldots, x_r \in \mathbb{F}_m$,

$$\sum_{j=1}^{r}(x_j + c) = rc + \sum_{j=1}^{r} x_j,$$

$$\sum_{j=1}^{r}(x_j + c)^2 = \sum_{j=1}^{r} x_j^2 + 2c \sum_{j=1}^{n} x_j + rc^2.$$

Since $\gamma_1, \gamma_2 \in \mathbb{F}_m$ and $\mathbb{F}_m$ is closed under addition and multiplication, the relevant equations give

$$V_r(\gamma_1, \gamma_2) = V_r(\gamma_1 + rc, \gamma_2 + 2c\gamma_1 + rc^2),$$
$$W_r(\gamma_1, \gamma_2) = W_r(\gamma_1 + rc, \gamma_2 + 2c\gamma_1 + rc^2).$$

When $p \mid r$, $\gamma_2 + 2c\gamma_1 + rc^2 = \gamma_2 + 2c\gamma_1$ and $\gamma_1 + rc = \gamma_1$. By Corollary 6, for $\gamma_1 \in \mathbb{F}_m^*$,

$$V_r(\gamma_1, \gamma_2) = \frac{1}{m} \sum_{c \in \mathbb{F}_m} V_r(\gamma_1, \gamma_2 + 2c\gamma_1) = \frac{1}{m} \sum_{b \in \mathbb{F}_m} V_r(\gamma_1, b)$$

$$= \frac{1}{m} S_{\mathbb{F}_m}(\mathbb{F}_m^r, \gamma_1) = \frac{1}{m^2}\binom{m}{r} - \frac{1}{m^2}\binom{m/p}{r/p}.$$

Similarly,

$$W_r(\gamma_1, \gamma_2) = \frac{1}{m} \sum_{c \in \mathbb{F}_m} W_r(\gamma_1, \gamma_2 + 2c\gamma_1) = \frac{1}{m} \sum_{b \in \mathbb{F}_m} W_r(\gamma_1, b) = \frac{1}{m}\binom{m}{r-1}.$$

When $p \nmid r$ and $c = -r^{-1}\gamma_1$, we have $\gamma_1 + rc = 0$ and

$$\gamma_2 + 2c\gamma_1 + rc^2 = \gamma_2 + 2(-r^{-1}\gamma_1)\gamma_1 + r(-r^{-1}\gamma_1)^2$$
$$= \gamma_2 - 2r^{-1}\gamma_1^2 + r^{-1}\gamma_1^2 = \gamma_2 - r^{-1}\gamma_1^2.$$

The result follows from Eq. (16). □

For convenience, we define the following function for $\gamma \in \mathbb{F}_m$:

$$Q_{n,j}(\gamma) := \frac{m[\![\gamma = 0]\!] - 1 + \epsilon\sqrt{m}\eta_m(-\gamma)}{2m}(\epsilon\sqrt{m})^{j-[\![p\nmid n]\!]} F_{n-j}(m, \epsilon\sqrt{m})$$

$$+ \frac{m[\![\gamma = 0]\!] - 1 - \epsilon\sqrt{m}\eta_m(-\gamma)}{2m}(-\epsilon\sqrt{m})^{j-[\![p\nmid n]\!]} F_{n-j}(m, -\epsilon\sqrt{m}), \quad (18)$$

where $\epsilon = 1$ if $m \equiv 1 \pmod 4$ and $\epsilon = i$ if $m \equiv 3 \pmod 4$. We have the following for $V_r(\gamma_1, \gamma_2)$ and $W_r(\gamma_1, \gamma_2)$ using Lemma 8 and Lemma 4.

Theorem 3. *Let $q = p^e$ for an odd prime p and positive integer e. Suppose $\mathbb{F}_m$ is a subfield of $\mathbb{F}_q$ and $\gamma_1, \gamma_2 \in \mathbb{F}_m$. If $p \mid r$, then*

$$V_r(\gamma_1, \gamma_2) = \frac{1}{m^2}\binom{m}{r} + \frac{m[\![\gamma_1 = 0]\!] - 1}{m^2}\binom{m/p}{r/p} + [\![\gamma_1 = 0]\!](-1)^r Q_{r,0}(\gamma_2),$$

$$W_r(\gamma_1, \gamma_2) = \frac{1}{m}\binom{m}{r-1} - [\![\gamma_1 = 0]\!](-1)^r Q_{r,1}(\gamma_2),$$

If $p \nmid r$, then

$$V_r(\gamma_1, \gamma_2) = \frac{1}{m^2}\binom{m}{r} + (-1)^r \eta_m(r) Q_{r,0}(\gamma_2 - r^{-1}\gamma_1^2),$$

$$W_r(\gamma_1, \gamma_2) = \frac{1}{m}\binom{m}{r-1} - (-1)^r \eta_m(r) Q_{r,1}(\gamma_2 - r^{-1}\gamma_1^2).$$

When γ_1, γ_2 are both in $\mathbb{F}_m$, we have the following for $N_{n,r,\mathbb{F}_m}(\gamma_1, \gamma_2)$ using Lemma 5 and Theorem 3. We note the identity

$$\binom{m}{n}\binom{n}{r} = \binom{m}{r}\binom{m-r}{n-r}. \tag{19}$$

Corollary 1. *Let $q = p^e$ for an odd prime p and positive integer e and $n \geq 2$. Suppose $\mathbb{F}_m$ is a subfield of $\mathbb{F}_q$ and $\gamma_1, \gamma_2 \in \mathbb{F}_m$. If $p \mid n$, then*

$$N_{n,r,\mathbb{F}_m}(\gamma_1, \gamma_2) = q^{n-2-r}\binom{m}{r}\sum_{j=0}^{n-2-r}(-q)^{-j}\binom{m-r}{j}$$
$$+ \frac{(-1)^{n-r}}{m^2}\binom{m}{r}\left(\binom{m-r}{n-r} - m\binom{m-r}{n-1-r}\right)$$
$$+ \frac{m[\![\gamma_1 = 0]\!] - 1}{m^2}(-1)^{n-r}\binom{n}{r}\binom{m/p}{n/p}$$
$$+ [\![\gamma_1 = 0]\!](-1)^r\binom{n-1}{r}Q_{n,1}(-2\gamma_2)$$
$$+ [\![\gamma_1 = 0]\!](-1)^r\binom{n}{r}Q_{n,0}(-2\gamma_2).$$

If $p \nmid n$, then

$$N_{n,r,\mathbb{F}_m}(\gamma_1, \gamma_2) = q^{n-2-r}\binom{m}{r}\sum_{j=0}^{n-2-r}(-q)^{-j}\binom{m-r}{j}$$
$$+ \frac{(-1)^{n-r}}{m^2}\binom{m}{r}\left(\binom{m-r}{n-r} - m\binom{m-r}{n-1-r}\right)$$
$$+ (-1)^r\binom{n-1}{r}\eta_m(n)Q_{n,1}((1-n^{-1})\gamma_1^2 - 2\gamma_2)$$
$$+ (-1)^r\binom{n}{r}\eta_m(n)Q_{n,0}((1-n^{-1})\gamma_1^2 - 2\gamma_2).$$

Proof. The result follows from Theorem 3, Lemma 5, and identity (19). □

When γ_1, γ_2 are not both in $\mathbb{F}_m$, we have the following for $N_{n,r,\mathbb{F}_m}(\gamma_1, \gamma_2)$ using Lemma 5 and Theorem 3.

Corollary 2. *Let $q = p^e$ for an odd prime p and positive integer e and $n \geq 2$. Suppose $\mathbb{F}_m$ is a subfield of $\mathbb{F}_q$, $\gamma_1, \gamma_2 \in \mathbb{F}_q$ not both in $\mathbb{F}_m$. If there exists $\sigma \in \mathbb{F}_q$ where $\gamma_1 - \sigma \in \mathbb{F}_m$ and $\gamma_1^2 - 2\gamma_2 - \sigma^2 \in \mathbb{F}_m$, then*

$$N_{n,r,\mathbb{F}_m}(\gamma_1, \gamma_2) = q^{n-2-r} \binom{m}{r} \sum_{j=0}^{n-2-r} (-q)^{-j} \binom{m-r}{j}$$
$$- \frac{(-1)^{n-r}}{m} \binom{m}{r} \binom{m-r}{n-1-r}$$
$$- [\![p \mid n-1]\!] \frac{m[\![\gamma_1 = \sigma]\!] - 1}{m^2} (-1)^{n-r} \binom{n-1}{r} \binom{m/p}{\frac{n-1}{p}}$$
$$+ [\![p \mid n-1, \gamma_1 = \sigma]\!] (-1)^r \binom{n-1}{r} Q_{n-1,0}(-2\gamma_2)$$
$$+ [\![p \nmid n-1]\!] (-1)^r \binom{n-1}{r} \eta_m(n-1) Q_{n-1,0}(y)$$

where $y = \gamma_1^2 - 2\gamma_2 - \sigma^2 - (n-1)^{-1}(\gamma_1 - \sigma)^2$. Otherwise,

$$N_{n,r,\mathbb{F}_m}(\gamma_1, \gamma_2) = q^{n-2-r} \binom{m}{r} \sum_{j=0}^{n-2-r} (-q)^{-j} \binom{m-r}{j}.$$

Proof. Let $y_1 = \gamma_1$ and $y_2 = \gamma_1^2 - 2\gamma_2$. Then since γ_1, γ_2 are not both in $\mathbb{F}_m$, y_1, y_2 are not both in $\mathbb{F}_m$. We have $V_k(y_1, y_2) = 0$. Moreover,

$$W_k(y_1, y_2) = \sum_{\sigma \in \mathbb{F}_q} V_{k-1}(y_1 - \sigma, y_2 - \sigma^2).$$

Suppose $y_1 - \sigma \in \mathbb{F}_m$, $y_2 - \sigma^2 \in \mathbb{F}_m$ and $\sigma \in \mathbb{F}_q$. Then since either $y_1 \notin \mathbb{F}_m$ or $y_2 \notin \mathbb{F}_m$, we have $\sigma \notin \mathbb{F}_m$. Then for $c \in \mathbb{F}_q$, $y_2 - (\sigma + c)^2 = y_2 - \sigma^2 - 2\sigma c - c^2$. Since $\sigma \in \mathbb{F}_q \setminus \mathbb{F}_m$, it follows that

$$\begin{cases} y_1 - (\sigma + c) \in \mathbb{F}_m \\ y_2 - (\sigma + c)^2 \in \mathbb{F}_m \end{cases} \Rightarrow \begin{cases} c \in \mathbb{F}_m \\ 2\sigma c + c^2 \in \mathbb{F}_m \end{cases} \Rightarrow c = 0.$$

Thus, $W_k(y_1, y_2) = V_{k-1}(y_1 - \sigma, y_2 - \sigma^2)$. The result follows from Theorem 3, Lemma 5 and Eq. (19). □

For $n \geq m+2$, the formulas for $N_{n,r,\mathbb{F}_m}(\gamma_1, \gamma_2)$ are simpler.

Theorem 4. *Suppose $q = p^e$ for an odd prime p and positive integer e, $\mathbb{F}_m$ is a subfield of $\mathbb{F}_q$, and $\gamma_1, \gamma_2 \in \mathbb{F}_q$. Then for $n \geq m+2$,*

$$N_{n,r,\mathbb{F}_m}(\gamma_1, \gamma_2) = q^{n-2-m}\binom{m}{r}(q-1)^{m-r}.$$

Proof. For all $k \geq m+1$ and $a, b \in \mathbb{F}_q$, we have from Equation (14) that

$$V_k(a,b) = \left|\left\{S \subseteq \mathbb{F}_m : |S| = k, \sum_{x \in S} x = a, \sum_{x \in S} x^2 = b\right\}\right| = 0. \tag{20}$$

Since $n \geq m+2$, we have

$$V_n(\gamma_1, \gamma_1^2 - 2\gamma_2) = 0.$$

Since $n - 1 \geq m + 1$, we have from Equations (15) and (20) that

$$W_n(\gamma_1, \gamma_1^2 - 2\gamma_2) = \sum_{y \in \mathbb{F}_m} V_{n-1}(\gamma_1 - y, \gamma_1^2 - 2\gamma_2 - y^2) = 0.$$

By Lemma 5, since $n - 2 \geq m$, it follows from the binomial theorem that

$$N_{n,r,\mathbb{F}_m}(\gamma_1, \gamma_2) = q^{n-2-r}\binom{m}{r}\sum_{j=0}^{n-2-r}(-q)^{-j}\binom{m-r}{j}.$$

$$= q^{n-2-r}\binom{m}{r}(1-q^{-1})^{m-r}$$

$$= q^{n-2-m}\binom{m}{r}(q-1)^{m-r}.$$

Hence, we obtain the result. $\square$

Finally, we give the complete formulas for $N_{n,r,\mathbb{F}_3}(\gamma_1, \gamma_2)$ when $q = 3$.

Corollary 3. *Suppose $q = 3$ and $\gamma_1, \gamma_2 \in \mathbb{F}_3$. For $n \geq 5$,*

$$N_{n,r,\mathbb{F}_3}(\gamma_1, \gamma_2) = \binom{3}{r}3^{n-5}2^{3-r}.$$

For $n \in \{2, 3, 4\}$, the values of $N_{n,r,\mathbb{F}_3}(\gamma_1, \gamma_2)$ are given in Table 1.

Table 1. Values of $N_{n,r,\mathbb{F}_3}(\gamma_1, \gamma_2)$ when $q = 3$

n	r	(0,0)	(0,1)	(0,2)	(1,0)	(1,1)	(1,2)	(2,0)	(2,1)	(2,2)
2	0	0	1	0	0	0	1	0	0	1
2	1	1	0	0	0	1	0	0	1	0
2	2	0	0	1	1	0	0	1	0	0
2	3	0	0	0	0	0	0	0	0	0
3	0	0	0	2	1	1	1	1	1	1
3	1	3	3	0	1	1	1	1	1	1
3	2	0	0	0	1	1	1	1	1	1
3	3	0	0	1	0	0	0	0	0	0
4	0	3	3	2	3	3	2	3	3	2
4	1	3	3	6	3	3	6	3	3	6
4	2	3	3	0	3	3	0	3	3	0
4	3	0	0	1	0	0	1	0	0	1

References

1. Ding, C., Yang, J.: Hamming weights in irreducible cyclic codes. Discret. Math. **313**(4), 434–446 (2013)
2. Gao, Z.: Counting polynomials over finite fields with prescribed leading coefficients and linear factors. Finite Fields Appl. **82**, Paper No. 102052, 23 (2022)
3. Gao, Z.: Counting polynomials over finite fields with prescribed leading coefficients and linear factors (2022). https://arxiv.org/pdf/2105.12845
4. Kosters, M.: The subset sum problem for finite abelian groups. J. Comb. Theory Ser. A **120**(3), 527–530 (2013)
5. Li, J., Wan, D.: On the subset sum problem over finite fields. Finite Fields Appl. **14**(4), 911–929 (2008)
6. Li, J., Wan, D.: Counting subset sums of finite abelian groups. J. Comb. Theory Ser. A **119**(1), 170–182 (2012)
7. Lidi, R., Neiderreiter, J.: Finite Fields. Cambridge, 2nd edn. (1997)
8. Nguyen, J.: Higher moments subset sums over finite fields. Ph.D. thesis, UC Irvine (2019)
9. Serre, J.: A Course on Arithmetic, vol. 7. Springer-Verlag, New York (1973)
10. Wang, W., Wang, L., Zhou, H.: Subset sums of quadratic residues over finite fields. Finite Fields Appl. **43**, 106–122 (2017)

Generalized Class Group Actions on Oriented Elliptic Curves with Level Structure

Sarah Arpin[1,2], Wouter Castryck[3(✉)], Jonathan Komada Eriksen[4], Gioella Lorenzon[3], and Frederik Vercauteren[3]

[1] Mathematics Institute, Universiteit Leiden, Leiden, The Netherlands
[2] The Quantum Software Consortium of the Netherlands, Leiden, The Netherlands
[3] COSIC, ESAT, KU Leuven, Leuven, Belgium
wouter.castryck@esat.kuleuven.be
[4] Department of Information Security and Communication Technology, Norwegian University of Science and Technology, Trondheim, Norway

Abstract. We study a large family of generalized class groups of imaginary quadratic orders O and prove that they act freely and (essentially) transitively on the set of primitively O-oriented elliptic curves over a field k (assuming this set is non-empty) equipped with appropriate level structure. This extends, in several ways, a recent observation due to Galbraith, Perrin and Voloch for the ray class group. We show that this leads to a reinterpretation of the action of the class group of a suborder $O' \subseteq O$ on the set of O'-oriented elliptic curves, discuss several other examples, and briefly comment on the hardness of the corresponding vectorization problems.

1 Introduction

A current trend in isogeny-based cryptography is to study elliptic curve isogenies that respect certain *level structure*, or additional information about the curves. This has two main catalysts. Firstly, the recently established rapid mixing properties of isogeny graphs of supersingular elliptic curves with a marked cyclic subgroup of order N (Γ_N^0-level structure) have led to improved security foundations, e.g., of the distributed generation of supersingular elliptic curves with unknown endomorphism ring [Arp22, BCC+23]; see [CL23b] for a generalization of these rapid mixing results to arbitrary level structure. Secondly, Robert's unconditional break of SIDH [Rob23a] has revealed that the problem of finding an isogeny between two elliptic curves with full Γ_N-level structure is dramatically easier than in the case of plain elliptic curves, at least for N smooth and large enough compared to the degree of the isogeny. The security of several recently proposed variants of SIDH [FMP23, BMP23] also reduces to leveled

---

This work was supported in part by the European Research Council (ERC) under the European Union's Horizon 2020 research and innovation programme (grant agreement ISOCRYPT - No. 101020788) and by CyberSecurity Research Flanders with reference number VR20192203.

© The Author(s), under exclusive license to Springer Nature Switzerland AG 2025
S. Petkova-Nikova and D. Panario (Eds.): WAIFI 2024, LNCS 15176, pp. 171–190, 2025.
https://doi.org/10.1007/978-3-031-81824-0_12

isogeny problems. Some of these can again be broken much more efficiently than in the unleveled case; see [FFP24] for a recent, systematic discussion.

In this paper we find explicit generalized class groups which act on the set of isomorphism classes of elliptic curves with various types of level structure. In doing so, we connect the study of isogenies between elliptic curves with level structure to the study of class group actions in the oriented framework of Colò–Kohel [CK20] and Onuki [Onu21]. Briefly recall that, for K an imaginary quadratic field, a K-orientation on an elliptic curve E over a field k, say of positive characteristic p, is an embedding $\iota : K \hookrightarrow \text{End}^0(E)$ into the endomorphism algebra of E (assuming that such an embedding exists). For an order O of K, such an orientation is a primitive O-orientation if $O = \iota^{-1}\text{End}(E)$. For a fixed order $O \subseteq K$, the set $\mathcal{E}\ell\ell_k(O)$ of primitively O-oriented elliptic curves up to isomorphism naturally comes equipped with a free and (essentially) transitive action of the class group cl_O by isogenies, see Sect. 2.4 for more details. Isogenies arising from this class group action are called horizontal. For suitable parameters, this is considered a cryptographic group action, underpinning constructions like CRS [Cou06, RS06], CSIDH [CLM+18, CD20] and SCALLOP [FFK+23, CL23a].

Level structures have sneaked up in the oriented setting before, although the situation is more diffuse than in the non-oriented case. E.g., for N any prime different from p that splits in O, it is well-known that horizontal N-isogenies automatically preserve the two eigenspaces of any generator σ of O acting on the N-torsion [BF23, FFP24]; this can be seen as level structure amounting to the specification of two independent subgroups of order N. In [CS21], Chenu and Smith study ideal class groups acting on supersingular elliptic curves $E/\mathbb{F}_{p^2}$ together with an N-isogeny to their Frobenius conjugate; this can be viewed as Γ_N^0-level structure, see also [XZQ23]. However, our research was triggered by a recent note due to Galbraith, Perrin and Voloch [GPV23] in the case of supersingular elliptic curves over $\mathbb{F}_p$, which is the setting of CSIDH: Just as $\text{cl}_{\mathbb{Z}[\sqrt{-p}]}$ acts on the set of supersingular elliptic curves over $\mathbb{F}_p$ with $\mathbb{F}_p$-rational endomorphism ring $\mathbb{Z}[\sqrt{-p}]$, its ray class group for modulus (N) acts on supersingular elliptic curves over $\mathbb{F}_p$ with full Γ_N-level structure. In fact, the observation that ray class groups act on oriented elliptic curves with level structure seems due to Còlo and Kohel [CK23, Rmk. on p. 3]. In [GPV23], the authors observe that although using supersingular elliptic curves over $\mathbb{F}_p$ equipped with such level structure yields a larger key space for the usual CSIDH parameters, the security of such an enhanced protocol immediately reduces to that of the original CSIDH protocol, so this does not provide an advantage. We generalize this observation and contrast it with class group actions of suborders as in [FFK+23].

Our goal is to analyze to what extent the observation by Galbraith et al. is part of a bigger story. Instead of starting from a type of level structure and trying to devise a corresponding class group action, we invert the viewpoint and start from the action of a *generalized class group*, subsequently finding the correct level structure in order to have a group action. In Sect. 2 we provide an overview of the relevant background on elliptic curves with level structure, orientations, and generalized class groups. Our main results are discussed in Sect. 3, where we study a large family of generalized class groups, study their properties, and show that they act freely and (essentially) transitively on oriented elliptic curves with suitable level structure. We discuss several interesting examples, one of which sheds a new light on actions by class groups of non-maximal orders. Finally, in Sect. 4, we discuss the hardness of the vectorization problem for our generalized class group actions.

2 Background

2.1 Elliptic Curves with Level Structure

In this section, fix an integer $N \geq 2$ and a field k such that $\operatorname{char} k \nmid N$, and let E be an elliptic curve over k. Let $E[N] \cong \mathbb{Z}/N\mathbb{Z} \times \mathbb{Z}/N\mathbb{Z}$ denote the N-torsion group of E.

Definition 1 (Level structure). *Let Γ be a subgroup of $\operatorname{GL}_2(\mathbb{Z}/N\mathbb{Z})$. A Γ-level structure on an elliptic curve E is a choice of isomorphism $\Phi : \mathbb{Z}/N\mathbb{Z} \times \mathbb{Z}/N\mathbb{Z} \cong E[N]$, up to pre-composition with an element of Γ. We denote this by the triple (E, Φ, Γ), just writing (E, Φ) when Γ is understood from context.*

Choosing such an isomorphism $\Phi : \mathbb{Z}/N\mathbb{Z} \times \mathbb{Z}/N\mathbb{Z} \cong E[N]$ amounts to specifying a basis $P, Q \in E[N]$ and considering it up to base change by matrices from the prescribed group $\Gamma \subseteq \operatorname{GL}_2(\mathbb{Z}/(N))$. An isogeny $\varphi : E_1 \to E_2$ respects the level structures P_1, Q_1 resp. P_2, Q_2 if and only if $\varphi(P_1) = Q_1$, $\varphi(P_2) = Q_2$, modulo the action of Γ. Commonly studied examples are

$$\Gamma_N^0 = \left\{ \begin{pmatrix} * & * \\ 0 & * \end{pmatrix} \right\}, \quad \Gamma_N^{0,0} = \left\{ \begin{pmatrix} * & 0 \\ 0 & * \end{pmatrix} \right\}, \quad \Gamma_N^1 = \left\{ \begin{pmatrix} 1 & * \\ 0 & * \end{pmatrix} \right\}, \quad \Gamma_N = \left\{ \begin{pmatrix} 1 & 0 \\ 0 & 1 \end{pmatrix} \right\}$$

where the level structure corresponds to specifying a cyclic subgroup of order N, two independent cyclic subgroups of order N, a point of order N, or a basis of $E[N]$ ("full level structure"), respectively.

Example 1. Fix an isomorphism $\Phi : (\mathbb{Z}/N\mathbb{Z} \times \mathbb{Z}/N\mathbb{Z}) \to E[N]$ and let $P = \Phi(1,0)$, $Q = \Phi(0,1)$. For any

$$h = \begin{pmatrix} a & b \\ 0 & d \end{pmatrix} \in \Gamma_N^0,$$

$\Phi \circ h$ sends $(1, 0) \mapsto aP$ and $(0, 1) \mapsto bP + dQ$. The basis element $(0, 1)$ can be sent anywhere in the group $E[N]$ via Γ_N^0, but the image of $(1, 0)$ is always in $\langle P \rangle$. In this sense Φ fixes a choice of cyclic subgroup $\langle P \rangle$.

2.2 Congruence Subgroups and Generalized Class Groups

Our main references for this and the next section are [Cox13,KL22]. For O an order in an imaginary quadratic number field K, a *modulus in O* is a non-zero integral ideal $\mathfrak{m} \subseteq O$. We denote by I_O the group of proper fractional ideals in K, which we recall are lattices $\mathfrak{a} \subseteq K$ such that $\{\alpha \in K \mid \alpha \mathfrak{a} \subseteq \mathfrak{a}\} = O$. Let P_O

be the subgroup of principal fractional ideals, i.e., fractional ideals of the form αO with $\alpha \in K \setminus \{0\}$. Then the class group of O is the quotient $\mathrm{cl}_O = I_O/P_O$.

It can be shown that for any choice of modulus $\mathfrak{m}$, every class in cl_O contains an ideal $\mathfrak{a} \subseteq O$ that is coprime with $\mathfrak{m}$, i.e., $\mathfrak{a} + \mathfrak{m} = O$ [Cox13, Cor. 7.17]. Equivalently, if we define $I_O(\mathfrak{m}) \subseteq I_O$ to be the subgroup generated by all proper integral ideals that are coprime with $\mathfrak{m}$, and we let $P_O(\mathfrak{m}) = P_O \cap I_O(\mathfrak{m})$,[1] then the natural map $I_O(\mathfrak{m})/P_O(\mathfrak{m}) \to \mathrm{cl}_O : [\mathfrak{a}] \mapsto [\mathfrak{a}]$ is an isomorphism.

A *ray* for modulus $\mathfrak{m}$ is a principal fractional ideal of the form

$$\alpha O \text{ with } \alpha \in K^* \text{ such that } \alpha \equiv 1 \bmod \mathfrak{m},$$

where a congruence $\alpha \equiv \beta \bmod \mathfrak{m}$ means that for any $\alpha_1, \alpha_2, \beta_1, \beta_2 \in O$ such that $\alpha = \alpha_1/\alpha_2$ and $\beta = \beta_1/\beta_2$ we have $\alpha_1\beta_2 - \alpha_2\beta_1 \in \mathfrak{m}$; see [KL22, Def. 4.2].

The rays form a subgroup $P_{O,1}(\mathfrak{m}) \subseteq P_O(\mathfrak{m})$ called the *ray group* for modulus $\mathfrak{m}$. Any group H such that $P_{O,1}(\mathfrak{m}) \subseteq H \subseteq I_O(\mathfrak{m})$ is then called a *congruence subgroup* for modulus $\mathfrak{m}$. The corresponding quotient $I_O(\mathfrak{m})/H$ is known as a *generalized class group*; such groups play a crucial role in the study of abelian extensions of K. In the extremal case $H = P_{O,1}(\mathfrak{m})$ one ends up with the ray class group $\mathrm{cl}_{O,1}(\mathfrak{m})$. It was observed by Galbraith et al. that $\mathrm{cl}_{\mathbb{Z}[\sqrt{-p}],1}(N\mathbb{Z}[\sqrt{-p}])$ acts freely on the set of supersingular elliptic curves with $\mathbb{F}_p$-rational endomorphism ring $\mathbb{Z}[\sqrt{-p}]$ equipped with full N-level structure [GPV23, Prop. 2.5], where N denotes an integer coprime to p; we will generalize this in Sect. 3, but first we discuss how class groups of non-maximal orders are related to generalized class groups of their superorders.

2.3 Class Groups of Suborders

Let $O' \subseteq O$ be orders in K. Then $O' = \mathbb{Z} + fO$ for a unique positive integer f called the *conductor* of O' relative to O. The following result shows that $\mathrm{cl}_{O'}$ can also be viewed as a generalized class group of O:

Theorem 1. *Let*

$$P_{O,\mathbb{Z}}(fO) = \{\alpha O \mid \alpha \in K^* \text{ and } \alpha \equiv g \bmod fO \text{ for some } g \in \mathbb{Z} \text{ coprime with } f\}.$$

Then the map

$$\mathrm{cl}_{O'} \to I_O(fO)/P_{O,\mathbb{Z}}(fO) : [\mathfrak{a}] \mapsto [\mathfrak{a}O],$$

where it can be assumed that $[\mathfrak{a}]$ is represented by an integral O'-ideal that is coprime with fO', is an isomorphism of groups.

Proof. This is the relative version of [Cox13, Prop. 7.22], with the same proof. □

The following classical exact sequence provides a foundational framework for working with generalized class groups of orders.

---

[1] Equivalently, $P_O(\mathfrak{m})$ is the subgroup of P_O generated by all principal integral ideals of O that are coprime with $\mathfrak{m}$; see [KL22, §4.3], or see [Cox13, Pf. of Prop. 7.19] for the case $\mathfrak{m} = fO$ with $f \in \mathbb{Z}_{>0}$.

Theorem 2 ([KL22, Theorem 5.4]). *Let K be an imaginary quadratic number field with ring of integers O_K and orders $O' \subseteq O \subseteq O_K$. Let $\mathfrak{f}$ be the conductor of O' relative to O. Then, the following sequence is exact:*

$$1 \longrightarrow O^\times/(O')^\times \longrightarrow (O/\mathfrak{f}O)^\times/(O'/\mathfrak{f}O)^\times \longrightarrow \mathrm{cl}_{O'} \longrightarrow \mathrm{cl}_O \longrightarrow 1.$$

Proof This is a generalization of the classical exact sequences with $O = O_K$ [Neu99, Theorem I.12.12]. The proof follows from [KL22, Thm. 5.4] by choosing $\mathfrak{m} = O, \mathfrak{m}' = O', \mathfrak{d} = \mathfrak{f}O'$ (notice that the roles of O and O' are reversed). □

2.4 Class Group Actions on Sets of Elliptic Curves

In this section, we assume to be working in characteristic $p > 0$ and let $k \subset \overline{\mathbb{F}_p}$. We describe the class group actions on certain sets of elliptic curves over finite fields. We begin following an approach of Waterhouse [Wat69], who provides such a group action for isomorphism classes of ordinary elliptic curves over finite fields. The endomorphism ring of an ordinary elliptic curve is isomorphic to an imaginary quadratic order O. The class group of the endomorphism ring is used to define a group action on the set of isomorphism classes of curves with isomorphic endomorphism rings. We write $\mathrm{End}(E)$ to denote the ring of all endomorphisms of E/k, defined over $\overline{k} = \overline{\mathbb{F}_p}$. When it arises, we write $\mathrm{End}_k(E)$ to specify the subring of endomorphisms of E defined over k.

Theorem 3 ([Wat69, Theorem 4.5]). *Let E be an ordinary elliptic curve over a field $k \subset \overline{\mathbb{F}_p}$ having endomorphism ring $\mathrm{End}(E) \cong O$. Then the class group of O acts freely and transitively on the set of elliptic curves over k with endomorphism ring isomorphic to O.*

The ideal class group of $\mathrm{End}(E)$ acts on the set of isomorphism classes of ordinary elliptic curves with endomorphism rings isomorphic to the order O in the following sense:

Definition 2. *Let E/k be an elliptic curve over a field $k \subset \overline{\mathbb{F}_p}$ with commutative endomorphism ring $\mathrm{End}(E)$. Take a proper integral ideal I of $\mathrm{End}(E)$ with $N(I)$ coprime to p. Define*

$$E[I] := \bigcap_{\alpha \in I} \ker \alpha.$$

As I is a finitely generated $\mathbb{Z}$-module, the set $E[I]$ is a finite group. This finite group defines an isogeny $\varphi_I : E \to E/E[I]$ with kernel $E[I]$. Define

$$I * E := E/E[I].$$

Each ideal class contains an integral ideal representative which is of norm coprime to p. The principal ideals are generated by a single endomorphism, and so act trivially. In [Wat69, §3], Waterhouse establishes that, in the case where E is ordinary, this is a free and transitive group action on the set of elliptic curves with endomorphism ring isomorphic to $\mathrm{End}(E)$. The proof goes through showing that ideals of $\mathrm{End}(E)$ satisfy certain properties qualifying them as *kernel ideals*.

In the supersingular case, the situation is more complicated. If E is supersingular, then $\operatorname{End}(E)$ is a noncommutative ring in a quaternion algebra. In particular, $\operatorname{End}(E)$ does not have a class *group* of (left or right) ideals. There is a partial remedy and a full remedy. The partial remedy: if $k = \mathbb{F}_p$, then $\operatorname{End}_k(E)$ is isomorphic to an order in the imaginary quadratic field $\mathbb{Q}(\sqrt{-p})$. In this case, one can work with the class group of $\operatorname{End}_k(E)$ and use Definition 2, just as in the ordinary case. The group action will be free and transitive (modulo a minor subtlety highlighted in the proof of [Sch87, Thm. 4.5]). However, if $k = \mathbb{F}_{p^n}$ for $n > 1$ or $k = \overline{\mathbb{F}}_p$, we find ourselves in need of additional framework: orientations on supersingular elliptic curves are the full remedy. The remainder of this section deals with the supersingular case.

Let $k \subset \overline{\mathbb{F}}_p$ and consider a supersingular elliptic curve E/k. By [Sil09], the endomorphism ring $\operatorname{End}(E)$ is isomorphic to a maximal order in the quaternion algebra $B_{p,\infty} := \operatorname{End}(E) \otimes_{\mathbb{Z}} \mathbb{Q}$ ramified precisely at p and ∞. Any non-scalar element $\alpha \in \operatorname{End}(E) \setminus \mathbb{Z}$ generates an imaginary quadratic order. Let K be an imaginary quadratic field in which p does not split. This condition gives the existence of an embedding of K into $B_{p,\infty}$ [Voi21, Prop. 14.6.7].

Definition 3. *A K-orientation on an elliptic curve E is an embedding*

$$\iota : K \hookrightarrow \operatorname{End}(E) \otimes_{\mathbb{Z}} \mathbb{Q}.$$

For an order O of K such an embedding is called a primitive O-orientation if $\iota(O) = \iota(K) \cap \operatorname{End}(E)$. The pair (E, ι) is called a primitively O-oriented supersingular elliptic curve. We denote by $\mathcal{E}\ell\ell_k(O)$ the set of primitively O-oriented elliptic curves over k.

Remark 1. When E is supersingular as above, a K-orientation ι maps into a four-dimensional $\mathbb{Q}$-algebra. In the case where the endomorphism ring of E is commutative, Definition 3 can still apply: the map ι defines an isomorphism $O \cong \operatorname{End}(E)$. Note that in both cases we call such a map a primitive O-orientation on E, to unify notation.

Definition 4 (K-oriented isogeny). *A K-oriented isogeny is an isogeny $\varphi : (E_0, \iota_0) \to (E_1, \iota_\varphi)$ between K-oriented elliptic curves such that $\varphi : E_0 \to E_1$ as an isogeny of elliptic curves and $\iota_\varphi(-) = \frac{1}{\deg \varphi} \varphi \circ \iota_0(-) \circ \widehat{\varphi}$.*

An isomorphism of elliptic curves is likewise a K-oriented isomorphism if it is of degree-1 and satisfies the above properties.

Via the Deuring lifting theorem, the theory of oriented supersingular elliptic curves is closely related to the theory of CM elliptic curves.

Definition 5. *There is an extension L' of the ring class field L of O and a prime $\mathfrak{p}$ above p in $O_{L'}$ such that every elliptic curve with CM by O has a representative defined over L' with good reduction at $\mathfrak{p}$. Let $\mathcal{E}\ell\ell_{L'}(O)$ denote the set of isomorphism classes of elliptic curves with endomorphism ring isomorphic to O and having good reduction over $\mathfrak{p}$. Let $\rho : \mathcal{E}\ell\ell_{L'}(O) \to \mathcal{E}\ell\ell_k(O)$ denote the reduction map modulo $\mathfrak{p}$.*

Definition 6. Let $(E, \iota) \in \mathcal{E}\ell\ell_k(O)$ and take an ideal $\mathfrak{a}$ of O. Define the (group-theoretic) intersection:
$$E[\iota(\mathfrak{a})] := \bigcap_{\alpha \in \mathfrak{a}} \ker(\iota(\alpha)).$$

Let $\varphi_{\mathfrak{a}}$ denote a K-oriented isogeny of (E, ι) with kernel $E[\iota(\mathfrak{a})]$. Such an isogeny $\varphi_{\mathfrak{a}} : (E, \iota) \to (E_{\mathfrak{a}}, \iota_{\mathfrak{a}})$ is unique up to K-oriented automorphism on the codomain. When $\#E[\iota(\mathfrak{a})] = N(\mathfrak{a})$, we define the action of $\mathfrak{a}$ on (E, ι) to be: $\mathfrak{a} * (E, \iota) = (E_{\mathfrak{a}}, \iota_{\mathfrak{a}})$.

Definition 6 is precisely the supersingular analogue of Definition 2 that we need. The principal ideals (β) of O are generated by endomorphisms of E, and thus act trivially on (E, ι) since $\beta \in O$ commutes with the image of ι in $\mathrm{End}(E)$. The following theorem of Onuki completes the picture by providing the supersingular analogue of Theorem 3.

Theorem 4 ([Onu21, Theorem 3.4]). *When p is not split in the imaginary quadratic field K and p is coprime to the conductor of the order O of K, then Definition 6 gives a free and transitive action of cl_O on $\rho(\mathcal{E}\ell\ell_{L'}(O))$.*

To understand when this gives a free and transitive action on the set $\mathcal{E}\ell\ell_k(O)$ of primitively O-oriented elliptic curves, we need to understand the relationship between the sets $\rho(\mathcal{E}\ell\ell_{L'}(O))$ and its superset $\mathcal{E}\ell\ell_k(O)$:

Corollary 1 ([Sch87, Theorem 4.5], [ACL+24, Theorem 4.4]). *If p is ramified in the imaginary quadratic field $O \otimes_{\mathbb{Z}} \mathbb{Q}$, then cl_O acts freely and transitively on the set of primitively O-oriented supersingular elliptic curves over $\overline{\mathbb{F}_p}$. If p is inert in the imaginary quadratic field $O \otimes_{\mathbb{Z}} \mathbb{Q}$, then cl_O has two orbits in the set of primitively O-oriented supersingular elliptic curves over $\overline{\mathbb{F}_p}$.*

Concretely, if p is not inert in the imaginary quadratic field K containing O, then cl_O acts freely and transitively on the set of isomorphism classes of primitively O-oriented elliptic curves $\mathcal{E}\ell\ell_k(O)$. If p is split in K, $\mathcal{E}\ell\ell_k(O)$ is a set of ordinary elliptic curves. If p is ramified, $\rho(\mathcal{E}\ell\ell_{L'}(O)) = \mathcal{E}\ell\ell_k(O)$ is a set of primitively O-oriented supersingular elliptic curves. If p is inert, cl_O acts on $\rho(\mathcal{E}\ell\ell_{L'}(O))$, which is again a set of primitively O-oriented supersingular elliptic curves.

3 Generalized Class Group Actions

Let K be an imaginary quadratic number field and let O be an order in K. Let $k \subset \overline{\mathbb{F}_p}$. In this section, if p is inert in K, we fix the orbit $\rho(\mathcal{E}\ell\ell_{L'}(O))$ of cl_O and call it $\mathcal{E}\ell\ell_k(O)$ by an abuse of notation.

Let $\mathfrak{m}$ be a modulus in O. Let H be a congruence subgroup for $\mathfrak{m}$ that is contained in $P_O(\mathfrak{m})$. Let
$$\mathrm{cl}_H = I_O(\mathfrak{m})/H$$

be the corresponding generalized class group. Because $H \subseteq P_O(\mathfrak{m})$ the map

$$\mathrm{cl}_H \times \mathcal{E}\ell\ell_k(O) \to \mathcal{E}\ell\ell_k(O) : ([\mathfrak{a}], E) \mapsto \varphi_\mathfrak{a}(E) = E/E[\mathfrak{a}]$$

remains a well-defined group action. However, if $H \subsetneq P_O(\mathfrak{m})$ then this no longer yields a free group action: the class of any ideal $\mathfrak{a} \in P_O(\mathfrak{m}) \setminus H$ is a non-trivial element acting trivially. This creates room for an action of cl_H on elements of $\mathcal{E}\ell\ell_k(O)$ equipped with extra data, i.e., with $\mathfrak{m}$-*level structure*, which we now define.

3.1 $\mathfrak{m}$-Level Structures

Our starting observation is:

Lemma 1. *Let O be an imaginary quadratic order and let $E \in \mathcal{E}\ell\ell_k(O)$. Suppose $\mathfrak{m} \subseteq O$ is a proper ideal of norm coprime with p. Then*

$$E[\mathfrak{m}] \cong O/\mathfrak{m}$$

as O-modules; in particular, they are also isomorphic as groups.

Proof. If O is a maximal order, then for $k \subseteq \mathbb{C}$ this is precisely [Sil94, Proposition II.1.4(b)], while for k finite one can use the Deuring lifting theorem to reduce to the case $k \subseteq \mathbb{C}$. To obtain the statement for arbitrary imaginary quadratic orders and over arbitrary base fields, we start by writing $\mathfrak{m} = NO + \alpha O$ with N the positive generator of $\mathfrak{m} \cap \mathbb{Z}$. We can assume that the norm of α is coprime with p. Indeed, this follows from the fact that $N(\mathfrak{m})$ equals the greatest common divisor of the norms of the elements of $\mathfrak{m}$ (because it is a proper ideal).

We rely on a result due to Lenstra [Len96, Lemma 2.1], stating that the lemma is true for principal ideals. Applying this to αO, we obtain an isomorphism

$$\phi : \frac{O}{\alpha O} \longrightarrow E[\alpha]$$

of O-modules. From this it readily follows that $O/\mathfrak{m} \cong E[\mathfrak{m}]$ as groups, because for any finite abelian group A we have $A/NA \cong A[N]$. However, our goal is to establish this isomorphism at the level of O-modules.

Proving this amounts to showing that for any δ such that $O = \mathbb{Z}[\delta]$ we can find a point $P \in E[\mathfrak{m}]$ such that

$$E[\mathfrak{m}] = \langle P, \delta(P) \rangle.$$

Indeed, this gives a well-defined map of O-modules given by

$$O/\mathfrak{m} \to E[\mathfrak{m}] : 1 \mapsto P,$$

which is clearly a surjective group homomorphism, hence bijective because we already know an isomorphism as groups.

We again use the fact that $\mathfrak{m}$ is proper, which implies that $\bar{\mathfrak{m}}\mathfrak{m} = N(\mathfrak{m})O$. We start with the curve $E_\mathfrak{m} := E/E[\mathfrak{m}]$. Note that $E_\mathfrak{m} \in \mathcal{E}\ell\ell_k(O)$ since the corresponding isogeny $\varphi_\mathfrak{m}$ is horizontal, so by abuse of notation we can view δ as an endomorphism of $E_\mathfrak{m}$. Again by Lenstra's theorem, there exists an isomorphism of O-modules

$$O/N(\mathfrak{m})O \to E_\mathfrak{m}[N(\mathfrak{m})].$$

Set $P \in E_\mathfrak{m}[N(\mathfrak{m})]$ to be the image of 1 under this isomorphism. This implies that $E_\mathfrak{m}[N(\mathfrak{m})] = \langle P, \delta(P) \rangle$. Then

$$E[\mathfrak{m}] = \langle \varphi_{\bar{\mathfrak{m}}}(P), \varphi_{\bar{\mathfrak{m}}}(\delta(P)) \rangle = \langle \varphi_{\bar{\mathfrak{m}}}(P), \delta(\varphi_{\bar{\mathfrak{m}}}(P)) \rangle,$$

where the first equality follows from the fact that $\varphi_{\bar{\mathfrak{m}}} = \widehat{\varphi_\mathfrak{m}}$, and the second follows from the fact that δ "commutes" with horizontal isogenies. Thus we have constructed our basis in the correct form. □

Remark 2. Lemma 1 covers all cases of interest to this paper, but it holds more generally: e.g., from the proof, it is immediate that the statement remains true if $\mathfrak{m}$ is principal and generated by a separable endomorphism, or whenever $\mathfrak{m}$ contains a separable element and $E[\mathfrak{m}]$ is cyclic. However, not all conditions can be dropped: one clearly runs into issues as soon as $\varphi_\mathfrak{m}$ is inseparable, while a more subtle counterexample is $O = \mathbb{Z}[\ell\sqrt{-1}]$ and $\mathfrak{m}$ the O-ideal generated by ℓ^2 and $\ell \cdot \ell\sqrt{-1}$ (where ℓ denotes any prime number different from p).

Applying Lemma 1 to $\mathfrak{m} = NO$ for some integer N coprime to char k, we recover the well-known fact that

$$E[N] \cong O/NO \cong \mathbb{Z}/N\mathbb{Z} \times \mathbb{Z}/N\mathbb{Z}$$

as groups, where upon writing $O = \mathbb{Z}[\sigma]$ for some generator σ, an instance of the last isomorphism is given by $1 \mapsto (1,0)$, $\sigma \mapsto (0,1)$. This motivates the following generalization of Definition 1.

Definition 7. *Let $\mathfrak{m} \subseteq O$ be a proper ideal coprime to char k. Let $\Gamma \subseteq \mathrm{GL}(O/\mathfrak{m})$ be a subgroup and let E be an elliptic curve primitively oriented by O. A Γ-level structure on E is then a group isomorphism*

$$\Phi : O/\mathfrak{m} \to E[\mathfrak{m}]$$

defined up to pre-composition with an element $\gamma \in \Gamma$ and post-composition with a K-oriented automorphism. We denote by Y_Γ the set of primitively O-oriented elliptic curves equipped with a Γ-level structure, up to K-oriented isomorphisms. If Γ consists of O-module automorphisms, then we denote by $Z_\Gamma \subseteq Y_\Gamma$ the subset for which the level structure is an isomorphism of O-modules.

The reason for highlighting the subset Z_Γ will become apparent in Sect. 3.2.

Remark 3. Considering Γ-level structures up to K-oriented automorphisms amounts to identifying (E, Φ) and $(E, \iota(u) \circ \Phi)$ for every $u \in O^\times$; here ι denotes the implicit embedding of O in $\mathrm{End}(E)$. However, in most cases this can be ignored because it is already taken care of by the Γ-level structure. E.g., this is true if $O^\times = \{\pm 1\}$ and Γ is closed under negation.

More concretely, in view of the lemma below, defining a Γ-level structure amounts to specifying a point P of order $a_{\mathfrak{m}}$ and a point Q of order $b_{\mathfrak{m}}$ such that $\frac{a_{\mathfrak{m}}}{b_{\mathfrak{m}}}P, Q$ is a basis of $E[b_{\mathfrak{m}}]$, considered up to "base changes" as specified by the subgroup Γ.

Lemma 2. *Let $\mathfrak{m} \subseteq O$ be a proper ideal. Then there exist unique $a_{\mathfrak{m}}, b_{\mathfrak{m}} \in \mathbb{Z}$ such that $\operatorname{char} k \nmid b_{\mathfrak{m}} \mid a_{\mathfrak{m}}$ and*

$$E[\mathfrak{m}] \cong \frac{\mathbb{Z}}{(a_{\mathfrak{m}})} \times \frac{\mathbb{Z}}{(b_{\mathfrak{m}})}$$

for all $E \in \mathcal{E}\ell\ell_k(O)$.

Proof. This is standard: every finite subgroup of E admits such a decomposition, and the independence of E follows because any two such curves are connected by a horizontal isogeny of norm coprime with $\mathfrak{m}$. □

In order to motivate the next sections, let us conclude by restating (a slightly extended version of) the observation made by Galbraith, Perrin, and Voloch [GPV23] in the context of CSIDH. Here one considers $\mathfrak{m} = NO$ and $\Gamma = \{\mathrm{id}\}$; for simplicity we will just write Y_N, Z_N instead of $Y_{\{\mathrm{id}\}}, Z_{\{\mathrm{id}\}}$. Putting an $\{\mathrm{id}\}$-level structure on a curve $E \in \mathcal{E}\ell\ell_k(O)$ just amounts to choosing a basis $P, Q \in E[N]$, i.e., a full level-N structure. Writing $O = \mathbb{Z}[\sigma]$, elements of $Z_N \subseteq Y_N$ correspond to bases of the form $P, \sigma(P)$; it is a consequence of Lemma 1 that such bases indeed exist.

Theorem 5. *Let N be a positive integer coprime to $\operatorname{char} k$. Then the ray class group*

$$\operatorname{cl}_{O,1}(NO) = I_O(NO)/P_{O,1}(NO)$$

acts freely on both Y_N and Z_N; in the latter case, the action is also transitive.

Proof. This is a special case of Theorem 6 below. □

3.2 A Family of Congruence Subgroups

The goal of this section (and of this paper) is to embed Theorem 5 in a more general story. We concentrate on congruence subgroups of the form

$$P_{O,\Lambda}(\mathfrak{m}) = \{\alpha O \mid \alpha \in K^\times \text{ and } \alpha \equiv \lambda \bmod \mathfrak{m} \text{ for some } \lambda \in \Lambda \text{ coprime to } N(\mathfrak{m})\},$$

where Λ is a multiplicatively closed subset of O. This covers the aforementioned congruence subgroups $P_{O,\mathbb{Z}}(fO)$ and $P_{O,1}(\mathfrak{m}) = P_{O,\{1\}}(\mathfrak{m})$ as special cases, yet it also introduces several interesting new examples.

Remark 4. Notice that Λ and $\pm\Lambda$ or more generally $O^\times \Lambda$ define the same congruence subgroup, as one can always change the generator α of a principal ideal accordingly. Thus it would make sense to impose $O^\times \subseteq \Lambda$. However, we refrain from doing this, in order to keep covering standard notation such as $P_{O,\mathbb{Z}}(fO)$; also the exact sequence from Proposition 1 is affected by this, see Remark 6.

Now, to such a congruence subgroup $P_{O,\Lambda}(\mathfrak{m})$ we can naturally associate the subgroup
$$\Gamma_{O,\Lambda}(\mathfrak{m}) = \{\,\mu_\alpha \mid \alpha O \in P_{O,\Lambda}(\mathfrak{m})\,\} = \{\,\mu_\lambda \mid \lambda \in O^\times \Lambda\,\} \subseteq \mathrm{GL}(O/\mathfrak{m})$$
where μ_α refers to the action of multiplication by α on $O/\mathfrak{m}$. By definition of $P_{O,\Lambda}(\mathfrak{m})$, this is a multiplicative subset of the finite group $\mathrm{GL}(O/\mathfrak{m})$, hence indeed a subgroup. Note that the μ_α's are O-module automorphisms, so both $Y_{\Gamma_{O,\Lambda}(\mathfrak{m})}$ and $Z_{\Gamma_{O,\Lambda}(\mathfrak{m})}$ are well-defined.

Theorem 6. *Let $\mathfrak{m} \subseteq O$ be a proper ideal, and let $H = P_{O,\Lambda}(\mathfrak{m})$ be as above. Then*
$$[\mathfrak{a}] \star (E, \Phi) = (\varphi_\mathfrak{a}(E), \varphi_\mathfrak{a} \circ \Phi) \tag{1}$$
is a well-defined free action of cl_H on $Z_{\Gamma_{O,\Lambda}(\mathfrak{m})}$. Moreover, this action is transitive. If $\Lambda \subseteq O^\times \mathbb{Z}$ then this extends to a free action of cl_H on $Y_{\Gamma_{O,\Lambda}(\mathfrak{m})}$.

Proof. Since $\deg \varphi_\mathfrak{a} = N(\mathfrak{a})$ is assumed coprime with $\mathfrak{m}$, it follows readily that the right-hand side of (1) is an element of $Y_{\Gamma_{O,\Lambda}(\mathfrak{m})}$. Using that $\varphi_\mathfrak{a}$ is K-oriented, we also see that it concerns an element of $Z_{\Gamma_{O,\Lambda}(\mathfrak{m})}$ as soon as (E, Φ) is.

Now assume $(E, \Phi) \in Z_{\Gamma_{K,\Lambda}(\mathfrak{m})}$ and let $\mathfrak{a} = \alpha O$ be the principal ideal generated by some $\alpha \in O$. Then
$$\varphi_\mathfrak{a} \circ \Phi = \Phi \circ \mu_\alpha \tag{2}$$
because Φ is an isomorphism of O-modules. It follows that Φ and $\varphi_\mathfrak{a} \circ \Phi$ define the same $\Gamma_{O,\Lambda}(\mathfrak{m})$-level structure on E if and only if $\alpha O \in P_{O,\Lambda}(\mathfrak{m})$. But this implies that the action is well-defined and free. As for the transitivity, it suffices to argue that if
$$\Phi_1, \Phi_2 : O/\mathfrak{m} \to E[\mathfrak{m}]$$
are two isomorphisms as O-modules, then there exists $\alpha \in O$ such that $\Phi_2 = \varphi_{\alpha O} \circ \Phi_1$. This is evident from the fact that we are dealing with free rank-1 modules over $O/\mathfrak{m}$.

Finally, we need to show that if $\Lambda \subseteq O^\times \mathbb{Z}$, then we still have a well-defined and free action on all of $Y_{\Gamma_{O,\Lambda}(\mathfrak{m})}$. By ignoring post-compositions with K-oriented automorphisms, we can in fact assume $\Lambda \subseteq \mathbb{Z}$. For this we need to show that
$$\varphi_{\alpha O} \circ \Phi = \Phi \circ \mu_{\alpha'}$$
for some $\alpha' O \in P_{O,\Lambda}(\mathfrak{m})$ if and only if $\alpha O \in P_{O,\Lambda}(\mathfrak{m})$. Since we are working modulo $\mathfrak{m}$, this amounts to saying that
$$\varphi_{\alpha O} \circ \Phi = \Phi \circ \mu_\lambda = [\lambda] \circ \Phi$$
for some $\lambda \in \Lambda$ if and only if $\alpha O \in P_{O,\Lambda}(\mathfrak{m})$; the last equality follows because Φ is a group homomorphism. If $\alpha O \in P_{O,\Lambda}(\mathfrak{m})$ then the existence of such a λ is clear. On the other hand, if such a λ exists then from [Sil09, Cor. III.4.11] it follows that $\alpha \equiv \lambda \bmod \mathfrak{m}$, as wanted. (Note that, in the above reasoning, we have used that we can ignore units, in view of Remark 3.) □

In general, the action of the generalized class group cl_H on $Y_{\Gamma_{O,\Lambda}(\mathfrak{m})}$ is far from transitive. E.g., recall from Theorem 5 that the ray class group acts freely on
$$Y_{\Gamma_N} = \{\,(E,P,Q)\mid E \in \mathcal{E}\ell\ell_k(O),\ P,Q \text{ basis of } E[N]\,\},$$
but when writing $O = \mathbb{Z}[\sigma]$, it is easy to see that if P happens to be an eigenvector of σ, this can never be "undone" by acting with a ray class. There are two natural ways to make the action more transitive:

- Restricting to a subset of $Y_{\Gamma_{O,\Lambda}(\mathfrak{m})}$; this is exactly what we did above when studying $Z_{\Gamma_{O,\Lambda}}$, which seems to be the most natural option.
- Further identifying elements of $Y_{\Gamma_{O,\Lambda}}$ by working with a bigger group $\Gamma \supseteq \Gamma_{O,\Lambda}$.

We now analyse the action of cl_H on a set defined by $\Gamma \supseteq \Gamma_{O,\Lambda}(\mathfrak{m})$. First, note that we are free to chose any such set, as the following lemma shows.

Lemma 3. *Assume $\Lambda \subseteq \mathbb{Z}$, let $H = P_{O,\Lambda}(\mathfrak{m})$ and consider the free action of cl_H on $Y_{\Gamma_{O,\Lambda}}$ from above. Then this descends to a well-defined action of cl_H on Y_Γ for any $\Gamma \supseteq \Gamma_{O,\Lambda}(\mathfrak{m})$.*

Proof. The set Y_Γ consists of equivalence classes of elements of $Y_{\Gamma_{O,\Lambda}}$. Thus, we only need to show that if $(E,\Phi) \sim (E,\Phi')$, i.e. $\Phi' = \Phi \circ T$ for some $T \in \Gamma$, then $(\varphi_\mathfrak{a}(E), \varphi_\mathfrak{a} \circ \Phi) \sim (\varphi_\mathfrak{a}(E), \varphi_\mathfrak{a} \circ \Phi')$ for all $[\mathfrak{a}] \in \mathrm{cl}_H$. But this is clearly true, since we still have $\varphi_\mathfrak{a} \circ \Phi' = \varphi_\mathfrak{a} \circ \Phi \circ T$. □

3.3 A Generalised Exact Sequence

Recall that $H \subseteq P_O(\mathfrak{m})$. Thus, the class group cl_H surjects onto cl_O, and as the action of cl_O is well understood, we aim to study the action of the kernel of this surjection. To do this, we start by slightly generalising the exact sequence from Theorem 2:

Proposition 1. *Let $\mathfrak{m}, \Lambda$ and O be as above. Let $H = P_{O,\Lambda}(\mathfrak{m})$. Then Λ defines a subgroup of $(O/\mathfrak{m})^\times$, defined as $\Delta := \phi(\Lambda) \cap (O/\mathfrak{m})^\times$, where ϕ denotes the natural surjection from O to $O/\mathfrak{m}$. Then, there is an exact sequence*
$$1 \to O^\times/(O^\times \cap (\Lambda + \mathfrak{m})) \to (O/\mathfrak{m})^\times/\Delta \to \mathrm{cl}_H \to \mathrm{cl}_O \to 1$$

Proof. The proof closely follows the proof from Cox [Cox13, Theorem 7.24], which proves the special case where $\Lambda = \mathbb{Z}$, $\mathfrak{m} = (f)$ and $O = O_K$.

We prove this from the right to left. The surjection $\pi : \mathrm{cl}_H \to \mathrm{cl}_O$ is obtained from the natural map sending $[\mathfrak{a}] \in I_O(\mathfrak{m})/P_{O,\Lambda}(\mathfrak{m})$ to the class of $\mathfrak{a}$ in $I_O(\mathfrak{m})/P_O(\mathfrak{m}) \cong \mathrm{cl}_O$. The kernel is therefore exactly $P_O(\mathfrak{m})/P_{O,\Lambda}(\mathfrak{m})$. Next, we show that there is a surjection
$$(O/\mathfrak{m})^\times/\Delta \to P_O(\mathfrak{m})/P_{O,\Lambda}(\mathfrak{m})$$
obtained by sending $[[\alpha]] \in (O/\mathfrak{m})^\times/\Delta$ to αO (we will use the notation $[\gamma]$ for elements of $(O/\mathfrak{m})^\times$, and $[[\gamma]]$ for elements of $(O/\mathfrak{m})^\times/\Delta$). The ideal αO is clearly

in $P_O(\mathfrak{m})$. Further, let $[\alpha] = [\beta][\delta]$, for some $\delta \in \Lambda$, i.e. α and β are in the same class of $(O/\mathfrak{m})^\times / \Delta$. Then, unraveling the definitions, there exists some $u \in O$ such that $u\alpha \equiv u\beta\delta \equiv 1 \pmod{\mathfrak{m}}$. Further, we can choose some $\delta' \in \Lambda$ such that $[\delta'] \equiv [\delta]^{-1}$. Thus, we have that

$$\alpha O \cdot u\beta\delta O = \beta O \cdot u\alpha\delta^{-1}O,$$

which shows that the map is a well defined group homomorphism, since $u\beta\delta O \in P_{O,1}(\mathfrak{m}) \subseteq P_{O,\Lambda}(\mathfrak{m})$, and $u\alpha\delta^{-1}O \in P_{O,\Lambda}(\mathfrak{m})$.

Next, we show that the map is surjective. Let $\gamma O \in P_O(\mathfrak{m})$. Obviously, if $\gamma \in O$, then $[[\gamma]]$ maps to γO. In general, γ can be written as $\gamma_1 \gamma_2^{-1}$ for $\gamma_1, \gamma_2 \in O$, which both are coprime to $\mathfrak{m}$. Let N be the norm of γ_2. Since N is coprime to $\mathfrak{m}$, there exists a $k \in \mathbb{Z}$ such that $kN \equiv 1 \mod \mathfrak{m}$. Since $N\gamma_2^{-1} = \overline{\gamma_2}$, we have that the class $[[\gamma_1][k\overline{\gamma_2}]]$ maps to $k\gamma_1\overline{\gamma_2}O = \gamma O \cdot kNO$, and since $kN \equiv 1 \mod \mathfrak{m}$, we have $kNO \in P_{O,1}(\mathfrak{m}) \subseteq P_{O,\Lambda}(\mathfrak{m})$, proving that the map is surjective.

Finally, assume $[[\alpha]] \in (O/\mathfrak{m})^\times / \Delta$ satisfies $\alpha O \in P_{O,\Lambda}(\mathfrak{m})$. Thus, we have that $\alpha O = \beta\gamma^{-1}O$, for some β, γ satisfying $[\beta'][\gamma]^{-1} \in \Delta$, and that $\alpha = \mu\beta\gamma^{-1}$ for some $\mu \in O^\times$. This in turn means that $[\alpha] = [\mu][\beta][\gamma]^{-1}$, and since $[\beta][\gamma]^{-1} \in \Delta$, we see that $[[\alpha]] = [[\mu]]$, i.e. $[[\alpha]]$ is in the image of the natural homomorphism $O^\times \to (O/\mathfrak{m})^\times / \Delta$, whose kernel is in turn exactly $O^\times \cap (\Lambda + \mathfrak{m})$. □

Remark 5. We can compare Proposition 1 with both the classical $O = O_K$ version of Theorem 2 and the formula for computing the size of the ray class group from [Coh12, Theorem 3.2.4]: Specialising to $\Lambda = \mathbb{Z}$ and $\mathfrak{m} = (f)$, and writing $O := \mathbb{Z} + fO_K$, we immediately see that

$$O_K^\times \cap (\Lambda + \mathfrak{m}) = O_K^\times \cap (\mathbb{Z} + fO_K) = O^\times,$$

and similarly, $\Delta = \phi(\mathbb{Z}) \cap (O_K/fO_K)^\times = (\mathbb{Z} + fO_K/fO_K)^\times = (O/fO_K)^\times$, recovering the exact sequence from Theorem 2.

The size of the ray class group can be computed from the exact sequence when $\Lambda = \{1\}$, in which case one finds that $\Delta = \{[1]\}$, and thus

$$\#\mathrm{cl}_H = h(O_K)\frac{\#(O_K/\mathfrak{m})^\times}{\frac{O_K^\times}{O_K^\times \cap (1+\mathfrak{m})}} = h(O_K)\frac{\#(O_K/\mathfrak{m})^\times}{[O_K^\times : O_{K,\mathfrak{m}}^\times]}$$

where $O_{K,\mathfrak{m}}^\times$ is the group of units congruent to 1 mod $\mathfrak{m}$. This same remark applies in the relative case, $O' \subseteq O \subseteq O_K$.

Remark 6. Recall from Remark 4 that switching from Λ to $O^\times \Lambda$ does not affect the congruence subgroup $P_{O,\Lambda}(\mathfrak{m})$, and therefore it does not change the generalized class group either. Also, it does not affect the subgroup $\Gamma_{O,\Lambda}(\mathfrak{m})$. However, it is interesting to observe that it can slightly reorganize the terms in the exact sequence from Proposition 1. Indeed, switching from Λ to $O^\times \Lambda$ has the effect of folding the exact sequence, which is of the form

$$1 \to G_1 \to G_2 \xrightarrow{f} G_3 \to G_4 \to 1, \quad \text{into} \quad 1 \to \frac{G_2}{\ker f} \to G_3 \to G_4 \to 1.$$

Since we know that cl_O acts freely on the set of primitively O-oriented curves, we use the surjection $\pi : cl_H \to cl_O$ from the exact sequence above, and study the action of $\ker \pi$. By Proposition 1, The elements of $\ker \pi$ are principal ideals, which can be identified by elements of $O/\mathfrak{m}$ up to multiplication by Δ and $O^\times$. In particular, they are endomorphisms, leaving the curve fixed, and acting on the different Γ-level structures, which in turn can be identified (by definition) with left cosets of $\Gamma \subset \mathrm{GL}(O/\mathfrak{m})$, up to K-oriented isomorphisms. This action is fairly easy to describe explicitly, as the following lemma shows.

Corollary 2. *Let $O, \mathfrak{m}, \Lambda \subseteq \mathbb{Z}$ and the map π be as before. Let $\Gamma \supseteq \Gamma_{O,\Lambda}(\mathfrak{m})$. Then, $\ker \pi$ acts on the set*

$$X_\Gamma := \{M\Gamma \mid M \in \mathrm{GL}(O/\mathfrak{m})\}/\sim$$

where $\sim$ is the equivalence relation obtained by identifying cosets up to left multiplication by μ_u for $u \in O^\times$.

Proof. As we have seen, $\ker \pi$ can be identified with elements $[\alpha] \in (O/\mathfrak{m})^\times$, up to multiplication by Λ and $O^\times$. We show that the natural action of sending the left coset $M\Gamma$ to $\mu_\alpha M\Gamma$ is well defined. This is clearly a well defined action by $(O/\mathfrak{m})^\times$, so it suffices to show that multiplication by elements of Λ and $O^\times$ act trivially. Since $\lambda \in \Lambda \subseteq \mathbb{Z}$, it is clear that

$$\mu_\lambda M\Gamma = M\mu_\lambda \Gamma = M\Gamma,$$

and further, for $u \in O^\times$, μ_u acts trivially by definition of X_Γ. □

3.4 Suborder Class Group Actions

As one of our main examples, let us concentrate on the case where $\Lambda = \mathbb{Z}$ and $\mathfrak{m} = fO$ for some prime number f different from $\mathrm{char}\, k$. Pick $\sigma \in O$ such that $O = \mathbb{Z}[\sigma]$. Write $H = P_{O,\mathbb{Z}}(fO)$ and observe that $\Gamma := \Gamma_{O,\mathbb{Z}}(fO) \subseteq \mathrm{GL}(O/fO)$ is just the group of multiplications μ_λ by an integer λ that is not divisible by f. Thus we have

$$Y_\Gamma = \{(E, P, Q) \mid E \in \mathcal{E}\ell\ell_k(O),\ P, Q \text{ basis of } E[f]\}/\sim$$

where $(E, P, Q) \sim (E, \lambda P, \lambda Q)$ for any scalar $\lambda \in (\mathbb{Z}/f\mathbb{Z})^\times$. The action of cl_H on Y_Γ is not transitive. Recall that there are two approaches towards turning this into a "more transitive" action:

- Instead of Y_Γ, we can act on Z_Γ, i.e., we can require that the isomorphism

$$\Phi : O/fO \to E[f]$$

  is an isomorphism of O-modules. This amounts to picking a basis of $E[f]$ of the form $P, \sigma(P)$. Note that it suffices to specify P in this case, and because of the scaling we are in fact specifying a cyclic subgroup $C \subseteq E$ of order f.

However, since $P, \sigma(P)$ must be a basis, the subgroups we thus obtain are those that are *not* eigenspaces of σ acting on $E[f]$. In other words, we can identify

$$Z_\Gamma = \{\, (E,C) \mid E \in \mathcal{E}\ell\ell_k(O),\ C \subseteq E \text{ kernel of descending } f\text{-isogeny}\,\}. \qquad (3)$$

Thanks to Theorem 6 we know that cl_H acts freely and transitively on this set.

- Alternatively, we can apply the idea of making the action of cl_H on Y_Γ more transitive by enlarging Γ. In this case it is natural to consider

$$\Gamma_N^0 = \left\{ \begin{pmatrix} * & * \\ 0 & * \end{pmatrix} \right\} \supseteq \Gamma,$$

where the last inclusion makes sense upon identification of $\mathrm{GL}(O/fO)$ with $(\mathbb{Z}/f\mathbb{Z})^2$. Thus by Lemma 3 we also have a natural action of cl_H on

$$Y_N^0 = Y_{\Gamma_N^0} = \{\, (E,C) \mid E \in \mathcal{E}\ell\ell_k(O),\ C \subseteq E \text{ cyclic subgroup of order } f\,\}.$$

However, unless f is inert, this action is neither free nor transitive: any eigenspace of σ acting on $E[f]$ is fixed by every element of cl_H. To turn this into a free and transitive action one has to discard the eigenspaces; as such one again arrives at Z_Γ.

Now let $O' = \mathbb{Z} + fO$ be a suborder of relative conductor f and recall from Theorem 1 that the natural map

$$\mathrm{cl}_{O'} \to \mathrm{cl}_H : [\mathfrak{a}] \mapsto [\mathfrak{a}O] \qquad (4)$$

is an isomorphism. In fact, more generally, it is easy to check that the exact sequence from Proposition 1 fits in an isomorphism of exact sequences

$$\begin{array}{ccccccccc}
1 & \to & \dfrac{O^\times}{O'^\times} & \to & \dfrac{(O/fO)^\times}{(O'/fO)^\times} & \to & \mathrm{cl}_{O'} & \to & \mathrm{cl}_O \to 1 \\
& & \downarrow & & \downarrow & & \downarrow & & \downarrow \\
1 & \to & \dfrac{O^\times}{O^\times \cap (\varLambda + fO)} & \to & \dfrac{(O/fO)^\times}{\varDelta} & \to & \mathrm{cl}_H & \to & \mathrm{cl}_O \to 1
\end{array}$$

where the vertical maps are the natural maps, and where the sequence on top is the exact sequence from Theorem 2.

Now recall from Sect. 2.4 that we have a free and transitive action of $\mathrm{cl}_{O'}$ on $\mathcal{E}\ell\ell_k(O')$. On the other hand, as we have just discussed, there is also a free and transitive action of cl_H on Z_Γ. Finally, we have the isomorphism (4) connecting $\mathrm{cl}_{O'}$ to cl_H, as well as a natural bijection

$$Z_\Gamma \to \mathcal{E}\ell\ell_k(O') : (E,C) \mapsto \pi(E,C) := E/C.$$

It can be argued that all these maps are compatible with each other:

Lemma 4. *For every ideal class $[\mathfrak{a}] \in \mathrm{cl}_{O'}$ we have*

$$[\mathfrak{a}] \star \pi(E,C) = \pi([\mathfrak{a}O] \star (E,C)),$$

where the left action is that of $\mathrm{cl}_{O'}$ on $\mathcal{E}\ell\ell_k(O')$, while the right action is that of cl_H on Z_Γ.

Proof. Write $E_1 = \pi(E, C)$ and let $E_1' = \varphi_{\mathfrak{a}}(E_1)$. Then $E \in \mathcal{E}\ell\ell_k(O)$ lies above E_1 via an ascending isogeny φ with kernel $E_1[f, f\sigma]$. Likewise, there is an elliptic curve $E' \in \mathcal{E}\ell\ell_k(O)$ above E_1', which is the codomain of an ascending isogeny φ' with kernel $E_1'[f, f\sigma]$. It is easy to check that we obtain a commuting diagram

$$\begin{array}{ccc} E & \xrightarrow{\varphi_{\mathfrak{a}O}} & E' \\ \varphi \uparrow & & \uparrow \varphi' \\ E_1 & \xrightarrow{\varphi_{\mathfrak{a}}} & E_1' \end{array}$$

showing that $[\mathfrak{a}O] \star E = E'$, an equality which refers to the action of cl_O on $\mathcal{E}\ell\ell_k(O)$; see also the proof of [Sut13, Lem. 6]. It is then immediate that $C = \ker \hat{\varphi}$ is mapped via $\varphi_{\mathfrak{a}O}$ to $C' := \ker \hat{\varphi}'$, from which the statement follows. □

Remark 7. From the commutativity of the above diagram it follows that

$$\hat{\varphi}' \circ \varphi_{\mathfrak{a}O} \circ \varphi = \hat{\varphi}' \circ \varphi' \circ \varphi_{\mathfrak{a}} = [f] \circ \varphi_{\mathfrak{a}},$$

showing that this is the horizontal isogeny corresponding to the ideal $f\mathfrak{a}$. Since the natural map $\mathrm{cl}_{O'} \to \mathrm{cl}_O$ is not injective, one can also wonder what ideal we end up with when first choosing an isogeny $\psi : E \to E'$ and considering the horizontal isogeny $\hat{\varphi}' \circ \psi \circ \varphi$. One particular case is where $E = E'$, as in the case of SCALLOP. In this case the ideals corresponding to $\hat{\varphi}' \circ \varphi$ have a particularly nice interpretation: they correspond to O'-ideals of norm f^2, of the form $\mathfrak{a}_{\alpha,\beta} = (f^2, f(\alpha + \beta\sigma))$, for some $\alpha + \beta\sigma \in O$ such that $N_{K/\mathbb{Q}}(\alpha + \beta\sigma) \not\equiv 0 \bmod f$. Indeed, there is in this case a free and transitive action of $\ker(\mathrm{cl}_{O'} \to \mathrm{cl}_O)$ on the set of all $E/C \in \mathcal{E}\ell\ell_k(O')$ over cyclic f-subgroups C that are not eigenspaces of σ acting on $E[f]$, which corresponds to the free and transitive action of $\mathrm{cl}_{O'}$ on Z_Γ as in Theorem 6. It can be proven that ideal classes $[\mathfrak{a}_{\alpha,\beta}]$ in $\mathrm{cl}_{O'}$ are in bijection with $(\alpha : \beta) \in \mathbb{P}^1(\mathbb{F}_f)$ such that $N_{K/\mathbb{Q}}(\alpha + \beta\sigma) \not\equiv 0 \bmod f$, and that they constitute all of $\ker(\mathrm{cl}_{O'} \to \mathrm{cl}_O)$ (see also [BLS12, Lemma 3.2]). Once we fix a subgroup $C = \ker \hat{\varphi}$, the action of such classes is explicitly given by: $[\mathfrak{a}_{\alpha,\beta}] \star \pi(E, C) = \pi(E, (\alpha + \beta\iota(\bar{\sigma}))(C))]$, where $\bar{\sigma}$ is the complex conjugate of σ. This follows from Lemma 4 and the fact that $[\mathfrak{a}_{\alpha,\beta}] = [(N_{K/\mathbb{Q}}(\alpha + \beta\sigma), f(\alpha + \beta\bar{\sigma}))]$, hence $[\mathfrak{a}_{\alpha,\beta}O] = [(\alpha + \beta\bar{\sigma})]$.

3.5 Further Examples

Cyclic Torsion. Assume that $O/\mathfrak{m}$ is cyclic, i.e., $b_\mathfrak{m} = 1$ in Lemma 2. Then we can observe

- that every $\alpha \in O$ is congruent to an integer mod $\mathfrak{m}$; in particular it can be assumed without loss of generality that $\Lambda \subseteq \mathbb{Z}$,
- every group isomorphism $\Phi : O/\mathfrak{m} \to E[\mathfrak{m}]$ is necessarily an isomorphism of O-modules, i.e., $Y_\Gamma = Z_\Gamma$ for any $\Gamma \subseteq \mathrm{GL}(O/\mathfrak{m})$.

As a more concrete example, take $\Lambda = \{1\}$ and let f be a prime number such that $fO = \mathfrak{m} \cdot \overline{\mathfrak{m}}$ for some prime ideal $\mathfrak{m} \subseteq O$. By Theorem 6 the ray class group

$I_O(\mathfrak{m})/P_{O,1}(\mathfrak{m})$ acts freely and transitively on the set $Z_{\Gamma_{O,1}(\mathfrak{m})}$ of O-oriented elliptic curves E/k equipped with an eigenvector of σ acting on $E[f]$. For instance, if $K = \mathbb{Q}(\sqrt{-p})$, f is odd and $p \equiv 1 \bmod f$ as in CSIDH, then we can take $\mathfrak{m} = (f, \sqrt{-p} - 1)$ and this consists of supersingular elliptic curves over $\mathbb{F}_p$ with a distinguished $\mathbb{F}_p$-rational point of order f (up to negation).

Scaling by n-th Powers. Let O be an order and $\mathfrak{m} = (f)$ an ideal such that $f \equiv 1 \pmod 4$ is prime in O. Let $\Lambda = \mathbb{Z}^2 := \{\alpha^2 \mid \alpha \in \mathbb{Z}\}$, and assume further that $O^\times = \{\pm 1\}$, such that $O^\times \subset \Lambda + f\mathbb{Z}$ (here we use $f \equiv 1 \pmod 4$). Observe that $H := P_{O,\Lambda}(\mathfrak{m})$ is thus the set of principal ideals generated by elements that are equivalent to integers that are squares mod $\mathfrak{m}$, and then it follows from Proposition 1, that $\#\mathrm{cl}_H = 2(f+1)h(O)$. As in the situation in Subsect. 2.4, we get the set $Z_{\Gamma_{O,\Lambda}(\mathfrak{m})}$ consists of elements of the form $(E, P, \sigma(P))$, where $P \in E[f]$, and where we identify $(E, P, \sigma(P)) \sim (E, Q, \sigma(Q))$ if and only if $P = [\lambda]Q$ for some $\lambda \in \mathbb{Z}$ that is a square mod f. Thus, the situation here is a fine-grained version the action of $\mathrm{cl}_{\mathbb{Z}+fO}$ on $(E, \langle P \rangle)$ from Subsect. 2.4: The slightly larger class group acts on the set that can be recognized as curves, together with one of two points of order f for each subgroup of order f, and where the two points differ by multiplication by a non-square.

This example easily generalizes to $\Lambda = \mathbb{Z}^n$ for any $n \mid f - 1$, such that $O^\times \subset \Lambda + f\mathbb{Z}$.

The Full Class Group. If $\Lambda = O$, then $P_{O,\Lambda}(\mathfrak{m}) = P_O(\mathfrak{m})$ is the group of all fractional principal ideals coprime to $\mathfrak{m}$, and we end up with the standard action of cl_O on $\mathcal{E}\ell\ell_k(O)$, which indeed naturally coincides with Z_Γ where $\Gamma = \Gamma_{K,O_K}(\mathfrak{m})$. Note that in general we do not have a well-defined action of cl_O on the larger set Y_Γ; indeed the condition $\Lambda \subseteq O^\times \mathbb{Z}$ from Theorem 6 is violated. Nevertheless it makes sense to study Y_Γ as a set; e.g., when $\mathfrak{m} = NO$ then it parametrizes O-oriented elliptic curves E together with a basis of $E[N]$, where two bases are identified if and only if they can be transformed into one another via an endomorphism in $\iota(O)$.

4 Security Reductions and Non-reductions

Although we do not have cryptographic applications in mind, it is natural to extend the central question of [GPV23] to our generalized setting: given

$$(E_1, \Phi_1), (E_2, \Phi_2) \in Z_{\Gamma_{O,\Lambda}(\mathfrak{m})},$$

how hard is it to find a generalized ideal class $[\mathfrak{a}] \in I_O(\mathfrak{m})/P_{O,\Lambda}(\mathfrak{m})$ such that $[\mathfrak{a}](E_1, \Phi_1) = (E_2, \Phi_2)$? This problem is known as the vectorization problem [Cou06], which quantum computers can solve in sub-exponential time $L_h(1/2)$, with h denoting the size of the generalized class group.

Unsurprisingly, the main conclusion from [GPV23, Alg. 2] also applies here: despite the generalized ideal class group being larger, there is an immediate reduction to the vectorization problem for cl_O, potentially at the cost of a discrete logarithm computation (which may be hard classically, but succumbs to Shor's algorithm quantumly). Indeed, after finding an ideal $\mathfrak{a} \in I_O(\mathfrak{m})$ such that $\varphi_\mathfrak{a} : E_1 \to E_2$, one can find $\alpha \in O$ such that $\alpha\mathfrak{a}$ moreover maps Φ_1 to Φ_2, via the computation of Weil pairings and discrete logarithms.

Remark 8. It is worth contrasting this with actions by class groups of suborders $O' \subseteq O$. From Sect. 3.4 we know that the action of $\mathrm{cl}_{O'}$ on $\mathcal{E}\ell\ell_k(O')$ is a generalized class group action in disguise. However, it would be wrong to apply the previous discussion and conclude that the corresponding vectorization problem reduces to that of cl_O acting on $\mathcal{E}\ell\ell_k(O)$ at the cost of discrete logarithm computations. There is again a reduction, but it proceeds by walking to $\mathcal{E}\ell\ell_k(O)$ via isogenies; in other words, the "disguise" is crucial for security. An extreme case is SCALLOP [FFK+23, CL23a], where cl_O is the trivial group, but here the isogenies are of very large prime degree, hence infeasible to compute.

Cases Where Vectorization Becomes Easier. In view of the attacks on SIDH, the extra level structure may in fact make the vectorization problem much easier. E.g., in the case of the ray class group for scalar modulus $\mathfrak{m} = NO$, an attacker has access to a basis P_1, Q_1 of $E_1[N]$ along with their images under the secret isogeny $\varphi_\mathfrak{a} : E_1 \to E_2$. Assuming we are given a bound on $\deg \varphi_\mathfrak{a} = N(\mathfrak{a})$ and assuming that N is large enough and smooth, we can recover $\deg \varphi_\mathfrak{a}$ and run the algorithm from [Rob23b] to solve the vectorization problem in classical polynomial time. Another interesting case is the generalized class group action from Sect. 3.4, where we have a free and transitive action on oriented elliptic curves together with the kernel of a descending f-isogeny, see (3). Thus, here one is given access to such a kernel $C_1 \subseteq E_1[f]$ along with its image $C_2 \subseteq E_2[f]$. But then one also knows that $\sigma(C_1)$ is connected to $\sigma(C_2)$ via the same unknown scalar: the vectorization problem becomes an instance of the M-SIDH problem [FMP23], which can again be broken in overstretched cases. Moreover, if f splits then we also know that the action preserves the eigenspaces of σ acting on the f-torsion; as such we have access to *four* subgroups together with their images, and we can reduce to the case of SIDH via [FP22, Lem. 1].

Supergroups of $P_O(\mathfrak{m})$. Finally, let us drop the overall assumption made at the beginning of Sect. 3, namely that $H \subseteq P_O(\mathfrak{m})$: what if, conversely, our congruence subgroup H is a supergroup of $P_O(\mathfrak{m})$? In this case the generalized class group $\mathrm{cl}_H = I_O(\mathfrak{m})/H$ naturally acts on subsets $\{ [\mathfrak{h}]E \mid \mathfrak{h} \in H \} \subseteq \mathcal{E}\ell\ell_k(O)$ of oriented elliptic curves that can be connected via an ideal in H. In theory, this gives a reduction from the vectorization problem for cl_O to that of the smaller group cl_H: first find the class connecting $\{ [\mathfrak{h}]E_1 \mid \mathfrak{h} \in H \}$ and $\{ [\mathfrak{h}]E_2 \mid \mathfrak{h} \in H \}$ and then solve a vectorization problem for $H/P_O(\mathfrak{m})$. If this were possible, then this could be converted into a Pohlig–Hellman type reduction for class group actions. But unfortunately (or fortunately), it is unclear how to work with the sets $\{ [\mathfrak{h}]E \mid \mathfrak{h} \in H \}$; e.g., when merely working with a representant, one lacks tools for equality testing (which amounts to deciding whether or not two O-oriented elliptic curves E_1, E_2 are connected via an ideal in H).

Acknowledgments. We thank the anonymous reviewers of the International Workshop on the Arithmetic of Finite Fields 2024 for their helpful comments. We also thank Benjamin Wesolowski for pointing us to [CK23]. When preparing the final version of this paper, we were informed that Derek Perrin has independently been working on related ideas [Per24].

References

[ACL+24] Arpin, S., Chen, M., Lauter, K.E., Scheidler, R., Stange, K.E., Tran, H.T.N.: Orientations and cycles in supersingular isogeny graphs. In: Research Directions in Number Theory: Women in Numbers V, pp. 25–86. Springer, Cham (2024)

[Arp22] Arpin, S.: Supersingular elliptic curve isogeny graphs. Ph.D. thesis, University of Colorado Boulder (2022)

[BCC+23] Basso, A., et al.: Supersingular curves you can trust. In: EUROCRYPT (2). LNCS, vol. 14005, pp. 405–437. Springer (2023)

[BF23] Basso, A., Fouotsa, T.B.: New SIDH countermeasures for a more efficient key exchange. In: ASIACRYPT (8). LNCS, vol. 14445, pp. 208–233. Springer (2023)

[BLS12] Bröker, R., Lauter, K., Sutherland, A.V.: Modular polynomials via isogeny volcanoes. Math. Comput. **81**(278), 1201–1231 (2012)

[BMP23] Basso, A., Maino, L., Pope, G.: FESTA: fast encryption from supersingular torsion attacks. In: Advances in Cryptology—ASIACRYPT 2023. Part VII. LNCS, vol. 14444, pp. 98–126. Springer, Singapore (2023)

[CD20] Castryck, W., Decru, T.: CSIDH on the surface. In: PQCrypto. LNCS, vol. 12100, pp. 111–129. Springer (2020)

[CK20] Colò, L., Kohel, D.: Orienting supersingular isogeny graphs. J. Math. Cryptol. **14**(1), 414–437 (2020)

[CK23] Colò, L., Kohel, D.: On the modular OSIDH protocol (2023). https://www.leonardocolo.com/documents/articles/mosidh.pdf

[CL23a] Chen, M., Leroux, A.: SCALLOP-HD: group action from 2-dimensional isogenies. IACR Cryptology ePrint Archive, p. 1488 (2023)

[CL23b] Codogni, G., Lido, G.: Spectral theory of isogeny graphs (2023)

[CLM+18] Castryck, W., Lange, T., Martindale, C., Panny, L., Renes, J.: CSIDH: an efficient post-quantum commutative group action. In: Advances in Cryptology—ASIACRYPT 2018. Part III. LNCS, vol. 11274, pp. 395–427. Springer, Cham (2018)

[Coh12] Cohen, H.: Advanced Topics in Computational Number Theory, vol. 193. Springer (2012)

[Cou06] Couveignes, J.M.: Hard homogeneous spaces. IACR Cryptology ePrint Archive, p. 291 (2006)

[Cox13] Cox, D.A.: Primes of the Form $x^2 + ny^2$: Fermat, Class Field Theory, and Complex Multiplication. Pure and Applied Mathematics: A Wiley Series of Texts, Monographs and Tracts. Wiley (2013)

[CS21] Chenu, M., Smith, B.: Higher-degree supersingular group actions. Math. Cryptol. **1**(1), 1–15 (2021)

[FFK+23] De Feo, L., et al.: SCALLOP: scaling the CSI-FiSh. In: Public-Key Cryptography - PKC 2023 Part I. LNCS, vol. 13940, pp. 345–375. Springer (2023)

[FFP24] De Feo, L., Fouotsa, T.B., Panny, L.: Isogeny problems with level structure. In: EUROCRYPT 2024. Springer (2024). (to appear)

[FMP23] Fouotsa, T.B., Moriya, T., Petit, C.: M-SIDH and MDSIDH: countering SIDH attacks by masking information. In: EUROCRYPT (5). LNCS, vol. 14008, pp. 282–309. Springer (2023)

[FP22] Fouotsa, T.B., Petit, C.: A new adaptive attack on SIDH. In: Topics on Cryptology – CT-RSA. LNCS, vol. 13161, pp. 322–344. Springer (2022)

[GPV23] Galbraith, S.D., Perrin, D., Voloch, J.F.: CSIDH with level structure. IACR Cryptology ePrint Archive, p. 1726 (2023)

[KL22] Kopp, G.S., Lagarias, J.C.: Class field theory for orders of number fields (2022)

[Len96] Lenstra, H.W.: Complex multiplication structure of elliptic curves. J. Number Theory **56**(2), 227–241 (1996)

[Neu99] Neukirch, J., Algebraic number theory, volume 322 of Grundlehren der mathematischen Wissenschaften [Fundamental Principles of Mathematical Sciences]. Springer, Berlin (1999). Translated from the 1992 German original and with a note by Norbert Schappacher, With a foreword by G. Harder

[Onu21] Onuki, H.: On oriented supersingular elliptic curves. Finite Fields Their Appl. **69**, 101777 (2021)

[Per24] Perrin, D.: Ordinary isogeny graphs with level structure. Article in preparation (2024)

[Rob23a] Robert, D.: Breaking SIDH in polynomial time. In: EUROCRYPT (5). LNCS, vol. 14008, pp. 472–503. Springer (2023)

[Rob23b] Robert, D.: Breaking SIDH in polynomial time. In: Hazay, C., Stam, M., (eds.) Advances in Cryptology - EUROCRYPT 2023 - 42nd Annual International Conference on the Theory and Applications of Cryptographic Techniques, Lyon, France, 23–27 April 2023, Proceedings, Part V. LNCS, vol. 14008, pp. 472–503. Springer (2023)

[RS06] Rostovtsev, A., Stolbunov, A.: Public-key cryptosystem based on isogenies. Cryptology ePrint Archive, Paper 2006/145 (2006). https://eprint.iacr.org/2006/145

[Sch87] Schoof, R.: Nonsingular plane cubic curves over finite fields. J. Combin. Theory Ser. A **46**(2), 183–211 (1987)

[Sil94] Silverman, J.H.: Advanced Topics in the Arithmetic of Elliptic Curves, vol. 151. Springer (1994)

[Sil09] Silverman, J.H.: The Arithmetic of Elliptic Curves. Graduate Texts in Mathematics, vol. 106, 2nd edn. Springer (2009)

[Sut13] Sutherland, A.V.: Isogeny volcanoes. In: ANTS-X. Open Book Series, vol. 1, pp. 507–530. MSP (2013)

[Voi21] Voight, J.: Quaternion Algebras. Graduate Texts in Mathematics, vol. 288. Springer, Cham (2021)

[Wat69] Waterhouse, W.C.: Abelian varieties over finite fields. Ann. Sci. Ecole Norm. Sup. **2**, 521–560 (1969)

[XZQ23] Xiao, G., Zhou, Z., Qu, L.: Oriented supersingular elliptic curves and Eichler orders (2023)

Differential Biases, c-Differential Uniformity, and Their Relation to Differential Attacks

Daniele Bartoli[1], Lukas Kölsch[2(✉)], and Giacomo Micheli[2]

[1] University of Perugia, Perugia, Italy
daniele.bartoli@unipg.it
[2] University of South Florida and Center for Cryptographic Research at USF, Tampa, USA
{koelsch,gmicheli}@usf.edu

Abstract. Differential cryptanalysis famously uses statistical biases in the propagation of differences in a block cipher to attack the cipher. In this paper, we investigate the existence of more general statistical biases in the differences. To this end, we discuss the c-differential uniformity of S-boxes, which is a concept that was recently introduced to measure certain statistical biases that could potentially be used in attacks similar to differential attacks. Firstly, we prove that a large class of potential candidates for S-boxes necessarily has large c-differential uniformity for all but at most B choices of c, where B is a constant independent of the size of the finite field q. This result implies that for a large class of functions, certain statistical differential biases are inevitable.

In a second part, we discuss the practical consequences of this result; in particular we discuss the practical possibility of designing a differential attack based on weaknesses of S-boxes related to their c-differential uniformity.

Keywords: Differential attack · Substitution-Permutation Network · Differential Uniformity

1 Introduction

1.1 Background on Block Ciphers and Differential Cryptanalysis

Block Ciphers. A block cipher is a symmetric encryption scheme that transforms an n-bit plaintext into an n-bit ciphertext using a secret key. Block ciphers (like AES) constitute the majority of all symmetric ciphers in use today, and are a staple of modern cryptography. A classical construction of block ciphers is iterative, meaning that the entire cipher is a sequence of (generally simple and almost identical) round functions. A standard choice for a round function is a *Substitution-Permutation Network (SPN)*, which consists of a *linear layer*, sometimes called the permutation part and a substitution layer (usually several so called S-boxes running in parallel), with a key addition in between the rounds.

Differential Cryptanalysis. One of the most important attacks against block ciphers is the so-called differential attack introduced in [7], which also serves as a basis for more advanced attacks like boomerang attacks [37], rectangle attacks [6], or differential-linear attacks [28].

Let us briefly recall the main idea behind the classical differential attack. The attack uses that differences of plain texts propagate with different probabilities through the encryption process.

A cipher vulnerable to a differential attack has a (strongly) non-uniform distribution of differences that can then be exploited by a differential attack. The concept of *difference* used here is usually the XOR (i.e. addition in $\mathbb{F}_2^n$). The main reason for this is that the most standard design for SPNs uses *key additions*, i.e. the key is simply added to the message in between all rounds.

Clearly, we have $(x+\Delta+k)-(x+k) = (x+\Delta)-x$, so the key addition process does not impact the propagation of differences at all. Further, the differences are also not impacted by the linear layer of SPNs, so the only part that has direct influence is the S-box layer, which makes both analysis and design considerably easier. For a function $F\colon \mathbb{F}_2^n \to \mathbb{F}_2^n$, the differential attack thus considers the following distribution of probabilities, where Δ_I, Δ_O are the differences of plain and cipher texts, respectively:

$$P_{\Delta_I,\Delta_O} = P_x(F(x+\Delta_I) = \Delta_O + F(x)). \tag{1}$$

The main idea behind the differential attack is then that, because the key does not impact the propagation of differences, differences can be broken down over all rounds, so *differential trails* $(\Delta_0, \ldots, \Delta_r)$ can be constructed which capture the probability that an input difference Δ_0 propagates through an r-round block cipher to an output difference Δ_r via intermediate differences Δ_i after the i-th round. With the tacit assumption of uniformity, the differential analysis of the cipher can thus be broken down to analysing its constitutive rounds, which, in the case of a successful attack, allows the attacker to guess the key given enough plaintext - ciphertext pairs. To combat these differential attacks, the probabilities in Eq. (1) should be as uniform as possible.

Differential Uniformities of S-Boxes. As described above, the resistance of an SPN relies on its non-linear part, i.e. the choice of the S-box. The behaviour of an S-box F with regard to differential attacks is measured by its *Difference Distribution Table (DDT)* and *differential uniformity* $\delta(F)$ which are defined as

$$\mathrm{DDT}_F[a,b] = \#\{x \in \mathbb{F}_2^n : F(x+a) + F(x) = b\}, \tag{2}$$

$$\delta_F = \max_{a \in \mathbb{F}_2^{n*}, b \in \mathbb{F}_2^n} \mathrm{DDT}_F[a,b]. \tag{3}$$

The lower the differential uniformity of F, the better is its resistance against a differential attack.[1] The best possible differential uniformity for S-boxes over binary finite fields is 2, those S-boxes are called almost perfect nonlinear (APN).

Alternative Differences and c-Differential Uniformities. The basic idea of the differential attack can be transferred to another group operation that substitutes the role of the XOR/addition. For example, *multiplicative differentials* were considered in [9]. Instead of investigating the propagation on pairs $(x, x + \Delta)$, this kind of differential attack considers pairs $(x, \alpha \cdot x)$, where the multiplication is the multiplication in the ring $\mathbb{Z}_n$. These kind of attacks were motivated by a number of ciphers that did not just use a simple key addition, but instead relied on modular multiplication as a primitive operation. The authors of [14] use this motivation to consider a notion that tracks some statistical behaviour of the propagation of differences and is clearly inspired by the definition of the differential uniformity in (3). The definition uses a more general setting (not restricting to characteristic 2), and uses the well known isomorphism (as vector spaces) of $\mathbb{F}_p^n$ and $\mathbb{F}_{p^n}$.

Definition 1. *[14, Definition 1] Let $F : \mathbb{F}_{p^n} \to \mathbb{F}_{p^n}$, and $c \in \mathbb{F}_{p^n}$. For $a, b \in \mathbb{F}_{p^n}$, we let the entries of the c-Difference Distribution Table (c-DDT) be defined by $_c\mathrm{DDT}_F[a, b] = \#\{x \in \mathbb{F}_{p^n} : F(x + a) - cF(x) = b\}$. We call the quantity*

$$_c\delta_F = \max\left\{_c\mathrm{DDT}_F(a, b) : a, b \in \mathbb{F}_{p^n}, \text{ and } a \neq 0 \text{ if } c = 1\right\}$$

the c-differential uniformity of F.

In characteristic 2, the case $c = 1$ clearly recovers the usual definitions of DDT and differential uniformity in (2) and (3). Again, the function F has minimal statistical bias if the c-differential uniformity is as low as possible. If a function F achieves the minimal possible c-differential unreiformity 1, we call it *perfect c-nonlinear (PcN)*.

Remark 1. Since our investigations in this paper are largely independent of the characteristic of the underlying finite field, following [14], we chose to discuss the c-differential uniformity and its related concepts in full generality in all positive characteristics. We want to note that while symmetric cryptography in odd characteristic remains a rarity, there are some recent developments in this direction [25].

The new notion of c-differential uniformity has led to much new research, counting more than a dozen papers in less than two years dedicated to it (see e.g. [20, 21, 39], and in particular the survey paper [31] and the references therein), usually focusing on determining the c-differential uniformity of functions $F \colon \mathbb{F}_{p^n} \to \mathbb{F}_{p^n}$ that have seen use or are possible candidates for S-boxes. Despite the considerable attention that the c-differentials have received, a practical analysis has so far been lacking.

---

[1] Of course, the choice of the linear layer is also important for the resistance against a differential attack to maximize the number of active S-boxes.

1.2 Our Contribution and Organization of the Paper

The study of the c-differential uniformity serves as an example of two important and much more general questions for the design of block ciphers:

- Given a function of a certain pattern, is it possible to give general conditions when statistical biases are *guaranteed*?
- If we know that a function has a statistical bias of a certain form, can we find a framework to analyze the possibility of a practical attack using this bias?

In this paper, we are going to give answers to these two questions using the example of biases encoded by c-differential uniformities.

We want to emphasize that the purposes of this paper is to show that certain differential biases are essentially inevitable, and to provide a general discussion on whether it is possible to mount attacks based on those biases. It remains an open question whether these biases (that provably exist) can be used to mount a specific attack on a cipher that is currently in use.

Broadly speaking, the contribution of this paper is thus twofold: In Sect. 2, we show that for a wide class of functions $F: \mathbb{F}_{p^n} \to \mathbb{F}_{p^n}$ the c-differential uniformity is actually the *worst* possible for almost all c. More precisely, we show in Theorem 5 that in characteristic 2 there is an *explicit* constant B, independent of the size of the finite field $\mathbb{F}_q$, such that for all c's outside a set of size B any function of odd degree whose first and second Hasse derivative is non-zero has worst possible c-differential uniformity. Note that functions with second Hasse derivative equal to zero are highly non-generic and can be completely characterized (see Remark 4). The proofs in this section use novel techniques for this line of research, relying on the theory of algebraic function fields, Galois theory, and algebraic geometry.

The results of Sect. 2 indicate that in many cases, some extreme statistical biases in the c-differences are actually inevitable. This result of course leads to the obvious question whether or how one can use those biases to mount an attack on a cipher.

In Sect. 3, we discuss the possibility of constructing such an attack. In particular, we investigate what kind of cipher could be theoretically vulnerable to an attack based on a statistical weakness in the c-differential uniformity and we relate the c-differential uniformity to a special kind of "regular" differential attack. To do this, we introduce a general form of differential uniformity using a (more arbitrary) binary operation ∘ replacing the usual XOR. We in particular investigate the interaction of these generalized differences with the linear layer and the key addition process of a standard SPN.

We finish our paper with a conclusion and some interesting practical and theoretical questions related to our results.

2 On the c-Differential Uniformity of Low Degree Polynomials

Whereas most of the papers that appeared so far in the literature on c-differential uniformities deal with monomials (see e.g. [4,14,20,32,35,38,40]), in this section, we investigate necessary constraints on polynomials with low (i.e. good) c-differential uniformity. We start with a basic overview over the tools we employ to get the result.

2.1 Tools and Notation

Global Function Fields and Galois Theory. We give a brief overview on function fields and Galois theory to the extent that is needed for this section. We broadly follow the notation from [34].

Let $\mathbb{F}_q(t)$ be the rational function field in the variable t over the finite field of order q. A function field F is an algebraic extension of $\mathbb{F}_q(t)$. The field of constants k_F of F is the subfield of elements of F that are algebraic over $\mathbb{F}_q$.

A valuation ring $\mathcal{O}$ is a subring of F that contains $\mathbb{F}_q$, is different from F, and such that if $x \notin \mathcal{O}$, then $1/x \in \mathcal{O}$. A place P is the unique maximal ideal of a valuation ring $\mathcal{O}$ and it is principal. The valuation ring attached whose maximal ideal is the place P is denoted by $\mathcal{O}_P$.

Fix a place P. Each element $z \in F \setminus \{0\}$ can be uniquely written as $z = t^n u$, where t is a prime element for a place P and $u \in \mathcal{O}_P$ is invertible. We associate to P a function $v_P : F \to \mathbb{Z} \cup \{\infty\}$, called valuation at P, defined as $v_P(z) = n$, if $z = t^n u \in F \setminus \{0\}$, and $v_P(z) = \infty$, if $z = 0$.

An inclusion $F \subseteq L$ of function fields is said to be a function field extension, and we will denote it by $L : F$. The degree of $L : F$ is simply the integer $[L : F] := \dim_F(L)$. If P is a place of F, there are $Q_1, \ldots, Q_\ell$ places of L above P, i.e. places of L that contain P. The relative degree of Q_i over P is the integer $f(Q_i \mid P) = [\mathcal{O}_{Q_i}/Q_i : \mathcal{O}_P/P]$.

The ramification index $e(Q_i \mid P) := e$ of Q_i over P is a natural number such that $v_{Q_i}(x) = e \cdot v_P(x)$, for all $x \in F$. We say that $Q_i \mid P$ is ramified if $e(Q_i \mid P) > 1$, and unramified if $e(Q_i \mid P) = 1$. The fundamental equality for function field extensions states $\sum_{i=1}^{\ell} f(Q_i \mid P) e(Q_i \mid P) = [L : F]$. A finite separable extension of fields $M \supseteq F$ is said to be Galois if $\mathrm{Aut}_F(M) := \{f \in \mathrm{Aut}(M) : f(x) = x \ \forall x \in F\}$ has size $[M : F]$. Let us recall that every splitting field M of a separable polynomial in $F[x]$ is a Galois extension of F. Moreover, every Galois extension of F can be seen as a splitting field of some polynomial in $F[x]$.

Definition 2. *Let $M : F$ be a Galois extension of function fields and let $G := \mathrm{Gal}(M : F)$ be its Galois group. Let further R be a place of M lying over a place P of F. Then we define the decomposition group as $D(R \mid P) = \{g \in G \mid g(R) = R\}$ and the inertia group as*

$$I(R \mid P) = \{g \in D(R \mid P) \mid v_R(g(s) - s) \geq 1 \quad \forall s \in \mathcal{O}_R\}.$$

The following result is useful to connect the action of the Galois group on the roots to the intermediate splitting of places. See [36, Satz 1].

Lemma 1. *Let $L : K$ be a finite separable extension of function fields, let M be its Galois closure and $G := \mathrm{Gal}(M : K)$ be its Galois group. Let P be a place of K and $\mathcal{Q}$ be the set of places of L lying above P. Let R be a place of M lying above P. Then we have the following:*

1. There is a natural bijection between $\mathcal{Q}$ and the set of orbits of $H = \mathrm{Hom}_K(L, M)$ under the action of the decomposition group $D(R \mid P) = \{g \in G \mid g(R) = R\}$.
2. Let $Q \in \mathcal{Q}$ and let H_Q be the orbit of $D(R \mid P)$ corresponding to Q. Then $|H_Q| = e(Q \mid P)f(Q \mid P)$ where $e(Q \mid P)$ and $f(Q \mid P)$ are ramification index and relative degree, respectively.
3. The orbit H_Q partitions further under the action of the inertia group $I(R \mid P)$ into $f(Q \mid P)$ orbits of size $e(Q \mid P)$.

It follows immediately that

Corollary 1. *If $L : F$ is ramified at P then $D(R \mid P)$ is non-trivial. In particular, $L : F$ is ramified at P if and only if $M : F$ is ramified at P.*

Notice that if $M : K$ is a Galois extension of function fields over $\mathbb{F}_q$, then $kM : kK$ is Galois for any extension k of $\mathbb{F}_q$ (including $k = \overline{\mathbb{F}}_q$). It is well known [5, Lemma 2.4], that the geometric Galois group is generated by the inertia groups. This can be seen by employing the results in [34, Section 3.8]: in fact, the fixed field in $\overline{\mathbb{F}}_q M$ of the subgroup G generated by the inertia groups has to be the base field $\overline{\mathbb{F}}_q K$ (this follows because such fixed field would be an extension field of $\overline{\mathbb{F}}_q K$ without ramification, which would force it to be $\overline{\mathbb{F}}_q K$), which by the fundamental theorem of Galois Theory, forces $G = \mathrm{Gal}(\overline{\mathbb{F}}_q M : \overline{\mathbb{F}}_q K)$.

Lemma 2. *Let $L : K$ be a finite separable extension of function fields, let M be its Galois closure and let $G := \mathrm{Gal}(\overline{\mathbb{F}}_q M : \overline{\mathbb{F}}_q K)$. Then G is generated by the inertia groups $I(R \mid P)$, i.e.*

$$G = \langle I(R \mid P) : P \text{ place of } K, \quad R \mid P, \quad R \text{ place of } M \rangle.$$

The following result can be deduced from [26].

Theorem 1. *[3, Theorem 3.3] Let p be a prime number, m a positive integer, and $q = p^m$. Let $L : F$ be a separable extension of global function fields over $\mathbb{F}_q$ of degree n, M be the Galois closure of $L : F$, and suppose that the field of constants of M is $\mathbb{F}_q$. There exists an explicit constant $C \in \mathbb{R}^+$ depending only on the genus of M and the degree of $L : F$ such that if $q > C$ then $L : F$ has a totally split place.*

The following corollary shows that we can deduce arithmetic information (splitting over $\mathbb{F}_q$) from geometric information (splitting over $\overline{\mathbb{F}}_q$, i.e. ramification).

Corollary 2. *Let p be a prime number, m a positive integer, and $q = p^m$. Let $L : F$ be a separable extension of global function fields over $\mathbb{F}_q$ of degree n, M be the Galois closure of $L : F$. Then if $\mathrm{Gal}(\overline{\mathbb{F}}_q M : \overline{\mathbb{F}}_q F) = S_n$, we have that $L : F$ has a totally split place (over $\mathbb{F}_q$).*

Proof. It is a standard fact that the Galois group of $\overline{\mathbb{F}}_q M : \overline{\mathbb{F}}_q F$ is equal to the Galois group of $M : k_F F$, where k_F is the constant field of M. Therefore, since $S_n = \mathrm{Gal}(M : k_F F) \leq \mathrm{Gal}(M : F) \leq S_n$, we have that $k_F = \mathbb{F}_q$ by the Galois correspondence, and the final claim follows using Theorem 1. □

The following result follows from [22, Proposition 1, Page 275].

Proposition 1. *Let G be a transitive subgroup of S_n generated by transpositions. Then $G = S_n$.*

Curves and Varieties over Finite Fields. As a notation, $\mathbb{P}^r(\mathbb{F}_q)$ and $\mathbb{A}^r(\mathbb{F}_q)$ denote the projective and the affine space of dimension $r \in \mathbb{N}$ over the finite field $\mathbb{F}_q$, q a prime power. In the cases $r = 2$ and $r = 3$ we denote by $\ell_\infty := \mathbb{P}^2(\mathbb{F}_q) \setminus \mathbb{A}^2(\mathbb{F}_q)$ and $H_\infty := \mathbb{P}^3(\mathbb{F}_q) \setminus \mathbb{A}^3(\mathbb{F}_q)$ the line and the plane at infinity respectively. A variety and more specifically a curve, i.e. a variety of dimension 1, is described by a set of equations $F_1 = F_2 = \cdots = F_s = 0$, for some $s \in \mathbb{N}$, with coefficients in a finite field $\mathbb{F}_q$. We say that a variety $\mathcal{V}$ is *absolutely irreducible* if there are no varieties $\mathcal{V}'$ and $\mathcal{V}''$ defined over the algebraic closure of $\mathbb{F}_q$ and different from $\mathcal{V}$ such that $\mathcal{V} = \mathcal{V}' \cup \mathcal{V}''$. If a variety $\mathcal{V} \subset \mathbb{P}^r(\mathbb{F}_q)$ is defined by homogeneous polynomials $F_i(X_0, \ldots, X_r) = 0$, for $i = 1, \ldots, s$, an $\mathbb{F}_q$-rational point of $\mathcal{V}$ is a point $(x_0 : \ldots : x_r) \in \mathbb{P}^r(\mathbb{F}_q)$ such that $F_i(x_0, \ldots, x_r) = 0$, for $i = 1, \ldots s$. A point is affine if $x_0 \neq 0$. The set of the $\mathbb{F}_q$-rational points of $\mathcal{V}$ is usually denoted by $\mathcal{V}(\mathbb{F}_q)$. We denote by the same symbol homogenized polynomials and their dehomogenizations, if the context is clear. For a more comprehensive introduction to algebraic varieties and curves we refer to [19, 24].

In this paper we will mostly make use of hypersurfaces, i.e. varieties in $\mathbb{P}^r(\mathbb{F}_q)$ of dimension $r - 1$, and specifically curves ($r = 2$) and surfaces ($r = 3$). Any hypersurface is defined by a unique (up to non-zero scalar multiplication) homogeneous $f(X_0, \ldots, X_r)$ polynomial in $r+1$ variables. For the sake of convenience, for a hypersurface $\mathcal{W} : f(X_0, \ldots, X_r) = 0$ we will also make use of its affine equation $f(1, X_1 \ldots, X_r) = 0$.

Singular points of algebraic curves and surfaces can be investigated via the so-called Hasse derivative; see also [24, Page 148].

Definition 3 ([23]). *Let $F(X_1, \ldots, X_r) \in \mathbb{F}_q[X_1, \ldots, X_r]$ be a polynomial. For any $\alpha_1, \ldots, \alpha_r \in \overline{\mathbb{F}}_q$, $F(X_1 + \alpha_1, \ldots, X_r + \alpha_r)$ can be written uniquely as*

$$F(X_1 + \alpha_1, \ldots, X_r + \alpha_r) = \sum_{(i_1, \ldots, i_r) \in \mathbb{N}^r} F^{(i_1, \ldots, i_r)}(\alpha_1, \ldots, \alpha_r) X_1^{i_1} \cdots X_r^{i_r},$$

for some polynomials $F^{(i_1, \ldots, i_r)}(X_1, \ldots, X_r) \in \mathbb{F}_q[X_1, \ldots, X_r]$. For a given multi-index $i = (i_1, \ldots, i_r) \in \mathbb{N}^r$, we define the i-th Hasse derivative of $F(X_1, \ldots, X_r)$ as the polynomial $F^{(i_1, \ldots, i_r)}(X_1, \ldots, X_r) \in \mathbb{F}_q[X_1, \ldots, X_r]$.

It can be seen that for any monomial $X_1^{j_1} \cdots X_r^{j_r}$ its $i = (i_1, \ldots, i_r)$-th Hasse derivative is

$$\binom{j_1}{i_1} \cdots \binom{j_r}{i_r} X_1^{j_1 - i_1} \cdots X_r^{j_r - i_r}$$

and vanishes if $i_k > j_k$ for some k; see also [18].

As a notation, if a polynomial f is univariate, we denote its derivatives as f', f'', ... Let $F(X, Y) \in \mathbb{F}_q[X, Y]$ be a polynomial defining an affine plane curve $\mathcal{C}$, let $P = (u, v) \in \mathbb{A}^2(\mathbb{F}_q)$ be a point in the plane, and write

$$F(X + u, Y + v) = F_0(X, Y) + F_1(X, Y) + F_2(X, Y) + \cdots,$$

where F_i is either zero or homogeneous of degree i.

The *multiplicity* of $P \in \mathcal{C}$, written as $m_P(\mathcal{C})$, is the smallest integer m such that $F_m \neq 0$ and $F_i = 0$ for $i < m$; the polynomial F_m is the *tangent cone* of $\mathcal{C}$ at P. A linear divisor of the tangent cone is called a *tangent* of $\mathcal{C}$ at P. The point P is on the curve $\mathcal{C}$ if and only if $m_P(\mathcal{C}) \geq 1$. If P is on $\mathcal{C}$, then P is a *simple* point of $\mathcal{C}$ if $m_P(\mathcal{C}) = 1$, otherwise P is a *singular* point of $\mathcal{C}$. A quick criterion to decide whether an affine point P is singular is the following: P is singular if and only if $F(P) = F^{(1,0)}(P) = F^{(0,1)}(P) = 0$.

It is possible to define in a similar way the multiplicity of an ideal point of $\mathcal{C}$, that is a point of the curve lying on the line at infinity.

Given two plane curves $\mathcal{A}$ and $\mathcal{B}$ and a point P on the plane, the *intersection number* $I(P, \mathcal{A} \cap \mathcal{B})$ of $\mathcal{A}$ and $\mathcal{B}$ at the point P can be defined by seven axioms. We do not include its precise and long definition here. For more details, we refer to [15] and [24] where the intersection number is defined equivalently in terms of local rings and in terms of resultants, respectively.

Concerning the intersection number, the following two classical results can be found in most of the textbooks on algebraic curves.

Lemma 3. *Let $\mathcal{A}$ and $\mathcal{B}$ be two plane curves. For any affine point P, the intersection number satisfies the inequality $I(P, \mathcal{A} \cap \mathcal{B}) \geq m_P(\mathcal{A}) m_P(\mathcal{B})$, with equality if and only if the tangents at P to $\mathcal{A}$ are all distinct from the tangents at P to $\mathcal{B}$.*

Theorem 2 (Bézout's Theorem). *Let $\mathcal{A}$ and $\mathcal{B}$ be two projective plane curves over an algebraically closed field $\mathbb{K}$, having no component in common. Let A and B be the polynomials associated with $\mathcal{A}$ and $\mathcal{B}$ respectively. Then $\sum_P I(P, \mathcal{A} \cap \mathcal{B}) = \deg A \cdot \deg B$, where the sum runs over all points in the projective plane $\mathbb{P}^2(\mathbb{K})$.*

Concerning surfaces in $\mathbb{P}^3(\mathbb{F}_q)$, i.e. variety of dimension 2 defined by a homogeneous polynomial $F(X, Y, Z, T) \in \mathbb{F}_q[X, Y, Z, T]$, the same definitions for singular points and multiplicity hold. In particular a point P is singular for a surface $\mathcal{S}: F(X, Y, Z, T) = 0$ if and only if

$$F(P) = F^{(1,0,0,0)}(P) = F^{(0,1,0,0)}(P) = F^{(0,0,1,0)}(P) = F^{(0,0,0,1)}(P) = 0.$$

The following is a simple result about the irreducibility of plane sections of absolutely irreducible surfaces. We include here its proof for the sake of clarity.

Proposition 2. *Let $\mathcal{S} \subset \mathbb{P}^3(\mathbb{F}_q)$ be an absolutely irreducible surface and consider a plane π. A singular point O for the curve $\pi \cap \mathcal{S}$ is either a singular point for $\mathcal{S}$ or π is the tangent plane to $\mathcal{S}$ at O. In particular, if $\pi \cap \mathcal{S}$ is reducible then either π is the tangent plane at some point of $\pi \cap \mathcal{S}$ or π contains a singular point of $\mathcal{S}$.*

Proof. Without loss of generality we can suppose that O is the origin and $\pi : Z = 0$ and that $\mathcal{C} := \pi \cap \mathcal{S} : F(X, Y) = 0$, for some polynomial $F \in \mathbb{F}_q[X, Y]$ with $F(0,0) = F^{(1,0)}(0,0) = F^{(0,1)}(0,0) = 0$. This means that the affine equation of $\mathcal{S}$

is of type $F(X,Y) + ZH(X,Y,Z) = 0$ for some $H(X,Y,Z) \in \mathbb{F}_q[X,Y,Z]$. Now, either there exists a constant term in $H(X,Y,Z)$ and thus O is nonsingular for $\mathcal{S}$ and π is the tangent plane at O to $\mathcal{S}$ or $H(X,Y,Z)$ possesses monomials of degree at least one and thus O is a singular point for $\mathcal{S}$.

The second part of the claim directly follows, observing that if $\mathcal{C} := \pi \cap \mathcal{S}$ is reducible then the curve $\mathcal{C}$ possesses singular points (possibly defined over $\overline{\mathbb{F}}_q$). □

Finally, we include here references for estimating on the number of $\mathbb{F}_q$-rational points of algebraic varieties over finite fields. The most celebrated result is the Hasse-Weil Theorem.

Theorem 3. *[Hasse-Weil bound for curves] Let $\mathcal{C} \subset \mathbb{P}^n(\mathbb{F}_q)$ be a projective absolutely irreducible non-singular curve of genus g defined over $\mathbb{F}_q$. Then*

$$q + 1 - 2g\sqrt{q} \leq \#\mathcal{C}(\mathbb{F}_q) \leq q + 1 + 2g\sqrt{q}. \tag{4}$$

If $\mathcal{C}$ is a non-singular *plane* curve, then $g = (d-1)(d-2)/2$, where d is the degree of the curve $\mathcal{C}$, and (4) reads

$$q + 1 - (d-1)(d-2)\sqrt{q} \leq \#\mathcal{C}(\mathbb{F}_q) \leq q + 1 + (d-1)(d-2)\sqrt{q}. \tag{5}$$

If the curve $\mathcal{C}$ is singular, there is some ambiguity in defining what an $\mathbb{F}_q$-rational point of $\mathcal{C}$ actually is. Clearly, if $\mathcal{C}$ is non-singular, then there is a bijection between $\mathbb{F}_q$-rational places (or branches) of the function field associated with $\mathcal{C}$ and $\mathbb{F}_q$-rational points of $\mathcal{C}$. In the singular case, this is no more true. We refer the interested readers to [24, Section 9.6] where other relations are investigated. We point out that actually the bound (5) holds even for singular (absolutely irreducible) curves; [1, Corollary 2.5].

Concerning algebraic varieties of dimension larger than one, the first estimate on the number of $\mathbb{F}_q$-rational points was given by Lang and Weil [27] in 1954.

Theorem 4. *[Lang-Weil Theorem] Let $\mathcal{V} \subset \mathbb{P}^N(\mathbb{F}_q)$ be an absolutely irreducible variety of dimension n and degree d. Then there exists a constant C depending only on N, n, and d such that*

$$\left| \#\mathcal{V}(\mathbb{F}_q) - \sum_{i=0}^{n} q^i \right| \leq (d-1)(d-2)q^{n-1/2} + Cq^{n-1}. \tag{6}$$

Although the constant C was not computed in [27], explicit estimates have been provided for instance in [8,11,16,17,30,33] and they have the general shape $C = f(d)$ provided that $q > g(n,d)$, where f and g are polynomials of (usually) small degree. We refer to [11] for a survey on these bounds.

2.2 Results on the c-Differential Uniformity

In what follows we will consider polynomials $f(x) \in \mathbb{F}_q[x] \setminus \mathbb{F}_q[x^p]$, $q = p^n$, which are not monomials. Note that the polynomials $f(x) \in \mathbb{F}_q[x^p]$ are not a good choice for an S-box because the map $x \mapsto x^p$ is linear.

The first result we show concerns the 2-transitivity of the geometric Galois group of $F = f(x+a) - cf(x) - t \in \mathbb{F}_q(t)[X]$.

Lemma 4. Let $q = p^n$, p a prime, and $f \in \mathbb{F}_q[x] \setminus \mathbb{F}_q[x^p]$, with $d = \deg(f)$, $p \nmid d(d-1)$, and suppose that $f(x)$ is not a monomial. Then, the number of $c \in \mathbb{F}_q$ for which there exists $a \in \mathbb{F}_q^*$ such that the geometric Galois group of $F = f(x+a) - cf(x) - t \in \mathbb{F}_q(t)[X]$ is not 2-transitive is bounded by an explicit constant B independent of q.

Proof. It is well known that the geometric Galois group of F is 2-transitive if and only if the curve $(F(X) - F(Y))/(X - Y) = 0$ is absolutely irreducible; see for instance [29, Theorem 6.11].

Consider the surface

$$\mathcal{W}_c : \frac{f(X+Z) - cf(X) - f(Y+Z) + cf(Y)}{X - Y} = 0.$$

Note that since $f(x)$ is not a monomial, $\mathcal{W}_c$ is actually a surface and not a curve. The intersection $\mathcal{W}_c \cap H_\infty$ with the hyperplane at infinity is the curve

$$\mathcal{C} : \frac{(X+1)^d - cX^d - (Y+1)^d + cY^d}{(X-Y)} = 0$$

and by [2, Theorems 4.1 and 4.2] whenever $\sqrt[d-1]{c} \notin N_{d-1} := \{(1-\xi^i)/(\xi^k - \xi^j) : i,j,k \in \{0,\ldots,d-2\}, i \neq 0, k \neq j\}$, where ξ is a primitive $(d-1)$-th root of unity in $\overline{\mathbb{F}}_q$, $\mathcal{C}_c$ is nonsingular and therefore absolutely irreducible.

This shows that for any $c \notin N_{d-1}$ the surface $\mathcal{W}_c$ is absolutely irreducible.

From now on we pick up $c \notin N_{d-1}$. We want to bound the total number of singular points of $\mathcal{W}_c$. First note that they are only affine since $\mathcal{C}_c$ is nonsingular. A point $P = (\alpha, \alpha, \gamma)$ is singular for $\mathcal{W}_c$ only if it is of multiplicity three for $\varphi(X, Y, Z) := f(X+Z) - cf(X) - f(Y+Z) + cf(Y) = 0$, since P is of multiplicity one for the denominator $X - Y = 0$. This happens only if

$$\varphi(P) = \varphi^{(1,0,0)}(P) = \varphi^{(0,1,0)}(P) = \varphi^{(0,0,1)}(P) = \varphi^{(2,0,0)}(P) = \varphi^{(0,2,0)}(P)$$
$$= \varphi^{(1,1,0)}(P) = \varphi^{(0,1,1)}(P) = \varphi^{(1,0,1)}(P) = \varphi^{(0,0,2)}(P) = 0.$$

In particular $\varphi^{(1,0,1)}(P) = f''(\alpha + \gamma) = 0 = f''(\alpha+\gamma) - cf''(\alpha) = \varphi^{(2,0,0)}(P)$ and thus $f''(\alpha + \gamma) = f''(\alpha) = 0$ and there are at most $(d-2)^2$ possibilities for (α, α, γ), since there are at most $d - 2 = \deg(f'')$ choices for α satisfying $f''(\alpha) = 0$ and each of them provides at most $d-2$ distinct elements γ such that $f''(\alpha + \gamma) = 0$.

Consider now singular points of $\mathcal{W}_c$ off $X - Y = 0$. In particular, they satisfy

$$\begin{cases} f(X+Z) - cf(X) - f(Y+Z) + cf(Y) = 0 \\ f'(X+Z) - cf'(X) = 0 \\ f'(Y+Z) - cf'(Y) = 0. \end{cases}$$

Such a system defines a variety $\mathcal{U}_c$ which is the intersection of three surfaces in $\mathbb{P}^3(\overline{\mathbb{F}}_q)$ and $\mathcal{U}_c$ is of dimension 0. To see this it is enough to observe that

$\mathcal{U}_c \cap H_\infty$ is precisely the set of singular points of $\mathcal{C}$ which is empty. This means that $\mathcal{U}_c$ cannot be of dimension larger than 0 otherwise $\#(H_\infty \cap \mathcal{U}_c)$ would be positive. This shows that for any $c \notin N_{d-1}$ the number of singular points of $\mathcal{W}_c$ is $O(1)$.

Now we bound the number of tangent planes to $\mathcal{W}_c$ of the type $\pi_z : Z = z$. Clearly, if π_z is tangent at $P = (\alpha, \beta, z) \in \mathcal{W}_c$ then in particular

$$\begin{cases} f(\alpha + z) - cf(\alpha) - f(\beta + z) + cf(\beta) = 0 \\ f'(\alpha + z) - cf'(\alpha) = 0 \\ f'(\beta + z) - cf'(\beta) = 0, \end{cases}$$

if $\alpha \neq \beta$ and

$$f''(\alpha + z) - cf''(\alpha) = 0 = f'(\alpha + z) - cf'(\alpha)$$

if $\alpha = \beta$. In the former case we already saw that the number of solutions (α, β, z) is finite. In the latter case the two plane curves

$$f''(X + Z) - cf''(Z) = 0 \text{ and } f'(X + Z) - cf'(Z) = 0$$

do not share any component since their intersections with the line ℓ_∞ are disjoint and thus by Bézout's Theorem the number of intersection points, and thus the number of (α, α, z) is $O(1)$. This shows that there is $O(1)$ of $z \in \overline{\mathbb{F}}_q$ such that π_z is tangent to $\mathcal{W}_c$. The claim now follows from Proposition 2. □

Remark 2. Note that in the lemma above the constant B can be taken as $(d-2)^3$.

As a byproduct of the lemma above we can easily deal with the case $_c\delta_F = 1$.

Corollary 3. *Let $q = p^n$, p a prime, and $f \in \mathbb{F}_q[x] \setminus \mathbb{F}_q[x^p]$, with $d = \deg(f)$, $p \nmid d(d-1)$, and suppose that $f(x)$ is not a monomial. Then, the number of $c \in \mathbb{F}_q$ for which f can be PcN is bounded by an explicit constant B independent of q.*

Proof. Consider again the surface

$$\mathcal{W}_c : \frac{f(X+Z) - cf(X) - f(Y+Z) + cf(Y)}{X - Y} = 0.$$

In the proof of Lemma 4 it has already been proved that $\mathcal{W}_c$ is absolutely irreducible whenever c does not belong to a specific set of values of size at most $(d-2)^3$. By Lang-Weil Theorem the number of affine $\mathbb{F}_q$-rational points of $\mathcal{W}_c$ is lower-bounded by

$$q^2 - A(d)q^{3/2},$$

where $A(d)$ is an absolute constant depending only on the degree of $\mathcal{W}_c$. The set of parallel planes $\pi_a : Z - a = 0$, $a \in \mathbb{F}_q$, partition the set of its affine $\mathbb{F}_q$-rational points and thus there exists at least an $\bar{a} \in \mathbb{F}_q^*$ such that $\#(\pi_{\bar{a}} \cap \mathcal{W}_c) \geq q - A(d)q^{1/2}$. This means that the curve

$$\mathcal{C}_{c,\bar{a}} : \frac{f(X+a) - cf(X) - f(Y+a) + cf(Y)}{X - Y} = 0$$

has at least $q - A(d)q^{1/2}$ affine points and thus, since the line $X - Y = 0$ is not a component of $\mathcal{C}_{c,\bar{a}}, \mathcal{C}_{c,\underline{a}}$ (this can be seen by observing that the homogeneous part of the highest degree in $f(X+a) - cf(X) - f(Y+a) + cf(Y)$ is a multiple of $X^d - Y^d$, and $p \nmid d$ by assumption) possesses at least $q - A(d)q^{1/2} - d$ affine $\mathbb{F}_q$-rational points off $X - Y = 0$ and thus $f(X+a) - cf(X)$ is not a permutation polynomial and f is not PcN. □

Remark 3. In the corollary above the constant B arises from Lang-Weil Theorem and thus it can be computed applying the results in [8,11,16,17,30,33]. For instance, applying [11, Theorem 7.1] one can see that B can be chosen as $\max\{6.3d^{13/3}, (d-2)^2\}$.

Our main result of this section is the following.

Theorem 5 (Main Result). *Let $q = p^n$, p a prime, and $f \in \mathbb{F}_q[x] \setminus \mathbb{F}_q[x^p]$, f not a monomial, with $d = \deg(f)$. Suppose that one of the following holds:*

1. *$p = 2$, d is odd, and both the Hasse derivatives f' and f'' do not vanish;*
2. *$p > 2$ and $d \not\equiv 0, 1 \pmod{p}$.*

The number of $c \in \mathbb{F}_q$ for which $_c\delta_F < \deg(f)$ is bounded by an explicit constant B independent of q.

Remark 4. – As it will be clear from the proofs of the ancillary results below, the constant B in the statement of Theorem 5 is smaller than $4d^2$.
– The condition on the Hasse derivatives in Theorem 5 is not very restrictive. Indeed, the polynomials that violate these conditions can be classified explicitly: The polynomials f in $\mathbb{F}_{2^n}[x]$ such that the first Hasse derivative is zero live in $\mathbb{F}_{2^n}[x^2]$, and the ones for which the second Hasse derivative is zero are exactly the ones of the form

$$x\left(\sum_{i=0}^{s} a_i x^i\right)^4 + \left(\sum_{i=0}^{s} b_i x^i\right)^4.$$

For instance, note that a sufficient condition for which both Hasse derivatives f' and f'' do not vanish is the existence of at least one monomial in $f(x)$ of degree $i \equiv 3 \pmod{4}$.
– Theorem 5 states that if $\deg(f)$ is small compared to the field size q, it is inevitable that there exist many $c \in \mathbb{F}_q$ such that $_c\delta_F = \deg(f)$, which is the maximal possible (and thus worst) c-uniformity. Note that $\deg(f)$ here refers to the degree of f as a polynomial and not to the algebraic degree (which is in the characteristic 2 case equivalent to the highest binary weight of a monomial of f). This means that this result even applies to functions with high algebraic degree since it is clearly possible that a function with high algebraic degree has comparatively low degree as a polynomial.

The proof of Theorem 5 involves the two surfaces

$$\mathcal{W}_1 : \frac{f'(X+Z)(f(Y)-f(X)) - (f(Y+Z) - f(X+Z))f'(X)}{(X-Y)Z} = 0,$$

$$\mathcal{W}_2 : \frac{f'(Y+Z)f'(X) - f'(X+Z)f'(Y)}{(X-Y)Z} = 0.$$

Note that at this first step we strongly need that the polynomial f is not a monomial, otherwise $\mathcal{W}_1$ and $\mathcal{W}_2$ are curves, being defined by a homogeneous polynomial in three variables, and our arguments do not apply.

Proposition 3. *Let $q = p^n$, $p \nmid d$. The two surfaces $\mathcal{W}_1$ and $\mathcal{W}_2$ do not share any component.*

Proof. Consider the two curves $\mathcal{C}_1 := \mathcal{W}_1 \cap H_\infty$ and $\mathcal{C}_2 := \mathcal{W}_2 \cap H_\infty$. Then

$$\mathcal{C}_1 : \frac{(X+1)^{d-1}(Y^d - X^d) - X^{d-1}((Y+1)^d - (X+1)^d)}{X - Y} = 0$$

and

$$\mathcal{C}_2 : \frac{(Y+1)^{d-1}X^{d-1} - (X+1)^{d-1}Y^{d-1}}{X - Y} = 0.$$

It is readily seen that $\mathcal{C}_2$ factorizes as $\prod_\xi \left((Y+1)X - \xi(X+1)Y\right) = 0$, where ξ runs over the set of the $d-1$-th roots of unity distinct from 1.

Let $\ell_\xi : (Y+1)X - \xi(X+1)Y = 0$ be one of the components of $\mathcal{C}_2$. In order to show that $\ell_\xi \not\subset \mathcal{C}_1$, consider

$$\begin{cases} (Y+1)X - \xi(X+1)Y = 0 \\ (X+1)^{d-1}(Y^d - X^d) - X^{d-1}((Y+1)^d - (X+1)^d) = 0. \end{cases}$$

Since $(Y+1)X - \xi(X+1)Y = 0$, $(Y+1)^{d-1}X^{d-1} = (X+1)^{d-1}Y^{d-1}$ and thus $(Y+1)^d X^{d-1} = (X+1)^{d-1}Y^{d-1}(Y+1)$. So

$$(X+1)^{d-1}(Y^d - X^d) - X^{d-1}((Y+1)^d - (X+1)^d) =$$
$$(X+1)^{d-1}\left(X^{d-1} - Y^{d-1}\right) \neq 0.$$

Since $\mathcal{C}_1$ and $\mathcal{C}_2$ do not share any component, so do the surfaces $\mathcal{W}_1$ and $\mathcal{W}_2$. □

Proposition 4. *Let $q > (2d-2)(2d-3) + 1$, $q = p^n$, $f(x) \in \mathbb{F}_q[x] \setminus \mathbb{F}_q[x^p]$ not a monomial, $p \nmid d = \deg(f)$. There exists a set $\Theta_c \subset \mathbb{F}_q$ of size at most $(2d-2)(2d-3)$ such that for any $c \in \mathbb{F}_q \setminus \Theta_c$ there exists at least an $a_c \in \mathbb{F}_q^*$ such that $f(x + a_c) - cf(x) = t$ has at most one multiple root in $\overline{\mathbb{F}}_q$ for any fixed $t \in \overline{\mathbb{F}}_q$.*

Proof. Consider the system

$$\begin{cases} f(x_1+a) - cf(x_1) = t \\ f(x_2+a) - cf(x_2) = t \\ f'(x_1+a) - cf'(x_1) = 0 \\ f'(x_2+a) - cf'(x_2) = 0 \\ a(x_1-x_2) \neq 0. \end{cases}$$

The solutions $(\overline{x_1}, \overline{x_2}, \overline{a}, \overline{c}, \overline{t})$ of this system correspond to values $\overline{a}, \overline{c}, \overline{t}$ for which there exist two multiple roots of $f(x+\overline{a}) - \overline{c}f(x) = \overline{t}$.

The above system is equivalent to

$$\begin{cases} f(x_1+a) - cf(x_1) = t \\ \frac{f(x_2+a) - f(x_1+a)}{f(x_2) - f(x_1)} = c \\ a(x_1 - x_2) \neq 0 \\ \frac{f'(x_1+a)(f(x_2) - f(x_1)) - (f(x_2+a) - f(x_1+a))f'(x_1)}{a(x_1 - x_2)} = 0 \\ \frac{f'(x_2+a)f'(x_1) - f'(x_1+a)f'(x_2)}{a(x_1 - x_2)} = 0. \end{cases} \quad (7)$$

The last two equations define the surfaces $\mathcal{W}_1$ and $\mathcal{W}_2$ considered in Proposition 3. Since such surfaces do not share any component, their intersection is of dimension one (i.e. union of curves) and of degree at most $(2d-2)(2d-3)$.

Thus, apart from a small (at most $(2d-2)(2d-3)$) number of a, the intersection $W_1 \cap W_2 \cap (Z = a)$ consists of at most $(2d-2)(2d-3)$ points (on the algebraic closure $\overline{\mathbb{F}}_q$). Since $q > (2d-2)(2d-3) + 1$, there exists $\overline{a} \in \mathbb{F}_q^*$ for which the total number of solutions (x_1, x_2, c, t) of System (7) is upper bounded by $(2d-2)(2d-3)$. Let Θ be the set of all the solutions of System (7) for such an $\overline{a}$ and consider $\Theta_c := \{c \in \mathbb{F}_q : \exists (\overline{x_1}, \overline{x_2}, c, \overline{t}) \in \Theta\}$. Clearly $\#\Theta_c \le \#\Theta \le (2d-2)(2d-3)$. Thus, for any $c \in \mathbb{F}_q \setminus \Theta_c$ we have that $f(x+\overline{a}) - cf(x) = t$ has at most one multiple root x_1 for any fixed $t \in \mathbb{F}_q$ and the claim follows. □

Proposition 5. *Let* $q > \max\{2d-4, (d-1)^2\}$, $q = p^n$, $f(x) \in \mathbb{F}_q[x] \setminus \mathbb{F}_q[x^p]$ *not a monomial,* $d = \deg(f)$, *and* $p \nmid d(d-1)$. *There exists a set* $\Theta_c' \subset \mathbb{F}_q$ *of size at most* $2d-4$ *such that for any* $c \in \mathbb{F}_q \setminus \Theta_c'$ *there exists at least an* $a_c \in \mathbb{F}_q^*$ *such that* $f(x+a_c) - cf(x) = t$ *has no root in* $\overline{\mathbb{F}}_q$ *of multiplicity larger than 2 for any fixed* $t \in \mathbb{F}_q$.

Proof. Consider the system

$$f(x+a) - cf(x) - t = f'(x+a) - cf'(x) = f''(x+a) - cf''(x) = 0, \quad a \neq 0.$$

Any solution $(\overline{x}, \overline{a}, \overline{t}, \overline{c})$ provides values $\overline{a}, \overline{t}, \overline{c}$ such that $f(x+\overline{a}) - \overline{c}f(x) = \overline{t}$ has a root of multiplicity at least three.

The system above is equivalent to

$$\begin{cases} f(x+a) - cf(x) = t \\ f'(x+a) - cf'(x) = 0 \\ a \neq 0 \\ \frac{f'(x)f''(x+a) - f'(x+a)f''(x)}{a} = 0. \end{cases} \quad (8)$$

The last equation in System (8) is a non-vanishing equation of degree $2d-4$ for any a. To see this, let $f(x) = x^d + \alpha x^{d-1} + \cdots$ and thus

$$\begin{aligned} f'(x) &= dx^{d-1} + \alpha(d-1)x^{d-2} + \cdots, \\ f''(x) &= d(d-1)x^{d-2} + \alpha(d-1)(d-2)x^{d-3} + \cdots, \end{aligned}$$

and the leading coefficient of $f'(x)f''(x+a) - f'(x+a)f''(x)$ is $-ad^2(d-1) \neq 0$.

Note that $f'(x+a) - cf'(x)$ and $f''(x)$ are non-vanishing polynomials. The number of a for which $f'(x+a)$ and $f'(x)$ share a factor is upperbounded by $(\deg(f'))^2 \leq (d-1)^2$. Let $\bar{a} \in \mathbb{F}_q^*$ be such that $f'(x+a)$ and $f'(x)$ do not share any factor.

For this fixed $\bar{a}$, System (8) admits at most $2d-4$ solutions $(\bar{x}, \bar{a}, \bar{t}, \bar{c})$ and the claim follows. □

Proposition 6. *Let $q = 2^h > \max\{2d-4, (d-1)^2\}$ and $f(x) \in \mathbb{F}_q[x] \setminus \mathbb{F}_q[x^2]$ not a monomial. Suppose that both the Hasse derivatives f' and f'' do not vanish. There exists a set $\Theta'_c \subset \mathbb{F}_q$ of size at most $2d-4$ such that for any $c \in \mathbb{F}_q \setminus \Theta'_c$ there exists at least an $a_c \in \mathbb{F}_q^*$ such that $f(x + a_c) - cf(x) = t$ has no root in $\overline{\mathbb{F}}_q$ of multiplicity larger than 2 for any fixed $t \in \mathbb{F}_q$.*

Proof. The argument is the same as in the proof of Proposition 5. Now

$$f'(x) = \alpha_r x^r + \alpha x^{r-2} + \cdots, \qquad f''(x) = \alpha_s x^s + \beta x^{s-1} + \cdots,$$

where $r+1$ is the largest degree not divisible by 2 and $s+2$ the largest degree congruent to 3 (mod 4). So, the leading coefficient of $f'(x)f''(x+a) - f'(x+a)f''(x)$ is $a\alpha_s\alpha_r \neq 0$ and the claim follows. □

Combining Propositions 4, 5, and 6 we have the following.

Proposition 7. *Let $q > (2d-1)(2d-3) + 1$, $q = p^n$, and $f(x) \in \mathbb{F}_q[x] \setminus \mathbb{F}_q[x^p]$ not a monomial, $d = \deg(f)$, with*

1. *$p \nmid d(d-1)$ if p is odd;*
2. *the Hasse derivatives f' and f'' do not vanish if $p = 2$.*

There exists a set $\Psi_c \subset \mathbb{F}_q$ of size at most $(2d-1)(2d-3)$ such that for any $c \in \mathbb{F}_q \setminus \Psi_c$ there exists at least an $a_c \in \mathbb{F}_q^*$ such that $f(x + a_c) - cf(x) = t$ has either all distinct roots or precisely one double root in $\overline{\mathbb{F}}_q$ for any fixed $t \in \mathbb{F}_q$.

We are now ready to prove the main theorem of this section.

Proof (Proof of Theorem 5). By Corollary 2, if for some $c \in \mathbb{F}_q$ there exists $a \in \mathbb{F}_q^*$ such that geometric Galois group of $F = f(x+a) - cf(x) - t \in \mathbb{F}_q(t)[X]$ is the symmetric group S_n then there exists a totally split place over $\mathbb{F}_q$ for the extension $\mathbb{F}_q(x,t) : \mathbb{F}_q(t)$, i.e. an element $\bar{t} \in \mathbb{F}_q$ for which $f(x+a) - cf(x) - \bar{t} = 0$ has $\deg(f)$ distinct solutions in $\mathbb{F}_q$. Thus, using Corollary 2, it is enough to prove that for a fixed value c not belonging to a subset of $\mathbb{F}_q$ of size at most $O(1)$, there exists at least one element $a_c \in \mathbb{F}_q^*$ such that the geometric Galois group of $F = f(x+a) - cf(x) - t \in \mathbb{F}_q(t)[X]$ is the symmetric group S_n.

Observe that thanks to Lemma 1 we have that the inertia groups of the Galois group of F are all isomorphic to C_2 (the cyclic group of order 2) since, for each $t \in \overline{\mathbb{F}}_q$, $f(x+a) - cf(x) - t$ possesses at most one double root.

Since G is generated by its inertia groups thanks to Lemma 2, $G = \mathrm{Gal}(F \mid \overline{\mathbb{F}}_q)$ is a transitive subgroup of S_n generated by transpositions (since F is irreducible for any fixed pair (a,c)). Proposition 1 now implies that $G = S_n$. Therefore we obtain that $G = \mathrm{Gal}(F \mid \mathbb{F}_q) = S_n$ as well, from which the claim follows thanks to Corollary 2. □

3 The Feasibility of Differential Attacks Based on the c-Differential Uniformity

The focus in the research on c-differential uniformity has so far been almost purely on determining the c-differential uniformity of specific functions. The actual use case has been mostly neglected, which is surprising given that a clear attack based on bad c-differential uniformities has not been presented yet.

While the "standard" differential uniformity measures the probability of a difference propagating through the S-box (indeed $DDT_F[a,b]$ measures the probability of a difference a turning into a difference b times 2^n, see [12, Section 3.4.1]), it is not immediately clear what kind of statistical bias the c-differential uniformity measures. Clearly, the distribution of the values of $F(x+a) - cF(x)$ is also a measure of differential bias (since it compares the inputs of x and $x+a$), but the output is for $c \neq 1$ not a usual difference itself, so the construction of a differential trail that tracks the propagation through several rounds of the cipher is not readily possible. The c-differential uniformity thus does not measure a propagation of usual differences.

Unlike the multiplicative differentials [9] mentioned as inspiration for the c-differential uniformity in [14], the c-differential uniformity also does not measure the propagation of multiplicative "differences" of the form $(x, \alpha x)$ through the cipher since, as mentioned, the input difference used is actually the regular addition. It is however possible to find a kind of "difference" such that the c-differential uniformity does measure the propagation of this "difference", as we present now.

3.1 A General Differential Attack

Instead of using the usual difference $a - b$ (which, in the case of the usual $\mathbb{F}_2^n$ setting, boils down to the XOR $a \oplus b$), other binary operation can of course be used, for instance in the case of the multiplicative differentials from [9] mentioned above, this is the modular multiplication. So let $\circ\colon \mathbb{F}_p^n \times \mathbb{F}_p^n \to \mathbb{F}_p^n$ be a binary operation. We can then consider the propagation of pairs of those generalized differences $(x, x \circ a)$ and, identical to the usual differential attack, we can attempt to use biases in the probabilities of the propagation of those differences.

Generalizing the usual differential uniformity of a function $F\colon \mathbb{F}_{p^n} \to \mathbb{F}_{p^n}$, we can define the $\circ$-differential uniformity and the $\circ$-DDT.

Definition 4. *Let* $F\colon \mathbb{F}_{p^n} \to \mathbb{F}_{p^n}$ *be a function. We define for all* $a, b \in \mathbb{F}_{p^n}$

$$_\circ\mathrm{DDT}_F[a, b] = \#\{x \in \mathbb{F}_{p^n} : F(x \circ a) = b \circ F(x)\}$$

and

$$_\circ \delta_F = \max_{x \circ a \neq x,\, b \in \mathbb{F}_{p^n}} {_\circ\mathrm{DDT}_F[a, b]}.$$

Clearly, for $\circ = +$ one recovers the usual DDT and differential uniformity. But, more interestingly, this framework actually also allows us to recover the c-differential uniformity. Indeed, setting $a \circ_c b := a + cb$ for all $a, b \in \mathbb{F}_{p^n}$ and a fixed $c \in \mathbb{F}_{p^n}$, we get that $F(x \circ_c a) = b \circ_c F(x)$ if and only if $F(x + ca) = b + cF(x) \Leftrightarrow F(x + ca) - cF(x) = b$. Since c is fixed, a simple transformation $a \mapsto a/c$ then immediately relates the c-DDT with the $\circ$-DDT for this choice of $\circ$, and the respective differential uniformities are identical. In this sense, the c-differential uniformity is just a new special case of differential uniformity for this specific choice of the binary operation $\circ$. The c-differential uniformity thus seems to be just a tool to measure the resistance of a cipher against a specific differential attack based on this operation. We want to note that the general form of differential as described in Definition 4 was analysed in a series of papers [10, 13] with the idea to find specific binary operations that can lead to efficient differential attacks (or, possibly, for a malicious designer, to a "hidden", non-public binary operation that serves as a trapdoor to a cipher resistant against the usual attacks). However, the authors in those papers only analyse a subclass of binary operations that excludes the specific binary operation $\circ$ that we identified as being related to the c-differential uniformity here.

3.2 The Differential Attack Based on Weak c-Differential Uniformity

Let us now consider a potential differential attack using this specific binary operation $\circ_c$ defined via $a \circ_c b = a + cb$ for all $a, b \in \mathbb{F}_{p^n}$ and $c \in \mathbb{F}_{p^n}^*$. As explained briefly in the introduction on the classic differential attack earlier, the differential attack is possible for many block ciphers since they use simple key addition as a primitive operation, which means that differences propagate in the same way regardless of the key. Moreover, differences propagate also unchanged through the linear layer. We now check if/when those two properties are satisfied by $\circ_c$.

We start with a consideration of the linear layer. The following result states that a linear layer applied to the input of the function does not change its c-differential uniformity, however a linear layer applied to the output generally does.

Theorem 6. *Let $F\colon \mathbb{F}_{p^n} \to \mathbb{F}_{p^n}$ be a function and $A\colon \mathbb{F}_{p^n} \to \mathbb{F}_{p^n}$ be an affine permutation, where $A = L + s$ and L is the linear part of A. Then*

$$_c\mathrm{DDT}_F[a,b] = {}_c\mathrm{DDT}_{F\circ A}[L(a), b]$$

and in particular ${}_c\delta_F = {}_c\delta_{F\circ A}$.

Moreover, if A is linear over $\mathbb{F}_{p^l}$ where $l = [\mathbb{F}_p(c) : \mathbb{F}_p]$, then

$$_c\mathrm{DDT}_F[a,b] = {}_c\mathrm{DDT}_{A\circ F}[a, A^{-1}(b)]$$

and ${}_c\delta_F = {}_c\delta_{A\circ F}$. However, generally ${}_c\delta_F \neq {}_c\delta_{A\circ F}$ if $c \neq 1$.

Proof. Clearly, $(F\circ A)(x+a) - c(F\circ A)(x) = b$ if and only if $F(L(x) + L(a) + s) - cF(L(x) + s) = b$, and the first results follows since L is a permutation.

On the other hand, $(A\circ F)(x+a) - c(A\circ F)(x) = b$ if and only if $F(x+a) - A^{-1}(cA(F(x))) = A^{-1}(b)$. If $A(cx) = cA(x)$, then $A^{-1}(cA(x)) = c$ and ${}_c\mathrm{DDT}_F[a,b] = {}_c\mathrm{DDT}_{A\circ F}[a, A^{-1}(b)]$. Writing A as a polynomial $A = \sum_{i=0}^{n-1} a_i x^{p^i} + s$, we get

$$A(cx) = \sum_{i=0}^{n-1} c^{p^i} a_i x^{p^i} + s, \qquad cA(x) = \sum_{i=0}^{n-1} ca_i x^{p^i} + cs.$$

Comparing coefficients yields $s = 0$ (unless $c = 1$) and $c^{p^i} = c$ for all i with $a_i \neq 0$. $c^{p^i} = c$ is equivalent to $c \in \mathbb{F}_{p^i}$, so we conclude that $A(cx) = cA(x)$ for all $x \in \mathbb{F}_{p^n}$ if and only if $a_i = 0$ for all i such that $c \notin \mathbb{F}_{p^i}$ which is equivalent to $A \in \mathrm{GL}(\mathbb{F}_{p^l})$.

If this condition does not hold, it is easy to construct examples by computer search that show ${}_c\delta_F \neq {}_c\delta_{A\circ F}$. □

Theorem 6 shows that unless one picks very specific linear layers or $c \in \mathbb{F}_p$ (which in the char2 case most interesting for applications only holds for $c = 1$, i.e. the classical differential uniformity), the c-differential uniformity is actually affected by the choice of the linear layer. In particular, unlike the classical differential attack, the resistance of the cipher cannot be broken down to the S-box level, since linear layers used in block ciphers are of course generally not linear over extension fields.

Let us now consider the interaction of the $\circ$-differences with the process of key addition. For the classical differences, the key addition does not impact any differences since

$$(x + \Delta) + k - (x + k) = \Delta \tag{9}$$

for any choice of x, k, Δ. For our binary operation $\circ_c$, let $\overline{\circ_c}\colon \mathbb{F}_{p^n} \times \mathbb{F}_{p^n} \to \mathbb{F}_{p^n}$, defined as $a\overline{\circ_c}b := a - cb$, be the "inverse" of $\circ_c$ in the sense that $(a\circ_c b)\overline{\circ_c}b = a$ for all $a, b \in \mathbb{F}_{p^n}$. We now consider the differences with respect to $\circ_c$ by substituting $+$ and $-$ of the classical differences in Eq. (9) with $\circ_c$ and $\overline{\circ_c}$ (while of course keeping the regular addition for the key addition). This yields

$$((x \circ_c \Delta) + k)\overline{\circ_c}(x + k) = x + c\Delta + k - c(x + k) = c\Delta - (c - 1)(x + k).$$

It is immediate to see that the differences now are neither independent from the subkey k nor from the message x if $c \neq 1$.

So, in closing, it seems that a differential attack based on $\circ_c$-differences (attempting to abuse high c-differential uniformity) has several practical challenges as both the linear layers and the key addition process used in the vast majority of block ciphers make such an attack considerably harder than an attack relying on the classical concept of differences. For block ciphers that do not use a simple key addition but possibly another primitive operation, differential attacks based on different binary operations as lined out in this section might be of practical interest.

Regardless, the c-differential uniformities still measure biases in the distribution of differences and it might still theoretically be possible to construct an attack different than the one considered here to abuse this bias. However, it seems to be clear that an analysis would be made considerably more difficult by the fact that linear layers play a more significant role than in the classical differential attack and its derivatives.

4 Conclusions and Open Problems

Our main contribution concerns the c-differential uniformity of polynomials and states that for a generic polynomial there exists only a thin set of instances of $c \in \mathbb{F}_q$ for which $_c\delta_F$ is not the worst possible. Our investigation involves techniques from both Algebraic Geometry in positive characteristic and Galois Theory and tells us that in order to avoid the differential biases encoded in the c-differential uniformity, polynomials need to respect specific constraints on their degree structure. More generally, we show that extreme statistical differential biases for a big class of functions are inevitable. While our analysis in Sect. 3 indicates that constructing attacks based on those biases is not easy, it remains open if other attacks exploiting c-differential uniformities are possible. An obvious question is if it is possible to extend the techniques we used in this paper to analyze other statistical biases, in particular it would be interesting to see if results of the form of Theorem 5 can be achieved. An interesting theoretical open question concerns the possibility to obtain similar results involving weaker constraints, for instance dropping the condition $p \nmid \deg(f)$ in Theorem 5. In Sect. 3, we discussed a *general* differential attack with an arbitrary binary operation replacing the XOR. In [13], the authors investigate a similar generalized attack (starting from a different motivation), and argue that for some specific binary operations, there are enough weak keys that allow the exploitation of certain biases. While the results are not applicable to the case of c-differential uniformities, it would be interesting if the operations investigated in [13] behave similarly to the c-differential uniformity as described in this paper.

References

1. Aubry, Y., Perret, M.: A Weil theorem for singular curves. In: Arithmetic. Geometry and Coding Theory (Luminy, 1993), pp. 1–7. de Gruyter, Berlin (1996)
2. Bartoli, D., Calderini, M.: On construction and (non)existence of c-(almost) perfect nonlinear functions. Finite Fields Appl. **72**, 16 (2021). Paper No. 101835. https://doi.org/10.1016/j.ffa.2021.101835
3. Bartoli, D., Micheli, G.: Algebraic constructions of complete m-arcs. Combinatorica **42**, 673–700 (2022)
4. Bartoli, D., Timpanella, M.: On a generalization of planar functions. J. Algebraic Combin. **52**(2), 187–213 (2020). https://doi.org/10.1007/s10801-019-00899-2
5. Bastioni, L., Micheli, G.: On complete m-arcs. J. Algebra **638**, 238–254 (2024)
6. Biham, E., Dunkelman, O., Keller, N.: The rectangle attack – rectangling the serpent. In: Pfitzmann, B. (ed.) Advances in Cryptology – EUROCRYPT 2001, pp. 340–357. Springer, Heidelberg (2001)
7. Biham, E., Shamir, A.: Differential cryptanalysis of des-like cryptosystems. J. Cryptol. **4**(1), 3–72 (1991). https://doi.org/10.1007/BF00630563
8. Bombieri, E.: Counting points on curves over finite fields (d'après S. A. Stepanov). In: Séminaire Bourbaki, 25ème année (1972/1973), Exp. No. 430, pp. 234–241. Lecture Notes in Math., vol. 383 (1974)
9. Borisov, N., Chew, M., Johnson, R., Wagner, D.: Multiplicative differentials. In: Daemen, J., Rijmen, V. (eds.) Fast Software Encryption, pp. 17–33. Springer, Heidelberg (2002)
10. Brunetta, C., Calderini, M., Sala, M.: On hidden sums compatible with a given block cipher diffusion layer. Discrete Math. **342**(2), 373–386 (2019). https://doi.org/10.1016/j.disc.2018.10.003. https://www.sciencedirect.com/science/article/pii/S0012365X18303376
11. Cafure, A., Matera, G.: Improved explicit estimates on the number of solutions of equations over a finite field. Finite Fields Appl. **12**(2), 155–185 (2006). https://doi.org/10.1016/j.ffa.2005.03.003
12. Carlet, C.: Boolean Functions for Cryptography and Coding Theory. Cambridge University Press, Cambridge (2021). https://doi.org/10.1017/9781108606806
13. Civino, R., Blondeau, C., Sala, M.: Differential attacks: using alternative operations. Des. Codes Crypt. **87**(2), 225–247 (2019). https://doi.org/10.1007/s10623-018-0516-z
14. Ellingsen, P., Felke, P., Riera, C., Stănică, P., Tkachenko, A.: C-differentials, multiplicative uniformity, and (almost) perfect c-nonlinearity. IEEE Trans. Inf. Theory **66**(9), 5781–5789 (2020). https://doi.org/10.1109/TIT.2020.2971988
15. Fulton, W.: Algebraic curves. Advanced Book Classics. Addison-Wesley Publishing Company, Advanced Book Program, Redwood City, CA (1989). An introduction to algebraic geometry, Notes written with the collaboration of Richard Weiss, Reprint of 1969 original
16. Ghorpade, S.R., Lachaud, G.: Number of solutions of equations over finite fields and a conjecture of Lang and Weil. In: Number Theory and Discrete Mathematics (Chandigarh, 2000), pp. 269–291. Trends Math., Birkhäuser, Basel (2002)
17. Ghorpade, S.R., Lachaud, G.: Corrigenda and addenda: Étale cohomology, Lefschetz theorems and number of points of singular varieties over finite fields [mr1988974]. Mosc. Math. J. **9**(2), 431–438 (2009). https://doi.org/10.17323/1609-4514-2009-9-2-431-438

18. Goldschmidt, D.M.: Algebraic Functions and Projective Curves. Graduate Texts in Mathematics, vol. 215. Springer, New York (2003). https://doi.org/10.1007/b97844
19. Hartshorne, R.: Algebraic Geometry. Graduate Texts in Mathematics, No. 52. Springer, New York-Heidelberg (1977)
20. Hasan, S.U., Pal, M., Riera, C., Stănică, P.: On the c-differential uniformity of certain maps over finite fields. Des. Codes Crypt. **89**(2), 221–239 (2021). https://doi.org/10.1007/s10623-020-00812-0
21. Hasan, S.U., Pal, M., Stănică, P.: The c-differential uniformity and boomerang uniformity of two classes of permutation polynomials. IEEE Trans. Inf. Theory **68**(1), 679–691 (2022). https://doi.org/10.1109/TIT.2021.3123104
22. Hashimoto, K.I., Miyake, K., Nakamura, H.: Galois Theory and Modular Forms, vol. 11. Springer (2003)
23. Hasse, H.: Theorie der höheren Differentiale in einem algebraischen Funktionenkörper mit vollkommenem Konstantenkörper bei beliebiger Charakteristik. J. Reine Angew. Math. **175**, 50–54 (1936). https://doi.org/10.1515/crll.1936.175.50
24. Hirschfeld, J.W.P., Korchmáros, G., Torres, F.: Algebraic Curves Over a Finite Field. Princeton Series in Applied Mathematics. Princeton University Press, Princeton (2008)
25. Kölbl, S., Tischhauser, E., Derbez, P., Bogdanov, A.: Troika: a ternary cryptographic hash function. Des. Codes Crypt. **88**(1), 91–117 (2020). https://doi.org/10.1007/s10623-019-00673-2
26. Kosters, M.: A short proof of a chebotarev density theorem for function fields. Math. Commun. **22**(2), 227–233 (2017)
27. Lang, S., Weil, A.: Number of points of varieties in finite fields. Am. J. Math. **76**, 819–827 (1954). https://doi.org/10.2307/2372655
28. Langford, S.K., Hellman, M.E.: Differential-linear cryptanalysis. In: Desmedt, Y.G. (ed.) Advances in Cryptology – CRYPTO '94, pp. 17–25. Springer, Heidelberg (1994)
29. Lidl, R., Mullen, G., Turnwald, G.: Dickson Polynomials. Pitman Monographs in Pure and Applied Mathematics. Addison-Wesley, Reading **65**, 15–53 (1993)
30. Lidl, R., Niederreiter, H.: Finite Fields, Encyclopedia of Mathematics and its Applications, vol. 20. Addison-Wesley Publishing Company, Advanced Book Program, Reading (1983). With a foreword by P. M. Cohn
31. Mesnager, S., Mandal, B., Msahli, M.: Survey on recent trends towards generalized differential and boomerang uniformities. Cryptogr. Commun. (2021). https://doi.org/10.1007/s12095-021-00551-6
32. Mesnager, S., Riera, C., Stănică, P., Yan, H., Zhou, Z.: Investigations on c-(almost) perfect nonlinear functions. IEEE Trans. Inf. Theory **67**(10), 6916–6925 (2021). https://doi.org/10.1109/TIT.2021.3081348
33. Schmidt, W.M.: Equations over Finite Fields. An Elementary Approach. Lecture Notes in Mathematics, vol. 536. Springer, Berlin-New York (1976)
34. Stichtenoth, H.: Algebraic Function Fields and Codes, vol. 254. Springer (2009)
35. Tu, Z., Zeng, X., Jiang, Y., Tang, X.: A class of APcN power functions over finite fields of even characteristic (2021). https://doi.org/10.48550/ARXIV.2107.06464. https://arxiv.org/abs/2107.06464
36. Van der Waerden, B.: Die Zerlegungs-und Trägheitsgruppe als Permutationsgruppen. Math. Ann. **111**(1), 731–733 (1935)
37. Wagner, D.: The boomerang attack. In: Knudsen, L. (ed.) Fast Software Encryption, pp. 156–170. Springer, Heidelberg (1999)

38. Wang, X., Zheng, D.: Several classes of PCN power functions over finite fields (2021). https://doi.org/10.48550/ARXIV.2104.12942. https://arxiv.org/abs/2104.12942
39. Zha, Z., Hu, L.: Some classes of power functions with low c-differential uniformity over finite fields. Des. Codes Crypt. **89**(6), 1193–1210 (2021). https://doi.org/10.1007/s10623-021-00866-8
40. Zha, Z., Hu, L.: Some classes of power functions with low c-differential uniformity over finite fields. Des. Codes Crypt. **89**(6), 1193–1210 (2021)

On the Walsh and Fourier-Hadamard Supports of Boolean Functions From a Quantum Viewpoint

Claude Carlet[1,2], Ulises Pastor-Díaz[3(✉)], and José M. Tornero[3]

[1] Department of Informatics, University of Bergen, 5005 Bergen, Norway
[2] Department of Mathematics, University of Paris 8, 93526 Saint-Denis, France
[3] Departament of Algebra, University of Sevilla, 41012 Sevilla, Spain
{upastor,tornero}@us.es

Abstract. In this paper, we focus on the links between Boolean function theory and quantum computing. In particular, we study the notion of what we call fully-balanced functions and analyse the Fourier–Hadamard and Walsh supports of those functions having such property. We study the Walsh and Fourier supports of other relevant classes of functions, using what we call balancing sets. This leads us to revisit and complete certain classic results and to propose new ones.

We complete our study by extending the previous results to pseudo-Boolean functions (in relation to vectorial functions) and giving an insight on its applications in the analysis of the possibilities that a certain family of quantum algorithms can offer.

Keywords: Boolean functions · Quantum computing · Walsh supports

1 Introduction

The main results of this paper deal with Boolean functions and do not require a knowledge in quantum computing and quantum algorithms, but they have been highly motivated by and have important applications in the analysis of the Generalised Phase Kick-Back quantum algorithm (a quantum algorithm inspired by the phase kick-back technique that the second and third authors introduced in [10] and which is used to distinguish certain classes of functions). For this reason, we will begin this introduction by giving some notions about this model of computation and its relation to Boolean functions, specially for those readers coming from a Boolean function background. However, for a more general and in-depth explanation, [6,9] can be consulted.

A quantum computer is made up of qubits, which are Hilbert spaces of dimension two. A state of the qubit is a vector $|\psi\rangle = \alpha|0\rangle + \beta|1\rangle$, where $\alpha, \beta \in \mathbb{C}$, and which satisfies the normalisation condition $|\alpha|^2 + |\beta|^2 = 1$. Here, $|0\rangle = \begin{pmatrix} 1 & 0 \end{pmatrix}^t$ and $|1\rangle = \begin{pmatrix} 0 & 1 \end{pmatrix}^t$ are the column vectors of the canonical basis, and α and β are called the amplitudes of the state. We are using the Dirac or so-called *bra-ket* notation. In this notation, vectors are represented by kets, $|\cdot\rangle$, their duals (in the linear algebra sense) are denoted by bras, $|\cdot\rangle^* = \langle\cdot|$, and the inner product is denoted by a bracket, $\langle\cdot|\cdot\rangle$. Systems of multiple qubits are constructed using the tensor product (more particularly, the Kronecker specialisation) of the individual qubit systems, and a state is said to be entangled if it does not correspond to a pure tensor in said product (that is, it cannot be written as the Kronecker product of vectors in the individual qubit systems). In particular, elements of the canonical basis (called computational basis in this context) are represented using elements of $\mathbb{F}_2^n$ inside the ket: $|\mathbf{x}\rangle_n = \bigotimes_{i=1}^n |x_i\rangle$, where $\mathbf{x} = (x_1, x_2, \ldots, x_n) \in \mathbb{F}_2^n$. A general state of a system of n qubits can be then written as $|\psi\rangle_n = \sum_{\mathbf{x} \in \mathbb{F}_2^n} \alpha_\mathbf{x} |\mathbf{x}\rangle_n$, where $\alpha_\mathbf{x} \in \mathbb{C}$ for all $\mathbf{x} \in \mathbb{F}_2^n$, satisfying the normalisation condition $\sum_{\mathbf{x} \in \mathbb{F}_2^n} |\alpha_\mathbf{x}|^2 = 1$. These qubit systems evolve by means of unitary matrices, and a quantum algorithm consists of the application of a unitary transformation to the first element of the computational basis (that is, the $|\mathbf{0}\rangle_n$ vector) in a system of n qubits and measuring the resulting state. We should recall that $|\mathbf{0}\rangle_n = \otimes_{i=1}^n |0\rangle$ is not the zero vector. Indeed, vectors in the computational basis of an n-qubit system will be labeled using the elements of $\mathbb{F}_2^n$, and $|\mathbf{0}\rangle_n$ will be just the first of them. The process of measuring a state, $|\psi\rangle_n = \sum_{\mathbf{x} \in \mathbb{F}_2^n} \alpha_\mathbf{x} |\mathbf{x}\rangle$, makes it collapse into one of the elements of the computational basis, say $|\mathbf{y}\rangle$, with probability $|\alpha_\mathbf{y}|^2$, which in turn would give us $\mathbf{y} \in \mathbb{F}_2^n$ as a result.

The Walsh transform (that we shall take normalised) has a deep relation to some quantum algorithms, like the Deutsch–Jozsa algorithm [5] or the Bernstein–Vazirani algorithm [1]. In fact, in the final superposition of these algorithms before measuring, the amplitude of a given state of the computational basis, $|\mathbf{z}\rangle_n$ is

$$\alpha_\mathbf{z} = \frac{1}{\sqrt{2^n}} \sum_\mathbf{x} (-1)^{f(\mathbf{x}) \oplus \mathbf{x} \cdot \mathbf{z}},$$

where $f : \mathbb{F}_2^n \to \mathbb{F}_2$ is the function used as an input in the algorithms. This, in particular, allows us to distinguish balanced functions from constant ones, as the Walsh transform of a balanced function—which coincides with the unnormalised vector of amplitudes in the final state of the GPK—takes value zero when evaluated at the zero point (which corresponds with amplitude $\alpha_\mathbf{0} = 0$), while a constant function takes value 2^n (which corresponds with amplitude $\alpha_\mathbf{0} = 1$). This implies that, in the constant situation, we will always obtain zero as the result of our measurement, as the zero state has amplitude—and thus probability—one in the superposition, while in the balanced situation we will always obtain a value different from zero, as the zero state has amplitude zero. The relevance of studying these specific classes of functions springs from the fact that they are completely distinguishable using this technique, but is also due to its implications in quantum complexity theory [2]. However, different classes of functions can be considered as long as they have different Walsh supports, so it is paramount to study the Walsh support—and thus the Fourier–Hadamard support, both of which will be formally defined in the following section—of

Boolean functions in order to determine whether we can use a quantum algorithm to distinguish them efficiently.

The technique used in these algorithms, called the phase kick-back, and its generalised version, the Generalised Phase Kick-Back or GPK [10], make this relation even more relevant. Indeed, if we use a vectorial function $F : \mathbb{F}_2^n \to \mathbb{F}_2^m$ as an input, then, after choosing $\mathbf{y} \in \mathbb{F}_2^m$, which in this context is called a marker, the amplitudes of the states of the canonical basis in the final superposition of the GPK algorithm are:

$$\alpha_{\mathbf{z}} = \frac{1}{\sqrt{2^n}} \sum_{\mathbf{x}} (-1)^{F(\mathbf{x}) \cdot \mathbf{y} \oplus \mathbf{x} \cdot \mathbf{z}}.$$

This is, once again, a normalised version of the Walsh transform (in this case of a vectorial function) and thus it seems clear that we can use the properties of the Walsh transform in distinguishing classes of functions using the GPK. What is more, we will see that, using the concepts of balancing set and constant set, we can link the Walsh transform of a vectorial function with the Fourier–Hadamard support of the pseudo-Boolean function determined by its image, as highlighted in Theorem 6. This, in turn, allows us to distinguish classes of functions as long as the Fourier–Hadamard support of the pseudo-Boolean functions associated with their images are different.

More particularly, in Sect. 7 we will give a closed formula for this relation for a particular class of functions. The aforementioned class of functions, which will be defined in Sect. 4, will be referred to as the class of fully balanced functions, and the relation that we have pointed out can be used to solve the problem of determining the image of a fully balanced function when it is given as a black box using the GPK. However, this is done in a different article [8], as it involves a quantum and algorithmic approach. Both the phase kick-back and the GPK have seen a number of applications both in quantum computing—showing that there are oracle separations between classical and quantum complexity classes—and in cryptography, where it has been used as a tool in proving the resistance of hash functions to quantum techniques [12].

Regarding the overview of this paper, Sect. 2 will deal with the preliminaries introducing the notation, the Fourier–Hadamard and Walsh transforms and Reed-Muller codes. Section 3 will introduce the concepts of $\mathbf{y}$-constant and $\mathbf{y}$-balanced sets, and determine their relation to Fourier–Hadamard and Walsh supports. In Sect. 4 we will present the concept of fully balanced sets, and we will give a novel proof to a known result regarding the nature of these sets, while Sect. 5 will follow this up by analysing the Fourier–Hadamard and Walsh supports of some specific functions using the newly introduced tools. Section 6 will be a quick review of the migration of the previously mentioned results to the context of multisets. Finally, Sect. 7 further expands on the application of the obtained results in quantum computing and their implications.

2 Preliminaries

2.1 Notation

Throughout the whole paper we will work with vectors in $\mathbb{F}_2^n$, which we will write in bold. In particular, $\mathbf{0}$ will denote the zero vector. A subset of $\mathbb{F}_2^n$ will be called a Boolean set. Regarding the different operations, we will use $\oplus$ when dealing with additions modulo 2, but for additions either in $\mathbb{Z}$ or in $\mathbb{F}_2^n$ we will make use of $+$. Furthermore, we denote by $\cdot$ the usual inner product in $\mathbb{F}_2^n$, $\mathbf{x} \cdot \mathbf{y} = \bigoplus_{i=1}^{n} x_i y_i$, for $\mathbf{x}, \mathbf{y} \in \mathbb{F}_2^n$.

We will refer to mappings $f : \mathbb{F}_2^n \to \mathbb{F}_2$ as Boolean functions, mappings $f : \mathbb{F}_2^n \to \mathbb{R}$ as pseudo-Boolean functions (in particular, a Boolean function can be seen as a pseudo-Boolean function) and mappings $F : \mathbb{F}_2^n \to \mathbb{F}_2^m$ as (n,m)-functions. Some important concepts regarding a Boolean function $f : \mathbb{F}_2^n \to \mathbb{F}_2$ are the following. Its support, $\mathrm{supp}(f) = \{\mathbf{x} \in \mathbb{F}_2^n \mid f(\mathbf{x}) = 1\}$. Its Hamming weight, denoted by $\mathrm{wt}(f)$, will be the number of vectors $\mathbf{x} \in \mathbb{F}_2^n$ such that $f(\mathbf{x}) = 1$. In other words, $\mathrm{wt}(f) = |\mathrm{supp}(f)|$. Its sign function is the integer-valued function $\chi_f(\mathbf{x}) = (-1)^{f(\mathbf{x})} = 1 - 2f(\mathbf{x})$. Note also that f, can always be expressed uniquely as follows:

$$f(\mathbf{x}) = \bigoplus_{\mathbf{u} \in \mathbb{F}_2^n} a_\mathbf{u} \mathbf{x}^\mathbf{u}, \text{ where } \mathbf{x}^\mathbf{u} = \prod_{i=1}^{n} x_i^{u_i}.$$

This expression is called the algebraic normal form, or $\mathrm{ANF}(f)$. The degree of this polynomial is called the algebraic degree of f. The derivative of f in the direction of $\mathbf{a} \in \mathbb{F}_2^n$ is defined as $D_\mathbf{a}(f)(\mathbf{x}) = f(\mathbf{x}) \oplus f(\mathbf{x} + \mathbf{a})$. Finally, we will say that a Boolean multiset is a pair $M = (\mathbb{F}_2^n, m)$ where $m : \mathbb{F}_2^n \to \mathbb{Z}_{\geq 0}$ is a pseudo-Boolean function that can take the value 0. For each $\mathbf{x} \in \mathbb{F}_2^n$ we will call $m(\mathbf{x})$ its multiplicity and we will denote by $S_M = \{\mathbf{x} \in \mathbb{F}_2^n \mid m(\mathbf{x}) > 0\}$ the support of M. However, we will also represent multisets by using set notation but repeating every element of a given multiset as many times as the multiplicity indicates. For a more general overview on multisets [11] can be consulted.

2.2 Fourier–Hadamard and Walsh Transforms

We will now give a quick summary on some results for Boolean and pseudo-Boolean functions, but for a more general reference, [3] can be consulted.

The Fourier–Hadamard transform of a pseudo-Boolean function $f : \mathbb{F}_2^n \to \mathbb{R}$ is the function:

$$\widehat{f}(\mathbf{u}) = \sum_{\mathbf{x} \in \mathbb{F}_2^n} f(\mathbf{x}) (-1)^{\mathbf{x} \cdot \mathbf{u}}.$$

We will call the Fourier–Hadamard support of f the set of $\mathbf{u} \in \mathbb{F}_2^n$ such that $\widehat{f}(\mathbf{u}) \neq 0$ and its Fourier–Hadamard spectrum the multiset of all values $\widehat{f}(\mathbf{u})$. It is important to underline the relation between the Fourier–Hadamard transform and linear functions. If we denote $l_{\mathbf{u}}(\mathbf{x}) = \mathbf{u} \cdot \mathbf{x}$ for $\mathbf{u} \neq \mathbf{0}$, we have:

$$\widehat{f}(\mathbf{u}) = \sum_{\mathbf{x} \in \mathbb{F}_2^n} f(\mathbf{x})(1 - 2\mathbf{x} \cdot \mathbf{u}) = \mathrm{wt}(f) - 2\,\mathrm{wt}(f \cdot l_{\mathbf{u}})$$

$$= \mathrm{wt}(f \oplus l_{\mathbf{u}}) - \mathrm{wt}(l_{\mathbf{u}}) = \mathrm{wt}(f \oplus l_{\mathbf{u}}) - 2^{n-1},$$

while $\widehat{f}(\mathbf{0}) = \mathrm{wt}(f)$. Given a Boolean function f, we can also calculate its Walsh transform:

$$W_f(\mathbf{u}) = \sum_{\mathbf{x} \in \mathbb{F}_2^n} (-1)^{f(\mathbf{x}) \oplus \mathbf{x} \cdot \mathbf{u}}.$$

We will analogously call the Walsh support of f the set of $\mathbf{u} \in \mathbb{F}_2^n$ such that $W_f(\mathbf{u}) \neq 0$ and its Walsh spectrum the multiset of all values $W_f(\mathbf{u})$. It is clear that the Walsh transform of a Boolean function f is the Fourier–Hadamard transform of its sign function, which implies by the linearity of the Fourier–Hadamard transform: $W_f = 2^n \delta_{\mathbf{0}} - 2\widehat{f}$, where $\delta_{\mathbf{0}}$ is the indicator of $\{\mathbf{0}\}$ and the Boolean function f is viewed here as a pseudo-Boolean function. In particular, if $\mathbf{u} \neq \mathbf{0}$, then we have $W_f(\mathbf{u}) = -2\widehat{f}(\mathbf{u})$, and thus any $\mathbf{u} \neq \mathbf{0}$ is in the Fourier–Hadamard support if and only if it is in the Walsh support. Regarding the zero vector we need to analyse two particular situations. When f is the zero function, i.e., $f(\mathbf{x}) = 0$ for all $\mathbf{x} \in \mathbb{F}_2^n$, then $\widehat{f}(\mathbf{0}) = 0$ but $W_f(\mathbf{0}) = 2^n$. On the other hand, if f is a balanced function, that is, $\mathrm{wt}(f) = 2^{n-1}$, then $\widehat{f}(\mathbf{0}) = 2^{n-1}$ but $W_f(\mathbf{0}) = 0$. In any other situation the Fourier–Hadamard and Walsh supports will be the same. Some important properties of the Fourier–Hadamard transform—which result in similar properties for the Walsh transform—are the inverse Fourier–Hadamard transform formula: $\widehat{\widehat{f}} = 2^n f$, and Parseval's relation:

$$\sum_{\mathbf{u} \in \mathbb{F}_2^n} \widehat{f}^{\,2}(\mathbf{u}) = 2^n \sum_{\mathbf{x} \in \mathbb{F}_2^n} f^2(\mathbf{x}),$$

which for Boolean functions turns into

$$\sum_{\mathbf{u} \in \mathbb{F}_2^n} \widehat{f}^{\,2}(\mathbf{u}) = 2^n |\mathrm{supp}(f)|,$$

and for the Walsh transform becomes

$$\sum_{\mathbf{u} \in \mathbb{F}_2^n} W_f^2(\mathbf{u}) = 2^{2n}.$$

The Walsh transform of a vectorial function $F : \mathbb{F}_2^n \to \mathbb{F}_2^m$ is the function $W_F : \mathbb{F}_2^n \times \mathbb{F}_2^m \to \mathbb{Z}$ defined as follows:

$$W_F(\mathbf{u}, \mathbf{v}) = \sum_{\mathbf{x} \in \mathbb{F}_2^n} (-1)^{\mathbf{v} \cdot F(\mathbf{x}) \oplus \mathbf{u} \cdot \mathbf{x}},$$

where $\mathbf{u} \in \mathbb{F}_2^n$ and $\mathbf{v} \in \mathbb{F}_2^m$.

2.3 Reed–Muller Codes

We will devote Sect. 4 to analysing the concept of fully balanced sets and its relation with minimum weight codewords in Reed–Muller codes. For a deeper analysis on these codes [7] or [3] can be consulted. Given a Boolean function $f : \mathbb{F}_2^m \to \mathbb{F}_2$, we can identify it with a vector of length 2^m by fixing an ordering—we will use the lexicographical ordering—in $\mathbb{F}_2^m$. Said vector, $\mathbf{f}$, is then the vector of evaluations of f for the chosen order. The Reed–Muller code of order r and length $n = 2^m$, noted $\mathcal{R}(r,m)$, is the set of vectors $\mathbf{f}$ where $f : \mathbb{F}_2^m \to \mathbb{F}_2$ is a Boolean function of algebraic degree at most r. Reed–Muller codes are linear codes with minimum distance—i.e., minimum weight among its non-zero vectors—2^{m-r} and dimension $1 + m + \binom{m}{2} + \ldots + \binom{m}{r}$.

3 On Balanced Sets and Fourier–Hadamard Supports

In the introduction we have pointed out the two ideas that motivate most of the results going forward. Firstly, that we can use the Walsh support—and thus the Fourier–Hadarmard support—of Boolean and vectorial functions to distinguish classes of them. Secondly, that we can use the Fourier–Hadamard support of pseudo-Boolean functions associated to the image of vectorial functions to differentiate between classes of functions using the GPK. It is to advance both of these ideas that we need to discuss constant and balancing sets. In this section, we introduce very simple concepts, on which we will build more complex and interesting ones. The fact that a hyperplane $H_\mathbf{x} = \{\mathbf{y} \in \mathbb{F}_2^n \mid \mathbf{x} \cdot \mathbf{y} = 0\}$ for $\mathbf{x}$ nonzero has 2^{n-1} elements can be stated in a way reminiscent of the Fourier–Hadamard transform. Given a vector space E in $\mathbb{F}_2^n$, and $l_\mathbf{y}$ a nontrivial linear form in E, then

$$\sum_{\mathbf{x} \in E} (-1)^{l_\mathbf{y}(\mathbf{x})} = 0.$$

We will say that $\mathbf{y}$—the vector which determines $l_\mathbf{y}$—balances $\mathbb{F}_2^n$.

Definition 1 (**y**-Balanced sets.). *Let $\mathbf{y} \in \mathbb{F}_2^n$ be a nonzero binary vector, we say that a nonempty set $S \subset \mathbb{F}_2^n$ is balanced with respect to $\mathbf{y}$ or $\mathbf{y}$-balanced if:*

$$|(S \cap H_\mathbf{y})| = \frac{|S|}{2}.$$

That is, S is halved by $\mathbf{y}$ with respect to the inner product.

We will also say that $\mathbf{y}$ balances S. If the size of S is odd, then it is clear that no vector balances S.

Remark 1. There are a few equivalent ways to define this notion.

Let $\mathbf{1}_S$ be the indicator vector of the set S and $\mathbf{l}_\mathbf{y}$ the vector associated with the linear form $l_\mathbf{y}$, then stating that $\mathbf{y}$ balances S is equivalent to saying that $\text{wt}(\mathbf{1}_S \mathbf{l}_\mathbf{y}) = |S|/2$, as $\mathbf{1}_S \mathbf{l}_\mathbf{y}$—the component-wise product of the two vectors, also known as the Hadamard product—is the indicator vector of $S \cap (\mathbb{F}_2^n \setminus H_\mathbf{y})$.

In another equivalent way, $\mathbf{y} \neq \mathbf{0}$ balances S if (and only if) $1_S \oplus l_\mathbf{y}$ is a balanced function—i.e., a function of weight 2^{n-1}—where 1_S is the indicator function of S and $l_\mathbf{y}$ the linear function determined by $\mathbf{y}$, since $\mathrm{wt}(1_S \oplus l_\mathbf{y}) = \mathrm{wt}(1_S) + \mathrm{wt}(l_\mathbf{y}) - 2\,\mathrm{wt}(1_S\, l_\mathbf{y})$ and $\mathrm{wt}(l_\mathbf{y}) = 2^{n-1}$.

A third way of defining this notion, and the one we will mostly focus on, is by means of the Fourier–Hadamard transform. The nonzero vector $\mathbf{y} \in \mathbb{F}_2^n$ balances a nonempty set S if and only if:

$$\widehat{1_S}(\mathbf{y}) = \sum_{\mathbf{x}\in\mathbb{F}_2^n} 1_S(\mathbf{x})(-1)^{\mathbf{x}\cdot\mathbf{y}} = \sum_{\mathbf{x}\in S}(-1)^{\mathbf{x}\cdot\mathbf{y}} = 0,$$

or equivalently $W_{1_S}(\mathbf{y}) = 0$.

In the same manner, we can define the idea of being $\mathbf{y}$-constant.

Definition 2 *($\mathbf{y}$-Constant sets.). Let $\mathbf{y} \in \mathbb{F}_2^n \setminus \{\mathbf{0}\}$ be a nonzero binary vector, we say that a nonempty set $S \subset \mathbb{F}_2^n$ is constant with respect to $\mathbf{y}$ or $\mathbf{y}$-constant if either*

$$S \subset H_\mathbf{y} \quad \text{or} \quad S \cap H_\mathbf{y} = \varnothing,$$

that is, the product $\mathbf{x} \cdot \mathbf{y}$ is constant for all $\mathbf{x} \in S$. We will say that every nonempty set S is $\mathbf{0}$-constant.

Remark 2. We can also express this idea by means of the Fourier–Hadamard transform: given a nonempty set S and its indicator function 1_S, it is $\mathbf{y}$-constant if and only if

$$\left|\widehat{1_S}(\mathbf{y})\right| = \left|\sum_{\mathbf{x}\in S}(-1)^{\mathbf{x}\cdot\mathbf{y}}\right| = |S|.$$

Analogously, given a set $B \subset \mathbb{F}_2^n$, we will say that S is B-balanced (B-constant) if it is $\mathbf{y}$-balanced ($\mathbf{y}$-constant) for every $\mathbf{y} \in B$. It is important to note that the definition of B-constant does not require that the product $\mathbf{x} \cdot \mathbf{y}$ be the same for every pair $\mathbf{x} \in S$, $\mathbf{y} \in B$, but rather constant for $\mathbf{x} \in S$ once a $\mathbf{y} \in B$ is fixed.

Definition 3 *(Balancing set and constant set.). Let $S \subset \mathbb{F}_2^n$ be a nonempty set, then we call its balancing set, denoted by $B(S)$, the set of all binary vectors $\mathbf{y} \in \mathbb{F}_2^n$ such that S is $\mathbf{y}$-balanced and its constant set, denoted by $C(S)$, the set of all binary vectors $\mathbf{y} \in \mathbb{F}_2^n$ such that S is $\mathbf{y}$-constant.*

The reason for introducing these two concepts is that they are tied to the result of the GPK algorithm when using $\mathbf{y}$ as a marker, as will be shown by Theorem 6. In the next remark we will explain the interest of defining the balancing and constant sets in this manner.

Remark 3. Given a nonzero Boolean function $f : \mathbb{F}_2^n \to \mathbb{F}_2$ with support $S = \mathrm{supp}(f)$, we have that its Fourier–Hadamard support is $\mathrm{supp}(\widehat{f}) = \mathbb{F}_2^n \setminus B(S)$. Following the relation $W_f = 2^n \delta_\mathbf{0} - 2\widehat{f}$, if f is not a balanced function, we have that its Walsh support is $\mathrm{supp}(W_f) = \mathrm{supp}(\widehat{f}) = \mathbb{F}_2^n \setminus B(S)$. However, if f is balanced, we have $\mathrm{supp}(W_f) = \mathrm{supp}(\widehat{f}) \setminus \{\mathbf{0}\} = \mathbb{F}_2^n \setminus (B(S) \cup \{\mathbf{0}\})$.

We will begin by considering the constant set problem. The following result follows from the Fourier–Hadamard transform formula.

Lemma 1. *Let $S \subset \mathbb{F}_2^n$ and $\mathbf{s} + S = \{\mathbf{s} + \mathbf{x} \mid \mathbf{x} \in S\}$ be the translation of S by $\mathbf{s} \in \mathbb{F}_2^n$, then $C(S) = C(\mathbf{s} + S)$.*

Indeed, it is clear that $\widehat{1_{\mathbf{s}+S}}(\mathbf{y}) = (-1)^{\mathbf{s} \cdot \mathbf{y}} \widehat{1_S}(\mathbf{y})$. Using this result we can simply consider that $\mathbf{0} \in S$ without loss of generality, which simplifies things, as then, if $\mathbf{y} \in \mathbb{F}_2^n$ makes S constant, it actually makes it 0 and we can find $C(S)$ by solving a system of linear equations. In this situation, since $\mathbf{y}$ belongs to $C(S)$ if and only if it is orthogonal to every element $\mathbf{x} \in S$, we have:

Lemma 2 *(Constant set.)*. *Let S be a set such that $\mathbf{0} \in S$, then*

$$C(S) = \bigcap_{\mathbf{x} \in S} H_{\mathbf{x}},$$

which is the linear subspace $\langle S \rangle^{\perp}$ of dimension $n - \mathrm{rk}(S)$, where $\mathrm{rk}(S)$ (the rank of S) is the dimension of $\langle S \rangle$—the linear space spanned by S—and $\langle S \rangle^{\perp}$ is the orthogonal space of S with respect to the $\cdot$ product.

Note that, still assuming that $\mathbf{0} \in S$, we have then $C(C(S)) = \langle S \rangle$. If $\mathbf{0} \notin S$, then it suffices to consider $S' = \mathbf{s} + S$ for some $\mathbf{s} \in S$. Taking now a look into the balancing set, the following properties are straightforward.

Lemma 3 *(Properties.)*. *Let S be a nonempty Boolean set, $B(S)$ be as in the previous definition and let $S_1, S_2 \subset \mathbb{F}_2^n$ both B-balanced and nonempty, then:*

(i) For all $A \subset B(S)$, S is A-balanced.
(ii) If S_1 and S_2 are such that $S_1 \cap S_2 = \emptyset$, then $S_1 \cup S_2$ is B-balanced.
(iii) If $S_1 \subset S_2$, then $S_2 \setminus S_1$ is B-balanced.
(iv) $S_1 \cap S_2$ is B-balanced if and only if $S_1 \cup S_2$ is B-balanced.
(v) $B(S) = B(\mathbb{F}_2^n \setminus S)$ for all $S \subset \mathbb{F}_2^n$.
(vi) $B(\mathbb{F}_2^n) = \mathbb{F}_2^n \setminus \{\mathbf{0}\}$.
(vii) $B(S) = B(\mathbf{s} + S)$ for all $\mathbf{s} \in \mathbb{F}_2^n$. (Invariance by translation).

We also have the following.

Lemma 4. *Let S be a nonempty Boolean set:*

(viii) Let $\mathbf{s} \in \mathbb{F}_2^n \setminus \{\mathbf{0}\}$ such that $S = \mathbf{s} + S$ and $\mathbf{y} \in \mathbb{F}_2^n$ with $\mathbf{y} \cdot \mathbf{s} = 1$, then $\mathbf{y} \in B(S)$.
(ix) Let S be B-balanced, then $\langle S \rangle$ is B-balanced.
(x) Let S_1 be B_1-balanced and S_2 be B_2-balanced, then $\langle S_1, S_2 \rangle$—the vector space generated by $S_1 \cup S_2$—is $(B_1 \cup B_2)$-balanced.

Proof. Property (viii) follows from the fact that $\widehat{1_{\mathbf{s}+S}}(\mathbf{y}) = (-1)^{\mathbf{s} \cdot \mathbf{y}} \widehat{1_S}$.

For property (ix), given $\mathbf{y} \in B$, we know that there is an $\mathbf{x} \in S$ such that $\mathbf{y} \cdot \mathbf{x} = 1$, and by property (viii) we get the result, as $\mathbf{x} + \langle S \rangle = \langle S \rangle$.

Property (x) follows from the same idea, as for any $i = 1, 2$; $\mathbf{y} \in B_i$ implies that there is an $\mathbf{x} \in S_i \subset \langle S_1 \cup S_2 \rangle$ such that $\mathbf{y} \cdot \mathbf{x} = 1$. □

Remark 4. Given a nonzero Boolean function $f : \mathbb{F}_2^n \to \mathbb{F}_2$ and let $S = \text{supp}(f)$. Then, the constant set and the balancing set of S, and incidentally also the Fourier-Hadamard support of f, have the following structure.

1. Both $C(S)$ and $\widehat{f}^{-1}(|S|)$ are vector spaces, with $C(S) = \widehat{f}^{-1}(|S|)$ if $\mathbf{0} \in S$.
2. If $\mathbf{0} \in S$, then each of the sets $\widehat{f}^{-1}(z)$ with $z \in \mathbb{Z}$ is either empty or a union of disjoint cosets of $C(S)$, this applies in particular to $B(S) = \widehat{f}^{-1}(0)$. If r is the rank of S, then the dimension of $C(S)$ will be $n - r$ and thus we will have 2^r of these cosets.
3. If $\mathbf{0} \notin S$, taking $g(\mathbf{x}) = f(\mathbf{x} + \mathbf{s})$ for some $\mathbf{s} \in S$ we have that $\widehat{g}(\mathbf{u}) = (-1)^{\mathbf{s} \cdot \mathbf{u}} \widehat{f}(\mathbf{u})$. This implies that now the sets $\widehat{f}^{-1}(z) \cup \widehat{f}^{-1}(-z)$ are the ones which are either empty or a union of cosets of $C(S)$, but the situation of $B(S)$ does not change.

Taking into consideration this remark, it makes sense to define the following concept.

Definition 4 *(Balancing index.).* Let $\varnothing \neq S \subset \mathbb{F}_2^n$, then we define its balancing index to be:
$$b(S) = \frac{|B(S)|}{|C(S)|}.$$

This index is always an integer, and it corresponds to the amount of disjoint cosets of $C(S)$ that conform $B(S)$, as we have seen in Remark 4.

The balancing index is clearly invariant by isomorphism, but this can be taken even further

Proposition 1. *Let $S \subset \mathbb{F}_2^n$ be a Boolean set and $\varphi : \langle S \rangle \to \mathbb{F}_2^m$ a monomorphism (i.e., an injective linear function). Then $b(S) = b(\varphi(S))$.*

Proof. We have seen that both the constant and the balancing set are invariant by translation, so we will suppose that S and $\varphi(S)$ include the $\mathbf{0}$ vector and that $\varphi(\mathbf{0}) = \mathbf{0}$ without loss of generality. Let r be the rank of S—and also of $\varphi(S)$–and let $\mathbf{s}_1, \ldots, \mathbf{s}_r$ be independent elements of S, then it is clear that they conform a basis of $\langle S \rangle$, and that the vectors $\varphi(\mathbf{s}_1), \ldots, \varphi(\mathbf{s}_r)$ conform a basis of $\langle \varphi(S) \rangle$.

We also know that both the constant set and the balancing set can be computed as the sets of solutions to certain families of systems of equations of the form $\{\mathbf{s}_i \cdot \mathbf{x} = b_i \mid i = 1, \ldots, r\}$, where $\mathbf{x}$ is the vector of unknowns. The vector whose i-th component is b_i will be noted as $\mathbf{b} \in \mathbb{F}_2^r$. If, for a certain $\mathbf{b} \in \mathbb{F}_2^r$, the solutions to the previous system balance S, then the solutions to the system $\{\varphi(\mathbf{s}_i) \cdot \mathbf{x} = b_i \mid i = 1, \ldots, r\}$ will also balance $\varphi(S)$, and the same will happen in the opposite direction. Indeed, if we take any $\mathbf{s} \in S$ such that $\mathbf{s} = \sum_{i=1}^{r} \alpha_i \mathbf{s}_i$ for some $\alpha_i \in \mathbb{F}_2$, we have that $\varphi(\mathbf{s}) = \sum_{i=1}^{r} \alpha_i \varphi(\mathbf{s}_i)$. Let $\mathbf{z} \in \mathbb{F}_2^n$ be a solution to the system $\{\mathbf{s}_i \cdot \mathbf{x} = b_i \mid i = 1, \ldots, r\}$ and $\mathbf{z}' \in \mathbb{F}_2^m$ a solution to $\{\varphi(\mathbf{s}_i) \cdot \mathbf{x} = b_i \mid i = 1, \ldots, r\}$, then

$$\mathbf{s} \cdot \mathbf{z} = \left(\sum_{i=1}^{r} \alpha_i \mathbf{s}_i\right) \cdot \mathbf{z} = \bigoplus_{i=1}^{r} \alpha_i b_i = \left(\sum_{i=1}^{r} \alpha_i \varphi(\mathbf{s}_i)\right) \cdot \mathbf{z}' = \varphi(\mathbf{s}) \cdot \mathbf{z}'.$$

We saw in Remark 4 that the balancing sets of S and $\varphi(S)$ were composed of cosets of $C(S)$. Although the dimension of $C(\varphi(S))$ does not have to be the same as that of $C(S)$, the amount of cosets that balance each of these sets is the same. To see this we just need to explicitly construct the bijection between the cosets of $C(S)$ and those of $C(\varphi(S))$ that we have hinted at before. Each of these cosets can be assigned to the vector $\mathbf{b} \in \mathbb{F}_2^r$ of independent terms in the system of equations whose solution is said coset. We just need to identify cosets of $C(S)$ and of $C(\varphi(S))$ which are assigned to the same $\mathbf{b}$. □

This allows us, by taking $r = m$, to consider only the cases where S is made of the zero vector, the n vectors of the canonical basis of $\mathbb{F}_2^n$ and $|S| - n - 1$ other linear combinations of these vectors when studying certain properties. Indeed, let $S \subset \mathbb{F}_2^n$ be a Boolean set with rank $r \leq n$ and such that $\mathbf{0} \in S$, and let $\mathbf{s}_1, \ldots, \mathbf{s}_r$ be independent elements of S. Then, we can consider the application $\varphi : \langle S \rangle \to \mathbb{F}_2^r$ linearly determined by $\varphi(\mathbf{s}_i) = \mathbf{e}_i$ for $i = 1, \ldots, r$, where $\mathbf{e}_i$ are the elements of the standard basis in $\mathbb{F}_2^r$. As this application satisfies the conditions presented above, we know that $b(S) = b(\varphi(S))$.

4 Fully Balanced Sets

Throughout the previous section, it could be glimpsed that, given an injective vectorial function $F : \mathbb{F}_2^n \to \mathbb{F}_2^m$, elements of the constant and balancing set of the image of F play a very relevant role in the outcome of the GPK when they are used as markers. This is the main motivation for defining a class of sets—and thus of functions—that are always either $\mathbf{y}$-constant or $\mathbf{y}$-balanced for any given $\mathbf{y} \in \mathbb{F}_2^m$. In this section, we will define the notion of fully balanced sets and take a look into its relation with minimal distance codewords in a Reed–Muller code. This new approach will be specially relevant in generalising the result to multisets, which will be done in Sect. 6. We will begin by recalling the following result that we can find, for instance, in [3].

Proposition 2. *Let E be a vector subspace of $\mathbb{F}_2^n$ and 1_E its indicator function, then:*
$$\widehat{1_E} = |E| 1_{E^\perp}.$$

In particular, this implies that $C(E) = E^\perp$ and $B(E) = \mathbb{F}_2^n \setminus E^\perp$. Another interesting remark is that if $r = \dim(E)$, then $b(E) = 2^r - 1$. Moreover, we will always have $B(S) \cup C(S) = \mathbb{F}_2^n$, which is the property we will use for our following definition.

Definition 5 (*Fully balanced.*). *We say that a nonempty set $S \subset \mathbb{F}_2^n$ is fully balanced if $B(S) \cup C(S) = \mathbb{F}_2^n$.*

Remark 5. Of course, this property is equivalent to saying that $\widehat{1_S}$ is valued in $\{-|S|, 0, |S|\}$, but there are actually many other ways to define it. For instance, if $r = \text{rk}(S)$, then S is fully balanced if $b(S) = 2^r - 1$, as $|C(S)| = 2^{n-r}$ and $|B(S)| = 2^n - 2^{n-r}$ due to Definition 4. However, the more intuitive one is that a nonempty S is fully balanced if for every $\mathbf{y} \in \mathbb{F}_2^n \setminus \{\mathbf{0}\}$ we have that $|S \cap H_\mathbf{y}|$ is either $|S|$, 0 (these two cases corresponding to $y \in C(S)$ by Definition 2) or $|S|/2$ (the case where $\mathbf{y}$ balances S using Definition 1).

To answer the question of whether there are any other fully balanced sets apart from affine spaces, we turn to [7], and more particularly to Lemma 6 in its Chap. 13. The result we present here is just the aforementioned one, but we have rewritten it so we do not explicitly assume that the size of S is a power of 2, which is one of the premises set in [7]. This statement is not actually used in their proof, so our result is not fundamentally new, but it seems important to know that the result is more general that as stated in [7] (and we give an original and simpler proof).

Theorem 1. *Let $S \subset \mathbb{F}_2^n$ be a nonempty Boolean set and $\mathbf{1}_S$ its indicator vector, then the following statements are equivalent.*

(i) S is fully balanced.
(ii) S is an affine space.
(iii) $\mathbf{1}_S$ is a minimum weight codeword in $\mathcal{R}(r,n)$ for some r.

Proof. The equivalence $(ii) \iff (iii)$ can be found in [7], and we have already taken a look into $(ii) \implies (i)$.

The implication $(i) \implies (ii)$ is also implicitly in [7, Chapter 13], when we read the proof of its Lemma 6 due to Rothschild and Van Lint; indeed, the proof given in [7] does not use in fact that $|S|$ is a power of two. We will instead present another proof of this implication using the Fourier–Hadamard transform properties.

Let $\mathbf{1}_S$ be the indicator function of S. As S is fully balanced, we know that $\widehat{\mathbf{1}_S}(\mathbf{u})$ is either 0, $|S|$ or $-|S|$. We will suppose without loss of generality that $\mathbf{0} \in S$, as both properties (being fully balanced and being an affine space) are preserved by translation. As we have seen, in this situation the value $-|S|$ is not possible, and $C(S)$ is the vector space of those $\mathbf{u}$ such that $\widehat{\mathbf{1}_S}(\mathbf{u}) = |S|$, so:

$$\widehat{\mathbf{1}_S} = |S| \mathbf{1}_{C(S)},$$

where $\mathbf{1}_{C(S)}$ is the indicator function of $C(S)$. Using now the inverse Fourier–Hadamard transform formula, we have:

$$\widehat{\widehat{\mathbf{1}_S}} = 2^n \mathbf{1}_S = |S| \widehat{\mathbf{1}_{C(S)}} = |S||C(S)| \mathbf{1}_{C(S)^\perp}$$

making also use of Proposition 2. As $\mathbf{1}_S$ is a Boolean function, then $|S||C(S)| = 2^n$ and $S = C(S)^\perp$, so S is a vector space. □

Remark 6. In the proof by Rothschild and Van Lint, they suppose that $|S|$ is a power of two and proceed by induction on n. Their proof can also be seen in terms of the Fourier–Hadamard transform, so we will briefly present it this way to show the differences between both approaches. For $n = 2$, the result is trivial, so we will suppose the result to be true for $n \leq k-1$ and prove it for $n = k$.

Let 1_S be the indicator function of S, we know that 1_S is fully balanced if and only if $\widehat{1_S}(\mathbf{u}) \in \{0, \pm|S|\}$ for all $\mathbf{u}$. If there is $\mathbf{s} \in C(S)$ which is not zero, then there is a hyperplane $H \subset \mathbb{F}_2^k$ such that $S \subseteq H$ (this H is $\{\mathbf{0}, \mathbf{s}\}^\perp$ if $\widehat{1_S}(\mathbf{s}) = |S|$ and its complement if $\widehat{1_S}(\mathbf{s}) = -|S|$). Taking now any hyperplane X of H, we know that $X = H \cap H'$ for some hyperplane H' in $\mathbb{F}_2^n$, and thus $S \cap X = S \cap H'$. This implies that $|S \cap X|$ verifies the induction hypothesis and we have our result. If $C(S) = \{\mathbf{0}\}$, then Rothschild and Van Lint show that $|S| = 2^n$, and thus $S = \mathbb{F}_2^n$ by a geometrical argument counting hyperplanes, but it is simpler to do it via an equivalent argument using Parseval's relation. Using $C(S) = \{\mathbf{0}\}$ we know that

$$\sum_{\mathbf{u} \in \mathbb{F}_2^n} \left(\widehat{1_S}(\mathbf{u})\right)^2 = \left(\widehat{1_S}(\mathbf{0})\right)^2 = |S|^2,$$

but it is also equal to $2^n |S|$ due to Parseval's relation, so $|S|$ must be either 0 or 2^n.

5 Some Fourier–Hadamard and Walsh Supports

We now move on to analysing the Fourier–Hadamard and Walsh supports of functions that have not yet been studied from the viewpoint of the Fourier support (recall that, for any nonzero function, $B(S)$ is the complement of the Fourier–Hadamard support). A common problem that we deal with in quantum computing, and that has implications in quantum complexity theory, is that of distinguishing classes of functions. Indeed, if we can isolate two classes of functions (for instance, balanced and constant functions) that cannot be distinguished from each other efficiently in one of the classical models of computation, and we are able to discriminate between them efficiently in the quantum model, the result would have interesting implications. We will show in Sect. 7 that the Walsh support of vectorial functions is tied to the Fourier support of the Boolean or pseudo-Boolean functions determined by their images. For that reason, knowing the Fourier–Hadamard support of Boolean functions is paramount in applying the GPK algorithm to distinguish classes of vectorial functions. Few results are known regarding possible Walsh supports: we know that $\mathbb{F}_2^n$ can be a Walsh support (for instance of any function having odd Hamming weight), as well as any singleton $\{\mathbf{a}\}$ (in this latter case, this is equivalent to the fact that the function is affine). A set of the form $\mathbb{F}_2^n \setminus \{\mathbf{a}\}$ can also be a Walsh support, a study of which can be found in [4] together with a general review in what is already known on the subject. We study now the Fourier support of a class of functions that has never been studied:

Theorem 2 *(Balancing independent sets.).* Let $S \subset \mathbb{F}_2^n$ with $|S|$ even, $\mathbf{0} \in S$ and $r = \mathrm{rk}(S) = |S| - 1$. Then

$$b(S) = \binom{r}{(r+1)/2},$$

and $B(S)$ is the disjoint union of $\binom{r}{(r+1)/2}$ affine spaces of dimension $n-r$ whose underlying vector space is $C(S) = \langle S \rangle^{\perp}$.

Proof. Let $\mathbf{s}_i \in S$; $i = 1, \ldots, r$, be the nonzero elements of S. For a certain $\mathbf{x} \in \mathbb{F}_2^n$ to balance S, it must satisfy that $\mathbf{x} \cdot \mathbf{s}_i = 0$ for $(r-1)/2$ of the nonzero elements of S and $\mathbf{x} \cdot \mathbf{s}_i = 1$ for the remaining $(r+1)/2$. Let $B_{r,t}$ be the set of vectors of weight t in $\mathbb{F}_2^r$, and take $t = (r+1)/2$. We know that, for every $\mathbf{x} \in \mathbb{F}_2^n$ that balances S, the vector obtained after computing the r possible $\mathbf{x} \cdot \mathbf{s}_i$ values must be in $B_{r,t}$. Next, for each element $\mathbf{b} \in B_{r,t}$, we will consider the system of equations given by $\{\mathbf{s}_i \cdot \mathbf{x} = b_i \mid i = 1, \ldots, r\}$, where b_i is the i-th component of $\mathbf{b}$ and $\mathbf{x}$ is the vector of unknowns. It is obvious that the solution to any of these systems balances S, as $w(\mathbf{b}) = t$. Furthermore, any vector that balances S must be a solution to one of these systems. The next step will be to analyse the solutions to these systems of equations. As the r non-trivial vectors of S are independent, we know that $n \geq r$, and in particular, the rank of any of the systems will be r and they will always have a subspace of dimension $n-r$ as solution. The final step is simply to remark that the pairwise intersections of such subspaces are empty and that there are precisely

$$\binom{r}{(r+1)/2}$$

possible such systems of equations. □

In particular, we have that $b(S) = \binom{r}{(r+1)/2}$. Let us see what this implies in terms of the Fourier–Hadamard and Walsh supports.

Remark 7. Let $f : \mathbb{F}_2^n \to \mathbb{F}_2$ be a Boolean function whose support $S = \mathrm{supp}(f)$ satisfies the conditions of Theorem 2. Then, if f is not balanced—which is always the case if $n > 3$—the Fourier–Hadamard and Walsh supports are $\mathrm{supp}(\hat{f}) = \mathrm{supp}(W_f) = \mathbb{F}_2^n \setminus B(S)$, and thus we have supports of size:

$$2^n - \binom{r}{(r+1)/2} 2^{n-r}.$$

If f is balanced then the Walsh support is of size

$$2^n - \binom{r}{(r+1)/2} 2^{n-r} - 1.$$

The next step in our quest could be to consider the next general situation. Let us study the case where all nontrivial elements of S are independent except for one.

Proposition 3. *Let $S \subset \mathbb{F}_2^n$ with $|S|$ even, $\mathbf{0} \in S$ and $r = \mathrm{rk}(S) = |S| - 2$. Denoting by $\mathbf{s}_1, \ldots, \mathbf{s}_r$ r independent elements of S and by $\mathbf{s}$ the remaining nonzero element of S such that*

$$\mathbf{s} = \sum_{i=1}^{r} \alpha_i \mathbf{s}_i, \text{ where } \alpha_i \in \mathbb{F}_2 \text{ for all } i \in \{1 \ldots, r\},$$

let $k = \sum_{i=1}^{r} \alpha_i$, where the sum is calculated in $\mathbb{Z}$, then:

$$b(S) = \begin{cases} 0 & \text{if } \varphi_1(k) > \varphi_2(k) \\ \sum_{i=\varphi_1(k)}^{\varphi_2(k)} \binom{k}{i} \binom{r-k}{(r/2)+e(i)-i} & \text{otherwise,} \end{cases}$$

where

$$\varphi(k) = (\varphi_1(k), \varphi_2(k)) = \begin{cases} (0, k) & \text{if } k < r/2 \\ (1, k) & \text{if } k = r/2 \\ \left(k - \frac{r}{2} + e\left(k - \frac{r}{2}\right), \frac{r}{2} + e\left(\frac{r}{2} + 1\right)\right) & \text{if } k > r/2, \end{cases}$$

and

$$e(x) = \frac{1 + (-1)^x}{2}.$$

Once again, $B(S)$ will be a disjoint union of $b(S)$ affine spaces with $C(S)$ as their underlying vector space.

Proof. As announced after Proposition 1, we will consider $n = r$, $\mathbf{e}_j \in S$ for each $j = 1, \ldots, r$ and $\mathbf{s} = 1^k \, 0^{r-k}$ without loss of generality. In this situation, k is the weight of $\mathbf{s}$. We know that $b(S)$ corresponds to the amount of affine spaces with $C(S)$ as their underlying vector space that balance S, but since we have $C(S) = \langle S \rangle^\perp = \{\mathbf{0}\}$, each of these affine spaces consists of a single vector and $|B(S)| = b(S)$. Thus, we only need to count the vectors $\mathbf{b} \in \mathbb{F}_2^r$ that balance S. Let $\mathbf{b} \in \mathbb{F}_2^r$ be any such vector, then $\mathbf{b} \cdot \mathbf{s}_j = \mathbf{b} \cdot \mathbf{e}_j = b_j$ for every $j = 1, \ldots, r$ and

$$\mathbf{b} \cdot \mathbf{s} = \sum_{j=1}^{k} b_j,$$

where b_j is the j-th component of $\mathbf{b}$. It is clear that for $\mathbf{b}$ to balance S it must either have weight $r/2$ and have an odd amount of ones among the first k positions, or have weight $(r/2) + 1$ and have an even amount of ones among the first k positions. At this point, the only thing that remains is to count such $\mathbf{b}$ vectors. Let i be the number of ones among the first k positions, then there are exactly $\binom{k}{i}$ ways for them to be distributed. If i is odd, then we want $\text{wt}(\mathbf{b}) = r/2$, so for each choice of $b_1, \ldots, b_k$

$$\binom{r-k}{(r/2)-i}$$

ways of choosing the remaining ones. If i is even, then we want $\text{wt}(\mathbf{b}) = (r/2)+1$, and then the remaining combinations will be

$$\binom{r-k}{(r/2)+1-i}.$$

A compact way to consider both possibilities at the same time is

$$\binom{r-k}{(r/2)+e(i)-i},$$

so the total number of combinations will be:

$$\sum_i \binom{k}{i}\binom{r-k}{(r/2)+e(i)-i}.$$

The only thing left to do is to analyse which are the possible values for i, which is the task of the φ function. Given $i \leq k \leq r$, the previous expression is nonzero if and only if $i \leq k$ and $r/2 + e(i) - i \leq r - k$, that is, $i \geq k - r/2 + e(i)$. If $k < r/2$, then it is trivial that i can range between 0 and k. If $k = r/2$, then it is almost the same situation, with the slight difference that i cannot be 0, as then there would have to be $(r/2) + 1$ ones in the last $r/2$ positions. However, if $k > r/2$, then the maximum range we can achieve is from $k - (r/2)$ to $r/2$. This is not always possible, as if $r/2$ is odd, then $i = (r/2) + 1$ is even and it would not balance our set. The same happens with our lower bound and the parity of $k - (r/2)$. If it were even, then it would be impossible to balance our set with $i = k - (r/2)$, so we must adjust the limits of our range in the φ function accordingly. □

As we can see, the formulae get incredibly complicated when the set lacks structure.

Remark 8. Let $f : \mathbb{F}_2^n \to \mathbb{F}_2$ be a Boolean function such that its support $S = \mathrm{supp}(f)$ satisfies the conditions of Proposition 3. Then, if f is not balanced—which again is always the case if $n > 3$—we have that both the Fourier–Hadamard and Walsh supports are a disjoint union of $b(S)$ affine spaces with $C(S)$ as their underlying vector space. Therefore, their total size will be $2^n - b(S)2^{n-r}$, where $b(S)$ is as shown in Proposition 3 and r is the rank of S.

We will focus now on functions whose support has a certain structure. There are many known ways of determining the Fourier–Hadamard and Walsh supports of Boolean functions by decomposing them. A summary of these situations can be found in [3], where, in particular, we can find that, given two pseudo-Boolean functions, ψ and φ, in $\mathbb{F}_2^n$, then, $\widehat{\varphi \otimes \psi} = \widehat{\varphi} \times \widehat{\psi}$. Also, we know that $\widehat{\varphi \times \psi} = \widehat{\varphi} \otimes \widehat{\psi}/2^n$, where $\otimes$ stands here for the Kronecker product. We will proceed to do a similar thing, but using a different construction based on the 0-kernel of the function. We will first highlight the idea with an example.

Remark 9. Let us consider a nonempty $S \subset \mathbb{F}_2^n$ for which there is an $\mathbf{s} \in \mathbb{F}_2^n$ such that $\mathbf{s} + S = S$ and $\mathrm{rk}(S) = r$. Then, any $\mathbf{y} \in \mathbb{F}_2^n \setminus H_\mathbf{s}$ balances S, but for any $\mathbf{y} \in H_\mathbf{s}$ to balance S we would need it to balance $S/\langle \mathbf{s} \rangle$. In particular, if $|S|$ is not a multiple of 4, then $B(S) = \mathbb{F}_2^n \setminus H_\mathbf{s}$ and $b(S) = 2^{r-1}$.

If we note as f the indicator function of S, then what we have here is an $\mathbf{s}$ such that $D_\mathbf{s}(f) = 0$, the set of all the elements that fulfill this property is known as the 0-kernel, $\mathcal{E}_0(f) = \{\mathbf{a} \in \mathbb{F}_2^n \mid D_\mathbf{a}(f) = 0\}$, and we will use it to generalise the previous result. Let us recall some of its properties first:

Proposition 4 *(Properties.)*. Let $f : \mathbb{F}_2^n \to \mathbb{F}_2$ be a Boolean function and $S = \mathrm{supp}(f)$:

(i) $\mathcal{E}_0(f)$ is a vector subspace.
(ii) If $\mathbf{0} \in S$, then $\mathcal{E}_0(f) \subset S$.
(iii) $\mathcal{E}_0(f(\mathbf{x})) = \mathcal{E}_0(f(\mathbf{x}+\mathbf{a}))$ for all $\mathbf{a} \in \mathbb{F}_2^n$.
(iv) $\mathcal{E}_0(f) = \bigcap_{\mathbf{x} \in S}(\mathbf{x}+S)$.

What we are trying to do here is to construct $B(S)$ from both E and the balancing set of the set of classes $\mathbf{x} + E$. So let us delve a little deeper into this idea.

Definition 6 *(Classes modulo $\mathcal{E}_0$.)*. Let $S \subset \mathbb{F}_2^n$ be nonempty, f its indicator function and $E = \mathcal{E}_0(f)$. We will consider the set of classes modulo E, $S/E = \{\mathbf{x} + E \mid \mathbf{x} \in S\}$ and define the balancing set of S/E as:

$$B(S/E) = B(S) \cap H, \text{ where } H = \bigcap_{\mathbf{s} \in E} H_\mathbf{s}.$$

Annoyingly, this does not help us much, since we use $B(S)$ as part of the definition, but the next result will give us an alternative way to compute $B(S/E)$.

Lemma 5. Let $S \subset \mathbb{F}_2^n$ be nonempty, f its indicator function and $E = \mathcal{E}_0(f)$, and let S_E be a complete set of representatives of S/E. Then:

$$B(S/E) = B(S_E) \cap H, \text{ where } H = \bigcap_{\mathbf{s} \in E} H_\mathbf{s}.$$

Proof. Let $\mathbf{y} \in H$, stating that $\mathbf{y}$ balances S is equivalent to stating that it balances the classes, as $\mathbf{y} \cdot \mathbf{s} = 0$ for every $\mathbf{s} \in F$ and thus $\mathbf{y} \cdot \mathbf{x}_1 = \mathbf{y} \cdot \mathbf{x}_2$ for $\mathbf{x}_1, \mathbf{x}'_1$ in the same class. The result follows from the fact that, in this situation, balancing a complete set of representatives is the same as balancing the classes. □

Now the next result becomes obvious.

Theorem 3. Let $S \subset \mathbb{F}_2^n$ be nonempty, f its indicator function and $E = \mathcal{E}_0(f)$, with $k = \dim(E)$ and $r = \mathrm{rk}(S)$. Then

$$B(S) = (\mathbb{F}_2^n \setminus H) \sqcup B(S/H),$$

where $H = \bigcap_{\mathbf{s} \in E} H_\mathbf{s}$ and $\sqcup$ stands for the disjoint union. Furthermore, $b(S) = 2^{r-k}(2^k - 1) + b(S_E)$, where S_E is a complete set of representatives of S/E.

Proof. As every element of $\mathbb{F}_2^n \setminus H$ balances S, the result just follows from:

$$B(S) = \left[B(S) \cap (\mathbb{F}_2^n \setminus H) \right] \sqcup \left[B(S) \cap H \right].$$

For the first part, $(B(S) \cap (\mathbb{F}_2^n \setminus H)) = \mathbb{F}_2^n \setminus H$, while the second is just the definition of $B(S/E)$. Regarding $b(S)$, the $2^{r-k}(2^k - 1)$ part comes from $\mathbb{F}_2^n \setminus H$. For the $b(S_E)$ part, if we consider without loss of generality $\mathbf{0} \in S$, $n = r$, $\mathbf{e}_1, \ldots, \mathbf{e}_k \in E$ and $\mathbf{e}_{k+1}, \ldots, \mathbf{e}_r \in S$, where the $\mathbf{e}_i$ are the vectors of the canonical base, then each of the vectors in $b(S_E)$ comes from a compatible system of equations of rank r, and thus there will be $b(S_E)$ affine spaces in $B(S/H)$. □

Corollary 1. *Let $f : \mathbb{F}_2^n \to \mathbb{F}_2$ such that $\mathrm{supp}(f) = S$ is as in Theorem 3, then $\mathrm{supp}(\widehat{f}) = H \cap \mathrm{supp}(\widehat{f}_{S/H})$, where $f_{S/H}$ is the indicator function of S/H.*

The hope is that using this results we can advance further in solving the Fourier–Hadamard support problem, and thus we will be able to use them in distinguishing classes of functions using quantum techniques.

6 The Fourier–Hadamard Transform for Multisets

Let us devote some time to consider the Fourier–Hadamard transform for multisets, as it will become useful for analysing the Walsh transform of vectorial functions $F : \mathbb{F}_2^n \to \mathbb{F}_2^m$ through their image multisets. The main motivation for doing so is to define the concept of fully balanced function and apply the already presented results to analyse what the GPK has to offer. For a general review on multisets we refer the reader to [11]. As we will see, most of the results we have presented remain unchanged, so we will mostly just give a small sketch of the proofs. However, we do present some new results. In particular, we answer the question of determining the possible Fourier–Hadamard supports that a pseudo-Boolean function can have using the constant and balancing sets of multisets.

Definition 7 *(Balanced multiset.).* *Let $M = (\mathbb{F}_2^n, m)$ be a nonempty Boolean multiset. We say that $\mathbf{y} \in \mathbb{F}_2^n$ balances M if:*

$$\widehat{m}(\mathbf{y}) = \sum_{\mathbf{x} \in S_M} m(\mathbf{x}) \cdot (-1)^{\mathbf{x} \cdot \mathbf{y}} = 0,$$

and we say that it makes M constant if:

$$|\widehat{m}(\mathbf{y})| = \left| \sum_{\mathbf{x} \in S_M} m(\mathbf{x}) \cdot (-1)^{\mathbf{x} \cdot \mathbf{y}} \right| = \sum_{\mathbf{x} \in S_M} m(x) = |M|,$$

where $|M| = \sum_{x \in S_M} m(x)$ is the cardinality of M. We define $C(M)$, $B(M)$ and $b(M)$ analogously.

In order to solve the problem of the constant set, we only need the following result.

Lemma 6. *Let M be a Boolean multiset, then $C(M) = C(S_M)$.*

We will now present results analogous to those of Sect. 3 for the balancing problem.

Proposition 5. *Let $M = (\mathbb{F}_2^n, m)$ be a nonempty Boolean multiset and $\mathbf{s} \in \mathbb{F}_2^n$. Let $\mathbf{s} + M = (\mathbb{F}_2^n, m')$ where $m'(\mathbf{x}) = m(\mathbf{s} + \mathbf{x})$, then $B(M) = B(\mathbf{s} + M)$.*

Proof. Let $\mathbf{y} \in B(M)$, then

$$\sum_{\mathbf{x} \in S_M} m'(\mathbf{s} + \mathbf{x}) \cdot (-1)^{(\mathbf{s}+\mathbf{x})\cdot\mathbf{y}} = (-1)^{\mathbf{s}\cdot\mathbf{y}} \sum_{\mathbf{x} \in S_M} m(\mathbf{x}) \cdot (-1)^{\mathbf{x}\cdot\mathbf{y}} = 0.$$

And thus $\mathbf{y} \in B(\mathbf{s} + M)$. As $\mathbf{s} + (\mathbf{s} + M) = M$ we have the result. □

Proposition 6. *Let M be a nonempty Boolean multiset and let $C(M)$ be its constant set. Then $B(M)$ is a disjoint union of affine spaces all of them with $C(M)$ as their underlying vector space.*

The reasoning is the same as the one for the set situation, and we also still have that the balancing index of a Boolean multiset is invariant by isomorphisms of $\mathbb{F}_2^n$. We will now extend the definition of fully balanced and the main results regarding this property.

Definition 8 (*Fully balanced multiset.*)**.** *Let M be a nonempty Boolean multiset. We say that it is a fully balanced multiset if $B(M) \cup C(M) = \mathbb{F}_2^n$.*

Theorem 4. *Let $M = (\mathbb{F}_2^n, m)$ be a nonempty Boolean multiset; then it is fully balanced if and only if S_M is an affine space and m is constant.*

Proof. It is clear that any such M is fully balanced. The converse implication can be proven either by induction or by following the sketch in the proof of Theorem 1 but replacing $|S|$ by $|M|$ and 1_S by m. If we did so, we would still have that if $m(\mathbf{0}) \neq 0$, then $\hat{m} = |M|1_{C(M)}$, just as before. As the inverse Fourier–Hadamard formula still applies, we would end up with:

$$m = \frac{|M||C(M)|}{2^n} 1_{C(M)^\perp},$$

having thus our result. □

The main problem that we have been trying to solve has been that of determining the possible Fourier–Hadamard supports—or, equivalently, balancing sets—that Boolean functions can have, but what happens when we consider multisets? The final result we will reach is the following

Theorem 5 (*Multiset balancing sizes.*)**.** *For any set S such that $\mathbf{0} \in S$ there is a multiset whose multiplicity function has S as its Fourier–Hadamard support.*

However, we will not prove this right away, and instead, we will first focus on a constructive method to create multisets that have increasingly smaller balancing sets.

Lemma 7. *Let $M_1 = (\mathbb{F}_2^n, m_1)$ be a nonempty multiset such that there is an $\mathbf{y} \in B(M_1)$, then there is another multiset $M_2 = (\mathbb{F}_2^n, m_2)$ satisfying $B(M_2) = B(M_1) \setminus \{\mathbf{y}\}$.*

Proof. Thanks to the inverse Fourier–Hadamard transform we know that for all $\mathbf{x} \in \mathbb{F}_2^n$

$$m_1(\mathbf{x}) = \frac{1}{2^n} \widehat{\widehat{m_1}}(\mathbf{x}).$$

If we want to change some values in $\widehat{m_1}$, the only two conditions that $\widehat{m_1}(\mathbf{x})$ must satisfy in order to determine a multiset are the following: first, it is clear that $\widehat{m_1}(\mathbf{x})$ must be nonnegative for all $\mathbf{x} \in \mathbb{F}_2^n$, and secondly, it must be a multiple of 2^n. If we set then

$$\widehat{m_2}(\mathbf{x}) = \begin{cases} \widehat{m_1} + 2^{n-1} & \text{if } \mathbf{x} = \mathbf{0} \text{ or } \mathbf{x} = \mathbf{y} \\ \widehat{m_1}(\mathbf{x}) & \text{otherwise,} \end{cases}$$

it is clear that

$$\widehat{\widehat{m_2}}(\mathbf{x}) = \begin{cases} \widehat{\widehat{m_1}}(\mathbf{x}) & \text{if } \mathbf{x} \cdot \mathbf{y} = 1 \\ \widehat{\widehat{m_1}}(\mathbf{x}) + 2^n & \text{if } \mathbf{x} \cdot \mathbf{y} = 0 \end{cases}$$

and thus both conditions are satisfied. As M_1 was nonempty, $\widehat{m_1}(\mathbf{0})$ was not 0, so $B(M_2) = B(M_1) \setminus \{\mathbf{y}\}$. $\square$

We will see this with an example as soon as we finish the proof of the general result.

Proof (Theorem 5). We know multisets in $\mathbb{F}_2^n$ whose Fourier–Hadamard support is just $\{\mathbf{0}\}$—for instance $\mathbb{F}_2^n$—so we only need to iterate the procedure of Lemma 7 to obtain the desired multiset, but we can actually just begin with the empty multiset. $\square$

7 Application in Quantum Computing

We will finally show some of the results that we have been hinting at when pointing out why does it seem relevant to study the Fourier–Hadamard and Walsh transforms for multisets and in the fully balanced situation. This will be done by tying down the Walsh transform of vectorial functions and some results in quantum computing. As we already mentioned in the introduction, the Walsh transform appears in quantum algorithms in which the phase kick-back technique [6] is applied. In particular, it first appeared in the Deutsch–Jozsa algorithm [5], where the final amplitude associated to a certain state $|\mathbf{x}\rangle_n$ when the algorithm is applied to a Boolean function $f : \mathbb{F}_2^n \to \mathbb{F}_2$ is shown to be $W_f(\mathbf{x})/2^n$.

The Walsh transform of vectorial functions also shows up in the generalised version of this algorithm [10]. Indeed, once again, the final amplitude associated to a certain state $|\mathbf{x}\rangle_n$ when the algorithm is applied to a vectorial function $F: \mathbb{F}_2^n \to \mathbb{F}_2^m$ using a marker $\mathbf{y} \in \mathbb{F}_2^m$ is shown to be $W_F(\mathbf{x}, \mathbf{y})/2^n$. We will show now the relation between fully balanced multisets and the Walsh transform of vectorial functions. To do so, we will first define some concepts.

Definition 9 (**y**-Balanced function.). *Let $F: \mathbb{F}_2^n \to \mathbb{F}_2^m$ be a vectorial function and let $\mathbf{y} \in \mathbb{F}_2^m$. We say that F is **y**-balanced if we have $F(\mathbf{x}) \cdot \mathbf{y} = 0$ for half of the vectors $\mathbf{x} \in \mathbb{F}_2^n$ and $F(\mathbf{x}) \cdot \mathbf{y} = 1$ for the other half.*

In the same way, we can define the idea of **y**-constant functions.

Definition 10 (**y**-Constant function.). *Let $F: \mathbb{F}_2^n \to \mathbb{F}_2^m$ be a vectorial function and let $\mathbf{y} \in \mathbb{F}_2^m$. We say that F is **y**-constant if for every vector $\mathbf{x} \in \mathbb{F}_2^n$ the result of $F(\mathbf{x}) \cdot \mathbf{y}$ is the same.*

As we have seen, these definitions can be translated in terms of the Fourier–Hadamard transform.

Proposition 7. *Let $F: \mathbb{F}_2^n \to \mathbb{F}_2^m$ be a vectorial function. Then, F is **y**-balanced if and only if:*

$$\sum_{\mathbf{x} \in \mathbb{F}_2^n} (-1)^{F(\mathbf{x}) \cdot \mathbf{y}} = 0.$$

*Similarly, F is **y**-constant if and only if:*

$$\left| \sum_{\mathbf{x} \in \mathbb{F}_2^n} (-1)^{F(\mathbf{x}) \cdot \mathbf{y}} \right| = 2^n.$$

The next result now becomes trivial.

Theorem 6. *Let $F: \mathbb{F}_2^n \to \mathbb{F}_2^m$ be a vectorial function and $\mathbf{y} \in \mathbb{F}_2^m$. Then, F is **y**-constant if and only if $|W_F(\mathbf{0}, \mathbf{y})| = 2^n$. Besides, F is **y**-balanced if and only if $W_F(\mathbf{0}, \mathbf{y}) = 0$.*

Following things up, we will now focus on the fully balanced situation.

Definition 11 (Fully balanced functions.). *Let $F: \mathbb{F}_2^n \to \mathbb{F}_2^m$ be a vectorial function, we say that it is fully balanced if for every $\mathbf{y} \in \mathbb{F}_2^m$, F is either **y**-balanced or **y**-constant.*

So we say that a function is fully balanced if the multiset of its image is fully balanced. We can likewise define the concepts of $C(F)$, $B(F)$ and $b(F)$ for a given vectorial function F. We will finish things up by highlighting a trivial consequence of this result that ties down the Walsh transform of a vectorial function and the balancing set of its image multiset. As a previous notation remark, we will refer to the image of F multiset as $I_F = (\mathbb{F}_2^m, m(\mathbf{x}))$ with $m(\mathbf{x}) = |F^{-1}(\mathbf{x})|$.

Corollary 2. Let $F : \mathbb{F}_2^n \to \mathbb{F}_2^m$ be a fully balanced vectorial function. Then,

$$\left|W_F(\mathbf{0}, \mathbf{v})\right| = 2^n \, 1_{C(I_F)}(\mathbf{v}),$$

where $\mathbf{v} \in \mathbb{F}_2^m$ and $1_{C(I_F)}$ is the indicator function of $C(I_F)$.

In other words, when we are dealing with fully-balanced functions, we can determine whether a marker $\mathbf{y}$ is in the balancing set or in the constant set of its image multiset by looking at GPK($\mathbf{y}$). As we highlighted in Sect. 2, the final superposition of GPK($\mathbf{y}$) applied to the function F has as amplitudes the values of W_F—in particular, the amplitude of $\mathbf{z}$ in GPK($\mathbf{y}$) is $\alpha_{\mathbf{z}} = W_F(\mathbf{z}, \mathbf{y})$—so if the result after measuring is zero, then $\mathbf{y}$ is in fact in the constant set, while if we get any other result, then $\mathbf{y}$ is in the balancing set. The idea now is that we can use the GPK algorithm to determine the dimension of the image of a given fully balanced function, and even compute said image, but this will be done on a separate publication [8].

Acknowledgments. The research of the first author is partly supported by the *Norwegian Research Council*, ID 314395. The research of the second and third authors is supported by the *Ministerio de Ciencia e Innovación* under Project PID2020-114613GB-I00 (MCIN/AEI/10.13039/501100011033).

Disclosure of Interests. The authors report that there are no competing interests to declare.

References

1. Bernstein, E., Vazirani, U.: Quantum complexity theory. SIAM J. Comput. **26**(5), 1411–1473 (1997)
2. Berthiaume, A., Brassard, G.: The quantum challenge to structural complexity theory. In: 1992 Proceedings of the 7th Annual Structure in Complexity Theory Conference, pp. 132–137 (1992). https://doi.org/10.1109/SCT.1992.215388
3. Carlet, C.: Boolean Functions for Cryptography and Coding Theory. Cambridge University Press, Cambridge (2021)
4. Carlet, C., Mesnager, S.: On the supports of the Walsh transforms of Boolean functions. Cryptology ePrint Archive, Paper 2004/256 (2004). https://eprint.iacr.org/2004/256
5. Deutsch, D., Jozsa, R.: Rapid solution of problems by quantum computation. Proc. Roy. Soc. Lond. Ser. A: Math. Phys. Sci. **439**, 553–558 (1992)
6. Kaye, P., Laflamme, R., Mosca, M.: An Introduction to Quantum Computing. OUP Oxford (2007)
7. MacWilliams, F.J., Sloane, N.J.A.: The Theory of Error Correcting Codes. Elsevier (1977)
8. Osorio-Castillo, J., Pastor-Díaz, U., Tornero, J.M.: Further applications of the generalised phase kick-back. arXiv preprint arXiv:2405.03850 (2024)
9. Ossorio-Castillo, J., Tornero, J.M.: Quantum computing from a mathematical perspective: a description of the quantum circuit model. arXiv preprint arXiv:1810.08277 (2018)

10. Ossorio-Castillo, J., Pastor-Díaz, U., Tornero, J.M.: A generalisation of the phase kick-back. Quantum Inf. Process. **22**(3), 143 (2023)
11. Syropoulos, A.: Mathematics of multisets. In: Calude, C.S., PǍun, G., Rozenberg, G., Salomaa, A. (eds.) WMC 2000. LNCS, vol. 2235, pp. 347–358. Springer, Heidelberg (2001). https://doi.org/10.1007/3-540-45523-X_17
12. Zhandry, M.: How to record quantum queries, and applications to quantum indifferentiability. In: Boldyreva, A., Micciancio, D. (eds.) CRYPTO 2019. LNCS, vol. 11693, pp. 239–268. Springer, Cham (2019). https://doi.org/10.1007/978-3-030-26951-7_9

Postquantum Cryptography

Efficient Batch Post-quantum Signatures with Crystals Dilithium

Nazlı Deniz Türe[1,2]($\boxtimes$) and Murat Cenk[3]

[1] Middle East Technical University, Ankara, Turkey
denizzture@gmail.com
[2] FAME CRYPT, Ankara, Turkey
[3] Ripple Labs Inc., Toronto, Canada

Abstract. Digital signatures ensure authenticity and secure communication. They are used to verify the integrity and authenticity of signed documents and are widely utilized in various fields such as information technologies, finance, education, and law. They are crucial in securing servers against cyberattacks and authenticating connections between clients and servers. Performing multiple signature generation simultaneously and efficiently is highlighted as a beneficial approach for many systems. This work focuses on efficient batch signature generation using Crystal Dilithium, NIST's post-quantum digital signature standard. One of the main operations of signature generation using Dilithium is the matrix-vector product with polynomial entries. So, the naive approach to generate m signatures where $m > 1$ is to perform m such multiplications. In this paper, we propose to use efficient matrix multiplications of sizes greater than four to generate m signatures. To this end, a batch algorithm that transforms the polynomial matrix-vector multiplication in Dilithium's structure into polynomial matrix-matrix multiplication is designed. The batch numbers and the sizes of the matrices to be multiplied based on the number of repetitions of Dilithium's signature algorithm are determined. Moreover, many efficient matrix-matrix multiplication algorithms, such as Strassen-like multiplications and commutative matrix multiplications, are analyzed to design the best algorithms that are compatible with the specified dimensions and yield improvements. Various multiplication formulas are derived for different security levels of Dilithium, and improvements up to 27.28%, 32.0%, and 30.31% in the arithmetic complexities are observed at three different security levels, respectively. The proposed batch Dilithium signature algorithm and the efficient multiplication algorithms are also implemented, and 34.22%, 17.40%, and 10.15% improvements on CPU cycle counts for three security levels are obtained.

Keywords: Batch Digital Signature Generation · Multiple Signing · Post-Quantum Cryptography · Commutative Matrix Multiplication · Crystals Dilithium · Digital Signature

---

This study is supported by TÜBİTAK 2244 - Industrial PhD Program.

© The Author(s), under exclusive license to Springer Nature Switzerland AG 2025
S. Petkova-Nikova and D. Panario (Eds.): WAIFI 2024, LNCS 15176, pp. 237–257, 2025.
https://doi.org/10.1007/978-3-031-81824-0_15

1 Introduction

Digital signatures are essential to provide data integrity, authenticity, and security. Digital signatures are widely used in instant messaging applications, financial transactions, education, legal documents, and many other digital documents. Moreover, they are crucial for servers such as TLS (Transport Layer Security) servers. The servers often face cyber attacks. To prevent unauthorized access, using digital signatures for both the server and user to authenticate each other is essential. RSA [12] is used as a digital signing algorithm on many major servers, such as TLS servers. Servers use digital signatures whenever a secure connection between a client and the server is set. So, it can be said that digital signatures are used thousands or even millions of times daily on a busy server. For example, TLS servers can establish multiple connections per second. The processes need to be fast so that the system's flow is not disrupted. For this reason, performing multiple signing operations at once is faster and more advantageous for systems. Benjamin [4] has developed a system based on ECDSA and RSA that provides multiple signings for TLS. Techniques such as ElGamal [9] have implemented multiple signing in previous studies ([1,7,11]). Fiat [10] and Tanwar et al. [19] present a batch algorithm that depends on the RSA system. Pavlovski et al. [14] propose an efficient batch signature generation via binary tree structures.

High-capacity quantum computers can be used to break many of the cryptographic algorithms in use today, as demonstrated in 1994 by P. Shor [16]. Since then, a lot of work has been done, and many developments have been achieved. For this reason, the National Institute of Standards and Technology (NIST) has organized a contest to standardize algorithms that provide security against attacks via quantum computers. As a result of the third round of evaluations [2], CRYSTALS Dilithium [8] is selected as the digital signing standard. Following this process, Dilithium's integrability into TLS 1.3 is demonstrated in [17]. However, research has shown that using post-quantum in the TLS server will cause performance degradation [13].

The Dilithium structure is examined to increase performance, and the operations with the highest arithmetic complexity are identified. The security of Dilithium is based on the Module-Learning with Error (M-LWE) problem. In Dilithium's signing algorithm, if y is a column vector that is computed using the secret key, the column vector w is obtained as $w = A \cdot y$, where A is the matrix that the signer and the verifier can compute. The entries of the column vectors and the matrix are the elements of the selected commutative polynomial ring. The highlight is that in Dilithium, matrix-vector multiplication is the operation with the highest arithmetic complexity and forms the skeleton of the key generation, signing, and verification algorithms.

This work focuses on efficient batch digital signature generation using the post-quantum algorithm Dilithium. Instead of multiplying a matrix and a column vector, it is possible to multiply the matrix A by another matrix whose columns are column vectors formed for each message. In the article [5], the application of the Strassen method to FrodoKEM [3]'s system containing matrix-matrix multiplication is the starting point of the idea of transforming the primary approach of Dilithium from matrix-vector multiplication to matrix-matrix multiplication for multiple signing purposes. In this way, matrix-matrix multiplication algorithms using the least number of multiplications, such as [6,18,20], and [15], in multiple signing can increase efficiency. Note that since the entries are large-size polynomials, reducing the number of multiplications contributes significantly to complexity.

This study proposes a design of a batch signature generation algorithm for Dilithium's various security levels (i.e., Dilithium 2, Dilithium 3, and Dilithium 5). Matrix-vector multiplication in the Dilithium structure is converted to matrix-matrix multiplication using the batch technique. Batch numbers are determined separately for each security level, according to the number of repetitions in the signing and the probability that the signature produced is valid. Suitable matrix-matrix multiplication techniques are chosen for the selected batch numbers (4, 5, and 4 for Batch Dilithium 2, Batch Dilithium 3, and Batch Dilithium 5, respectively). Since the matrix and vector entries are polynomials from the ring $R_q = \mathbb{Z}_q[x]/(x^n + 1)$, multiplication is much more expensive than addition. Reducing the number of multiplications is a priority to increase efficiency when selecting these algorithms. Therefore, matrix multiplication algorithms are chosen based on the cost metric minimizing multiplications. It is observed that using [15] for Dilithium 2, Dilithium 3, and Dilithium 5 is more functional. Since Batch Dilithium 2 contains the multiplication of two square matrices of size $(2^d \times 2^d)$, it is possible to apply [6,18,20] to it recursively. However, note that multiplication in [15] cannot be used recursively since it requires entries to be commutative, and matrix multiplications are non-commutative. The multiplication formulas are derived from the most efficient matrix-matrix multiplication techniques based on batch numbers, which determine the size of the matrices to be multiplied. Improvements of up to 28.1%, 33.3%, and 31.5% are observed in the arithmetic complexities of multiplication operations at three different security levels, respectively. The batch system is also implemented for Dilithium's three security levels. Speed comparisons are made between the batch and the reference implementations for 20 random messages, and improvements in CPU cycle counts up to 34.22%, 17.40%, and 10.15% are observed for Dilithium 2, Dilithium 3, and Dilithium 5, respectively.

The organization of the remainder of the paper is as follows: In Sect. 2, related notations and functions are described, and the Crystals Dilithium algorithm is introduced, followed by the presentation of the new algorithm compatible with batch processing in Sect. 3. Then, arithmetic complexity analysis and the improvements observed via implementations for Dilithium's different security levels are detailed in Sect. 4. Section 5 summarizes this research's accomplishments.

2 Preliminaries and Notations

Notations and functions that are used in the remaining sections are given in Table 1.

Table 1. Notations, Functions and their definitions

Notation	Definition
A	Matrix (Bold Upper Case Letter)
$A[i][j]$	The entry in the ith row and j^{th} column of A
v	Column vector (Bold Lower Case Letter)
$v[i]$	i^{th} entry of v
R	Commutative ring $\mathbb{Z}[X]/(x^n+1)$
R_q	Commutative ring $\mathbb{Z}_q[x]/(x^n+1)$
$R_q^{k \times l}$	$k \times l$ matrices whose entries are from R_q
R_q^k	Column vectors of length k whose entries are from R_q
$\leftarrow S_\eta^l \times S_\eta^k$ and $\leftarrow S_\eta^l$	Uniform sampling
Function	Definition
NTT	Number Theoretic Transform described in [8]
NTT^{-1}	Inverse operation of Number Theoretic Transform
H	Hash Function (SHAKE-256)
$HighBits$ and $LowBits$	Decomposing operations defined in [8]
$\|z\|_\infty$	$\|z \mod {}^{\pm}q\|$ defined in [8]
$\|$	Concatenation Operation

2.1 Crystals Dilithium

Crystals Dilithium is an M-LWE (Module-Learning with Errors) based digital signature algorithm selected as a post-quantum signing standard in the NIST's standardization process. The key generation, signing, and verification algorithms of Dilithium are provided in Algorithms 1, 2 and 3, respectively.

Dilithium's key generation mechanism is explained in Algorithm 1. In the first stage, ρ, ρ' and K are generated using the 256-bit long ζ value. The parameter ρ is used to create the public polynomial matrix $\hat{A}$, and ρ' is used to produce the hidden polynomial vectors s_1 and s_2. In step 5, t is calculated using the generated polynomial matrix and vectors. Finally, the key pair is created through sub and hash functions.

Algorithm 1. Dilithium Key Generation [8]

Output: Public key pk, Secret key sk
1: $\zeta \leftarrow \{0,1\}^{256}$
2: $(\rho, \rho', K) \in \{0,1\}^{256} \times \{0,1\}^{512} \times \{0,1\}^{256} := H(\zeta)$
3: $\hat{A} \in R_q^{k \times l} := ExpandA(\rho)$ ▷ A is generated in NTT representation as $\hat{A}$
4: $(s_1, s_2) \in S_\eta^l \times S_\eta^k := ExpandS(\rho')$
5: $t := NTT^{-1}(\hat{A} \cdot NTT(s_1)) + s_2$
6: $(t_1, t_0) := Power2Round_q(t, d)$
7: $tr \in \{0,1\}^{256} := H(\rho\|t_1)$
8: return $pk = (\rho, t_1)$, $sk = (\rho, K, tr, s_1, s_2, t_0)$

Algorithm 2. Dilithium Signing [8]

Input: Secret key $sk = (\rho, K, tr, s_1, s_2, t_0)$, message M
Output: Signature σ
1: $\hat{A} \in R_q^{k \times l} := ExpandA(\rho)$ ▷ A is generated in NTT representation as $\hat{A}$
2: $\mu \in \{0,1\}^{512} := H(tr\|M)$
3: $\kappa := 0$, $(z, h) = \perp$
4: $\rho' \in \{0,1\}^{512} := H(K\|\mu)$
5: $\hat{s}_1 := NTT(s_1)$
6: $\hat{s}_2 := NTT(s_2)$
7: $\hat{t}_0 := NTT(t_0)$
8: while $(z, h) = \perp$ do
9: $\quad y \in \tilde{S}_{\gamma_1}^l := ExpandMask(\rho', \kappa)$
10: $\quad w := NTT^{-1}(\hat{A} \cdot NTT(y))$
11: $\quad w_1 := HighBits_q(w, 2\gamma_w)$
12: $\quad \tilde{c} \in \{0,1\}^{256} := H(\mu\|w_1)$
13: $\quad c \in B_\tau := SampleInBall(\tilde{c})$
14: $\quad z := y + NTT^{-1}(\hat{c}\hat{s}_1)$ ▷ $\hat{c} = NTT(c)$
15: $\quad r_0 := LowBits_q(w - NTT^{-1}(\hat{c} \cdot \hat{s}_2), 2\gamma_2)$
16: $\quad$ if $\|z\|_\infty \geq \gamma_1 - \beta$ or $\|r_0\|_\infty \geq \gamma_2 - \beta$ then
17: $\quad\quad (z, h) := \perp$
18: $\quad$ else
19: $\quad\quad h := MakeHint_q(-NTT^{-1}(\hat{c} \cdot \hat{t}_0), w - cs_2 + NTT^{-1}(\hat{c} \cdot \hat{t}_0), 2\gamma_2)$
20: $\quad\quad$ if $\|ct_0\|_\infty \geq \gamma_2$ or the # of 1's in h is greater than ω then
21: $\quad\quad\quad (z, h) := \perp$
22: $\quad \kappa := \kappa + l$
23: return $\sigma = (\tilde{c}, z, h)$

The Dilithium signature algorithm is described in Algorithm 2. The public polynomial matrix $\hat{A}$ is obtained using the ρ generated by the signature process. The official document is signed using the private key and $\hat{A}$ with the assistance of NTT. The signature phase's correctness is checked. This process needs to be repeated if the requirements are not satisfied. In Dilithium's official document [8], the probability that the signature is established correctly is described in detail, and the probability that the entire signature will be correct is calculated as $e^{-256 \cdot \beta \cdot k / \gamma_2}$.

In Algorithm 3, the validity of the signature is checked using the public key.

Algorithm 3. Dilithium Verification [8]

Input: Public key $pk = (\rho, t_1)$, message M, signature σ
Output: Valid or Not
1: $\hat{A} \in R_q^{k \times l} := ExpandA(\rho)$ ▷ A is generated in NTT representation as $\hat{A}$
2: $\mu \in \{0,1\}^{512} := H(H(\rho\|t_1)\|M)$
3: $c := SampleInBall(\tilde{c})$
4: $w_1' := UseHint_q(h, NTT^{-1}(\hat{A} \cdot NTT(z) - NTT(c) \cdot NTT(t_1 \cdot 2^d)))$
5: return $[\|z\|_\infty < \gamma_1 - \beta]$ and $[\tilde{c} = H(\mu \| w_1')]$ and [# of 1's in h is $\leq \omega$]

The parameters of Dilithium are presented in Table 2 where (k, l) is the size of the public matrix $\hat{A}_{k \times l}$. The entries of the matrix $\hat{A}$ are the polynomials from the ring $R_q = \mathbb{Z}_q[x]/(x^n + 1)$.

Table 2. Parameter Sets for Dilithium [8]

Security Level	Algorithm	n	(k,l)	q
2	Dilithium 2	256	(4,4)	8380417
3	Dilithium 3	256	(6,5)	8380417
5	Dilithium 5	256	(8,7)	8380417

The matrix-vector multiplications in the 5th step of the Algorithm 1, the 10th step of the Algorithm 2, and the 4th step of the Algorithm 3 are the most costly operations for Dilithium since all the elements in the matrices and column vectors are polynomials from the commutative ring $\mathbb{Z}_q[X]/(x^{256}+1)$ where q is in Table 2. To illustrate this, consider an example where $\boldsymbol{A}$ belongs to $R_q^{6\times 5}$ and $\boldsymbol{s'}$ belongs to R_q^5. To calculate $\boldsymbol{A} \cdot \boldsymbol{s'}$, 30 polynomial multiplications are needed, with all the polynomials coming from $\mathbb{Z}_q[X]/(x^{256}+1)$. Signing six different messages for six different users would require 180 polynomial multiplications. Signing multiple messages can be performed more efficiently by reducing the number of multiplications. The proposed algorithm, which ensures multiple messages can be signed at once and more efficiently, is explained in the next section.

3 Dilithium Signature Algorithm for Batch Operations

This section explains the proposed batch version of Dilithium's signing algorithm.

3.1 Batch Crystals Dilithium Signing

The batch Dilithium Signature Algorithm, that is based on the Dilithium Signature Algorithm specified in Algorithm 2 and allows signing more than one message at a time, is explained in Algorithm 4. As an example, batch numbers of the proposed algorithm are selected as 4, 5, and 4 for Dilithium 2, 3, and 5, respectively. How batch numbers can be selected is explained in detail in Sect. 3.3.

Algorithm 4. Batch Dilithium Signing for m Different Messages

Input: Secret key $sk = (\rho, K, tr, s_1, s_2, t_0)$, messages M_i, where $i = 0, 1, \ldots, m-1$
Output: Signatures σ_i, where $i = 0, 1, \ldots, m-1$
1: $\hat{A} \in R_q^{k \times l} := ExpandA(\rho)$ ▷ A is generated in NTT representation as $\hat{A}$
2: for $i = 0, 1, \ldots, m-1$ do
3:      $\mu_i \in \{0,1\}^{512} := H(tr\|M_i)$
4:      $\kappa_i := 0$, $(z'_i, h'_i) = \perp$
5:      $\rho'_i \in \{0,1\}^{512} := H(K\|\mu_i)$
6: $waitList_i = i \quad \forall i = 0, 1, 2, \ldots, m-1$
7: $\hat{s}_1 := NTT(s_1)$
8: $\hat{s}_2 := NTT(s_2)$
9: $\hat{t}_0 := NTT(t_0)$
10: while Length of $waitList \geq p$ do ▷ Number of batch $p = 4, 5, 4$ for Dilithium 2, 3, 5.
11:     $status_i = 0 \quad \forall i = 0, 1, 2, \ldots, m-1$
12:     for $i = 0, 1, \ldots, p-1$ do
13:         $y_i \in \tilde{S}_{\gamma_1}^l := ExpandMask(\rho'_{waitList_i}, \kappa_{waitList_i})$
14:     $\hat{Y} := (l \times p)$ matrix, whose columns are $NTT(y_i)$'s, where $i = 0, 1, 2, \ldots, p-1$
15:     $\hat{W} := \hat{A} \cdot \hat{Y}$, $\hat{W} : (k \times p)$ matrix, whose columns are $\hat{w}_i$'s, where $i = 0, 1, 2, \ldots, p-1$
16:     $W := (k \times p)$ matrix, whose columns are $NTT^{-1}(\hat{w}_i)$'s, where $i = 0, 1, 2, \ldots, p-1$
17:     for $i = 0, 1, \ldots, p-1$ do
18:         $w'_i := HighBits_q(i, 2\gamma_2)$
19:         $\tilde{c}_{waitList_i} \in \{0,1\}^{256} := H(\mu_{waitList_i} \| w'_i)$
20:         $c_{waitList_i} \in B_\tau := SampleInBall(\tilde{c}_{waitList_i})$
21:         $\hat{c}_{waitList_i} = NTT(c_{waitList_i})$
22:         $z'_i := y_i + NTT^{-1}(\hat{c}_{waitList_i} \cdot \hat{s}_1)$
23:         $r_i := LowBits_q(w_i - NTT^{-1}(\hat{c}_{waitList_i} \cdot \hat{s}_2), 2\gamma_2)$
24:         if $\|z'_i\|_\infty \geq \gamma_1 - \beta$ then
25:             $status_i = status_i + 1$
26:         if $\|r_i\|_\infty \geq \gamma_2 - \beta$ then
27:             $status_i = status_i + 1$
28:         $h'_i := MakeHint_q(-NTT^{-1}(\hat{c}_{waitList_i} \cdot \hat{t}_0), w_i - c_{waitList_i} \cdot s_2 + NTT^{-1}(\hat{c}_{waitList_i} \cdot \hat{t}_0), 2\gamma_2)$
29:         if $\|ct_0\|_\infty \geq \gamma_2$ then
30:             $status_i = status_i + 1$
31:         if The # of 1's in h'_i is greater than ω then
32:             $status_i = status_i + 1$
33:         $\kappa_{waitList_i} = \kappa_{waitList_i} + l$
34:     for $i = 0, 1, \ldots, p-1$ do
35:         if $!status_i$ then
36:             $z_{waitList_i} = z'_i$
37:             $h_{waitList_i} = h'_i$
38:             Delete $waitList_i$ from the list
39: return $\sigma_i = (\tilde{c}_i, z_i, h_i)$ where $i = 0, 1, \ldots, m-1$

The inputs of the batch signing algorithm are the user's secret key $sk = (\rho, K, tr, s_1, s_2, t_0)$ and m different messages. These messages are indexed as $M_0, M_1, \ldots, M_{m-1}$. First, the matrix $\hat{A}$ is produced using ρ as in Algorithm 2. Then, μ_i, κ_i, and ρ'_i values corresponding to each message M_i are computed. κ_i values start with 0, and (z'_i, h'_i) are initialized to $\perp$. A $waitList = 0, 1, 2, \ldots, m-1$ list representing indices of messages waiting to be signed is defined. The first p indices in the list represent which messages are in the signing phase at the same time. At the end of the while loop, the indices of messages with appropriate signatures are deleted from this list. Just before the signature phase, s_1, s_2, and t_0, which are the secret key elements, are converted to their NTT formats: $\hat{s}_2$, $\hat{s}_2$, and $\hat{t}_0$. Once the preliminary preparations are completed, the signing loop begins. The candidate signatures produced must satisfy certain conditions. The status of these conditions is controlled by $status_i$.

Note that y_i's are computed using the ρ_i''s and κ_i's of the relevant messages and brought to their NTT format. The produced $\hat{y}_i$'s form the columns of the $\hat{Y}$ matrix. The batch operation of the algorithm is carried out in the 15th step. This step, which is matrix-vector multiplication in the original version, is turned into matrix-matrix multiplication. Each column of the matrix $\hat{W}$ obtained by multiplication is converted to its normal form with NTT^{-1}. These are called w_i's and are used for p messages in the signing phase. In Steps 17–33, the operations performed for a single message in the original algorithm are performed for p messages. The candidate signatures generated for each message are checked with "if" statements, and as a result, the $status_i$'s are updated. Between stages 34 and 38, candidate signatures that meet all conditions are determined to be real signatures, and the indices of properly signed messages are deleted from the $waitList$. p messages waiting in the queue in the $waitList$ enter the loop with updated κ_i values (κ_i values are updated in step 33). The loop continues until less than p elements are left in the $waitList$. Messages that cannot be signed with Algorithm 4 are signed individually with the Original Dilithium Signature Algorithm (Algorithm 2). Thus, all messages are signed properly.

Example: Let M_0, M_1, M_2, M_3, M_4, M_5, M_6, M_7 ($m = 8$) messages to be signed with Dilithium 2. The number of batch p is four and $\kappa = [\kappa_0, \kappa_1, \kappa_2, \kappa_3, \kappa_4, \kappa_5, \kappa_6, \kappa_7]$ is initialized to $[0,0,0,0,0,0,0,0]$ as in Algorithm 4, line 4. The signatures are generated in the while loop and checked to determine whether they comply with the required conditions. Let the generated signatures $(\tilde{c}_3, z_3, h_3)$, $(\tilde{c}_1, z_1, h_1)$, $(\tilde{c}_5, z_5, h_5)$, $(\tilde{c}_6, z_6, h_6)$, and $(\tilde{c}_0, z_0, h_0)$ are valid in order. For example, the change of κ and $waitList$ according to the number of loops and valid signatures generated is shown in Table 3.

Table 3. Change of κ and $waitList$ according to the number of loops and valid signatures generated.

# of Loop	Valid	$\kappa = [\kappa_0, \kappa_1, \kappa_2, \kappa_3, \kappa_4, \kappa_5, \kappa_6, \kappa_7]$	$waitList$
0 (start)	–	[0,0,0,0,0,0,0,0]	[0,1,2,3,4,5,6,7]
1	$(\tilde{c}_3, z_3, h_3)$	[4,4,4,4,0,0,0,0]	[0,1,2,4,5,6,7]
2	$(\tilde{c}_1, z_1, h_1)$	[8,8,8,4,4,0,0,0]	[0,2,4,5,6,7]
3	$(\tilde{c}_5, z_5, h_5)$	[12,8,12,4,8,4,0,0]	[0,2,4,6,7]
4	$(\tilde{c}_6, z_6, h_6)$	[16,8,16,4,12,4,4,0]	[0,2,4,7]
5	$(\tilde{c}_0, z_0, h_0)$	[20,8,20,4,16,4,4,4]	[2,4,7]

First, the matrix $\hat{A}$ is generated via Algorithm 4 and μ_i, ρ_i' are computed for $i = 0, 1, 2, \ldots, 7$. Four messages are handled simultaneously since $p = 4$. The four messages that are processed are determined by the first four components of $waitList$. At the beginning, $waitList = [0, 1, 2, 3, 4, 5, 6, 7]$. Since the first four components are $[0, 1, 2, 3]$, the first while loop tries to generate signatures for the messages M_0, M_1, M_2, M_3. According to Table 3, the signature candidates produced at the end of the first loop are $(\tilde{c}_0, z_0, h_0)$, $(\tilde{c}_1, z_1, h_1)$, $(\tilde{c}_2, z_2, h_2)$, $(\tilde{c}_3, z_3, h_3)$. Only one of these four signature candidates satisfies the conditions and is valid, which is $\sigma_3 = (\tilde{c}_3, z_3, h_3)$. The corresponding κ values are updated to [4,4,4,4,0,0,0,0]. Since the signature generated for M_3 is valid, three is removed from the $waitList$ and $waitList = [0, 1, 2, 4, 5, 6, 7]$. In the second round, signature candidates are produced for M_0, M_1, M_2, M_4. Assume that only $(\tilde{c}_1, z_1, h_1)$ is correct among the four signatures generated at the end of the second round. Then, $\kappa = [8, 8, 8, 4, 4, 0, 0, 0]$ and 1 is deleted from $waitList$ and the list becomes $waitList = [0, 2, 4, 5, 6, 7]$. This structure continues until the number of elements of $waitList$ is less than p. Thus, at the end of the Batch Dilithium 2 Signing algorithm, σ_3, σ_1, σ_5, σ_7, and σ_0 are produced, respectively.

The messages M_2, M_4, and M_7 that cannot be signed with the batch algorithm are signed individually using the original Dilithium 2 Signing Algorithm (Algorithm 2). One step further, it is possible to initialize κ values of the original Dilithium 2 signing algorithm to the values obtained at the end of the batch algorithm since some κ values are tried for these messages, and no correct results are obtained. Finally, all messages are signed by combining Algorithms 4 and 2.

3.2 Making the Batch Algorithm More Efficient Compared to the Naive Approach

The process in step 15 of Algorithm 4 is emphasized to ensure that the batch algorithm is more efficient than the naive one. The matrix-vector multiplication, which is done separately for each message in the original algorithm, has now turned into matrix-matrix multiplication. Due to the structure of Dilithium, the elements of these matrices are not integers but polynomials of degree 255. For this reason, although the matrix sizes are very small, one of the most expensive phases of the algorithm is the multiplication of these matrices. If these multiplications are done using the schoolbook method, the batch version does not seem to have an advantage over the original version. However, many matrix-matrix multiplication algorithms are available in the literature that can be used depending on the dimensions and properties of the matrices. If appropriate algorithms are chosen, the number of polynomial multiplications required by this operation can be reduced, and generating multiple signatures at once is more efficient than generating them one by one.

The Importance of Commutativity Property and Choosing The Proper Efficient Matrix-Matrix Multiplication Algorithms. The dimensions of the matrices to be multiplied in the 15th step of Algorithm 4 according to the determined batch numbers and different security levels of Dilithium signing are shown in Table 4.

Table 4. Batch numbers and matrix sizes according to different security levels.

Security Level	Algorithm	Batch Number	Size of $\hat{A}$	Size of $\hat{Y}$	Size of $\hat{W}$
2	Dilithium 2	4	(4×4)	(4×4)	(4×4)
3	Dilithium 3	5	(6×5)	(5×5)	(6×5)
5	Dilithium 5	4	(8×7)	(7×4)	(8×4)

Considering the matrix dimensions and structure of Dilithium, many matrix-matrix multiplication algorithms can be used. Dilithium's ring, $R = \mathbb{Z}_{8380417}[X]/(X^{256}+1)$, is commutative. For this reason, the multiplication of polynomials, which are the entries of matrices, is commutative. Thanks to this feature, the product of any two entries, a_{ij} and y_{kl} has the equality $a_{ij} \cdot y_{kl} = y_{kl} \cdot a_{ij}$. By taking advantage of this feature, a fast commutative matrix-matrix multiplication algorithm detailed in [15] can be used for all security levels. This algorithm works more efficiently than the Strassen [18] method or the non-commutative methods in the literature. Non-commutative methods can also be preferred when the matrix multiplication is performed recursively for larger-size matrices.

Strassen-like multiplications such as Strassen method, Cenk&Hassan's method [6], and Winograd's Multiplication [20] are known to be the best recursive matrix-matrix multiplications.

In this work, we derive the Batch Dilithium 2 algorithm, $(4 \times 4) \cdot (4 \times 4)$ matrix-matrix multiplication formulas using Strassen's Method and the fast commutative method. Matrix multiplication formulas for $(6 \times 5) \cdot (5 \times 5)$ and $(8 \times 7) \cdot (7 \times 4)$ are obtained for Batch Dilithium 3 and Dilithium 5 using the fast commutative method.

3.3 Probability Computations and Choosing the Batch Sizes

Dilithium's signing algorithm requires specific requirements to be satisfied by signature candidates. The signature algorithm uses if statements to verify these conditions. Lines 24, 26, 29, and 31 in the batch algorithm (Algorithm 4) and 13, 17 in the classical algorithm (Algorithm 2) have these criteria. Step 13 is more dominant in the condition checks in Algorithm 2. Because of this, the following formula-which is defined in the [8]-is used to determine the probability that the signature created is valid (for step 13):

$$\approx e^{-256 \cdot \beta (l/\gamma_1 + k/\gamma_2)}. \tag{1}$$

The values of the variables included in the formula and probabilities that a generated signature candidate will satisfy the requirements based on the various Dilithium security levels are given in Table 5. Signatures that do not satisfy the conditions are reproduced with the while loop.

Table 5. Variables required to compute the probability

Variable	Dilithium 2	Dilithium 3	Dilithium 5
γ_1	2^{17}	2^{19}	2^{19}
γ_2	95232	261888	261888
(k, l)	(4,4)	(6,5)	(8,7)
β	78	196	120
Probability	≈ 0.24	≈ 0.196	≈ 0.26

Depending on the batch number, the probability of at least one of the p signatures produced by Dilithium 2, Dilithium 3, and Dilithium 5 being valid are calculated as $1-(1-0.24)^p$, $1-(1-0.196)^p$, and $1-(1-0.26)^p$, respectively.

The probability that at least one of the p signatures is valid increases with the batch number. When calculating the batch number p, two scenarios need to be taken into account. The first is the probability that when p number of messages are signed, at least one of these signatures is correct. The second is the rate of improvement that the chosen p will provide. Based on these two scenarios, the user may determine the priorities and select an appropriate batch number. Moreover, while choosing the batch number, the formulas to be derived based on the matrix sizes and the possible matrix-matrix multiplication algorithms should also be considered.

In Table 6, the probabilities that at least one of the p signatures is valid based on some eligible batch numbers (p) are given for Dilithium 2, 3, and 5.

Table 6. Probabilities of at least one of p signatures being correct according to batch number (p) for Dilithium 2, 3, and 5.

	Probability		
p (#of batch)	Dilithium 2	Dilithium 3	Dilithium 5
3	0.561	0.480	0.595
4	0.666	0.582	0.700
5	0.746	0.664	0.778
6	0.807	0.730	0.836
7	0.854	0.783	0.878
8	0.889	0.825	0.910
9	0.915	0.860	0.933
10	0.936	0.887	0.951
11	0.951	0.909	0.964
12	0.963	0.927	0.973
13	0.972	0.941	0.980
14	0.979	0.953	0.985
15	0.984	0.962	0.989
16	0.988	0.970	0.992
17	0.991	0.975	0.994
18	0.993	0.980	0.996
19	0.995	0.984	0.997
20	0.996	0.987	0.998

Figures 1, 2, and 3 show how the probability of at least one of p signatures being correct according to the batch number p for Dilithium 2, 3, and 5, respectively. Table 6 and Figs. 1, 2, and 3 show that the probability converges to 1 as the number of batches increases.

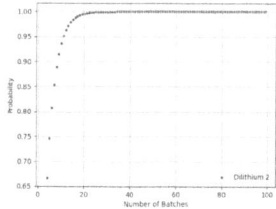 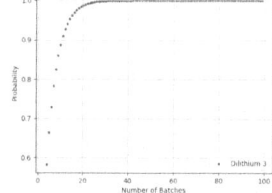 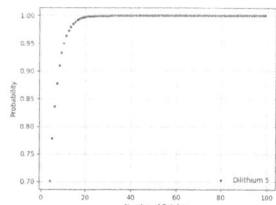

Fig. 1. Change in probability of at least one of p signatures generated with Dilithium 2 being correct according to batch number (p).

Fig. 2. Change in probability of at least one of p signatures generated with Dilithium 3 being correct according to batch number (p).

Fig. 3. Change in probability of at least one of p signatures generated with Dilithium 5 being correct according to batch number (p).

The number of polynomial multiplication operations that need to be performed according to the selected batch number is given in Table 7. For the batch method, [15] is used as the efficient matrix-matrix multiplication method.

In Dilithium 2, via [15] the number of multiplication operations required to multiply matrices $\hat{A}^{k \times l}$ and $\hat{Y}^{l \times p}$ can be computed as $l(kp+k+p-1)/2$ where $k=4$ and $l=4$. Similarly, in Dilithium 3 and 5, the number of multiplication operations required can be computed as $(l(kp+k+p-1)+k-1)/2$ for p is even and $l(kp+k+p-1)/2$ for p is odd, where $k=6$, $l=5$ for Dilithium 3, and $k=8$, $l=7$ for Dilithium 5.

Table 7. Required number of multiplications for a single signature assuming one of the p signatures is correct

p (# of Batch)	Dilithium 2		Dilithium 3		Dilithium 5	
	Batch	Classical	Batch	Classical	Batch	Classical
3	36	48	65	90	119	168
4	46	64	85	120	154	224
5	56	80	100	150	182	280
6	66	96	120	180	217	336
7	76	112	135	210	245	392
8	86	128	155	240	280	448
9	96	144	170	270	308	504
10	106	160	190	300	343	560
11	116	176	205	330	371	616
12	126	192	225	360	406	672
13	136	208	240	390	434	728
14	146	224	260	420	469	784
15	156	240	275	450	497	840
16	166	256	295	480	532	896
17	176	272	310	510	560	952
18	186	288	330	540	595	1008
19	196	304	345	570	623	1064
20	206	320	365	600	658	1120

Table 8. Improvement Rates (%) for 20 Signatures

p (#of Batch)	Dilithium 2 Impr (%)	Dilithium 3 Impr (%)	Dilithium 5 Impr (%)
3	22.50	25.00	26.25
4	23.91	24.79	26.56
5	24.00	26.67	28.00
6	23.44	25.00	26.56
7	22.50	25.00	26.25
8	21.33	23.02	24.38
9	20.00	22.22	23.33
10	18.56	20.17	21.31
11	17.05	18.94	19.89
12	15.47	16.88	17.81
13	13.85	15.38	16.15
14	12.19	13.33	14.06
15	10.50	11.67	12.25
16	8.79	9.64	10.16
17	7.06	7.84	8.24
18	5.31	5.83	6.15
19	3.55	3.95	4.14
20	1.78	1.96	2.06

For example, let 20 messages be signed with Batch Dilithium. The improvement rates that will be obtained by reducing the number of multiplications are given in Table 8. As can be observed from the table, the best improvement rate is obtained by selecting p as 5. According to improvement percentages and probability calculations, p can be selected by taking into account the available efficient matrix-matrix multiplication algorithms. Let C be the number of multiplications required for the classical method, B be the number of multiplications required for the efficient matrix-matrix multiplication algorithm used in the batch method, m be the number of messages to be signed, and p be the number of batches. The improvement rate is calculated with the following formula:

$$(Cm - (B(m-(p-1)) + C(p-1)))100/(Cm) \qquad (2)$$

As observed in Table 5, the probabilities of signatures being valid for Dilithium 2, 3, and 5 are approximately 1/4, 1/5, and 1/4, respectively. So, one of the four signatures produced by Dilithium 2, one of the five for Dilithium 3, and one of the four for Dilithium 5 are expected to be valid with a high probability. Therefore, the probability that at least one of the four signatures produced by Dilithium 2 is valid is $1 - (1 - 0.24)^4 \approx 0.67 = 67\%$. Similarly, the probability that at least one of the five signatures produced by Dilithium 3 is valid is 66%, and the probability that at least one of the four signatures produced by Dilithium 5 is valid is 70%. As an example, batch numbers are selected as 4, 5, and 4 for Dilithium2, 3, and 5, respectively.

4 Results

The results of the batch application are examined under two main headings. These are improvements in the arithmetic complexity and the improvements observed via the implementations.

4.1 Matrix Multiplications and Arithmetic Complexity Analysis

The batch method must be used together with efficient matrix-matrix multiplication algorithms in order to be effective. For this reason, the batch numbers to be used in the Batch Dilithium Algorithm for each security level are determined by the method explained in Sect. 3.3. The dimensions of the matrices in the 15th step of the Algorithm 4 according to each security level of Dilithium are given in Table 4. Since the entries of the matrices are polynomials and the multiplications of those entries are highly costly compared to their additions, the aim is to reduce the number of multiplications. The matrix-matrix multiplication algorithms in the literature with the minimum number of multiplications are generally obtained by using the Strassen-like multiplications recursively. However, since the sizes of the matrices in our work are small, they do not require recursions. So, the commutative matrix multiplication algorithms that have better multiplicative complexity than the non-commutative multiplication algorithms can also be used for our purposes since the entries are polynomials from the commutative ring R_q. Therefore, it is advantageous to use the commutative matrix-matrix multiplication algorithm described in [15].

Let (n_1, n_2, n_3) matrix multiplication be the product of a matrix of dimension $(n_1 \times n_2)$ and a matrix of dimension $(n_2 \times n_3)$. We need an efficient $(4, 4, 4)$ matrix multiplication for Dilithium 2, $(6, 5, 5)$ matrix multiplication for Dilithium 3, and $(8, 7, 4)$ for Dilithium 5. For the $(4, 4, 4)$ multiplications, the best choice seems to be Strassen-like recursive multiplications [6,18,20] with 49 multiplications and the commutative matrix multiplication method by [15] with 46 multiplications. For $(6, 5, 5)$ matrix multiplication in batch Dilithium 3 can be performed with 100 multiplications, and $(8, 7, 4)$ matrix multiplication in batch Dilithium 5 can be performed with 154 multiplications using [15]. We derive all explicit formulas and use them in our implementation.

Batch Dilithium 2. Assume that m messages are signed independently with Dilithium 2 using the classical approach. Each message needs 16 multiplications. However, due to the failure rate in the algorithm, $16 \cdot 4 = 64$ multiplications are expected to sign the message on average. Therefore, a total of $64m$ multiplications are needed for m messages.

In the batch method in Algorithm 4, step 15 is performed for four signatures. Since the probability that a signature candidate is valid for Dilithium 2 is approximately 0.24, one of the four signatures can be assumed to be valid. This process ends when $m - (p - 1)$ messages are signed. Thus, the batch operation requires a total of $46 \cdot (m - (p - 1)) = 46 \cdot (m - 3)$ multiplication operations. The remaining $p - 1 = 3$ messages from the batch process are signed individually with classical Dilithium 2. Step 7 in the Algorithm 2 requires 16 multiplications to sign each message. If we consider that the loop is repeated four times to generate a valid signature, the number of multiplications needed for a message is calculated as $16 \cdot 4 = 64$. Thus, the number of multiplications required for three messages is $3 \cdot 64 = 192$, and the number of multiplications required for batch Dilithium 2 is approximately $46 \cdot (m - 3) + 192 = 46m + 54$. The number of multiplications required for the classical and batch methods, together with the improvement rates obtained with batch Dilithium 2, is presented in Table 9 for the selected number of messages up to 100. The improvement rate is calculated as $(64m - (46m + 54)) \cdot 100/64m$, and as the number of messages increases, this rate converges to 28.1%.

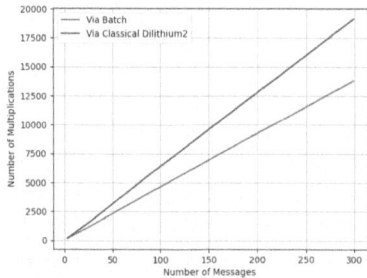 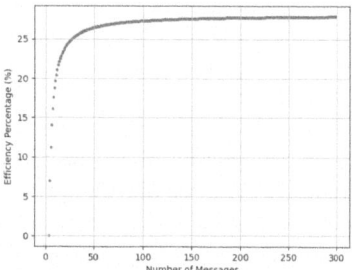

Fig. 4. Variation of the number of multiplications required according to the number of messages for Batch Dilithium 2 and Classical Dilithium 2

Fig. 5. Variation of the improvement rate (%) provided by the batch algorithm compared to the classical version depending on the number of messages for Dilithium 2

Variation of the number of multiplications required according to the number of messages in Batch Dilithium 2 and Classical Dilithium 2 is shown in Fig. 4. Variation of the improvement rate (%) provided by the batch algorithm compared to the classical version depending on the number of messages can be observed in Fig. 5.

Table 9. Variation of the number of multiplications required for batch and classical use of Dilithium 2, and the improvement rates provided by batch method with [15] according to the number of messages

# of Messages m	Batch ($p=4$) $46m+54$	Classic $64m$	Improvement (%)
4	238	256	7.0
5	284	320	11.25
6	330	384	14.1
7	376	448	16.1
8	422	512	17.58
9	468	576	18.85
10	514	640	19.69
20	974	1280	23.91
40	1894	2560	26.02
80	3734	5120	27.07
100	4654	6400	27.28

Batch Dilithium 3. If the classical method is used to sign m messages using Dilithium 3, it takes 30 multiplications for a message. But, because of the failure rate, $30 \cdot 5 = 150$ multiplications are done on average. For m messages, a total of $150m$ multiplications are expected.

The batch signature generation is performed for five signatures in Algorithm 4 at step 15. It is presumed that at least one of the five signatures is valid since the probability that a signature candidate is valid for Dilithium 3 is approximately 0.196. As a result of calculations similar to those in batch Dilithium 2, the batch operation requires a total of $100 \cdot (m-4)$ multiplication operations. When the remaining four messages are signed using the classical method, the total number of multiplications required for batch Dilithium 3 is $100m+200$. According to the number of messages, the number of multiplications required for the classical or batch method, and the improvement rates obtained with batch Dilithium 3 are provided in Table 10. The improvement rate is computed using $(150m - (100m+200)) \cdot 100/150m$. As the number of messages increases, this rate converges to 33.3%. Variation of the number of multiplications required according to the number of messages in Batch Dilithium 3 and Classical Dilithium 3 is shown in Fig. 6. Variation of the improvement rate (%) provided by the batch algorithm compared to the classical version depending on the number of messages can be observed in Fig. 7.

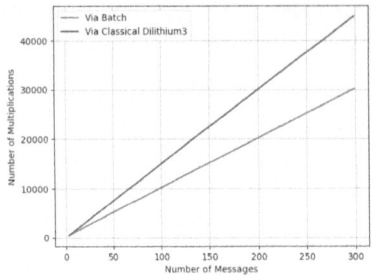
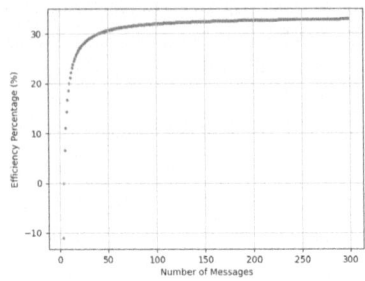

Fig. 6. Variation of the number of multiplications required according to the number of messages for Batch Dilithium 3 and Classical Dilithium 3

Fig. 7. Variation of the improvement rate (%) provided by the batch algorithm compared to the classical version depending on the number of messages for Dilithium 3

Table 10. Variation of the number of multiplications required for batch and classical use of Dilithium 3, and the improvement rates provided by batch method with [15] according to the number of messages

# of Messages m	Batch ($p=5$) $100\,m + 200$	Classic $150\,m$	Improvement (%)
5	700	750	6.7
6	800	900	11.11
7	900	1050	14.29
8	1000	1200	16.67
9	1100	1350	18.52
10	1200	1500	20.0
20	2200	3000	26.67
40	4200	6000	30.0
80	8200	12000	31.67
100	10200	15000	32.0

Batch Dilithium 5. Assume the classical method is used to sign m messages separately using Dilithium 5. It takes 56 multiplications for each message. Performing $56 \cdot 4 = 224$ multiplications is done under the assumption that a valid signature requires four repetitions. For m messages, a total of $224m$ multiplications must be done.

Similar to Batch Dilithium 2, 15th step of the Algorithm 4 is performed for four signatures. Since the probability that a signature candidate is valid for Dilithium 5 is approximately 0.26, it is assumed that at least one of the four signatures will be valid. As a result of calculations similar to those in batch Dilithium 2 and 3, the batch operation requires a total of $154 \cdot (m-3)$ multiplication operations. When the remaining three messages are signed using the classical method, the total number of multiplications required for batch Dilithium

5 is $154m + 210$. According to the number of messages, the number of multiplications required for the classical or batch method, and the improvement rates obtained with batch Dilithium 5 are presented in Table 11.

Table 11. Variation of the number of multiplications required for batch and classical use of Dilithium 5, and the improvement rates provided by batch method with [15] according to the number of messages

# of Messages m	Batch ($p=4$) $154m+210$	Classic $224m$	Improvement (%)
4	826	896	7.81
5	980	1120	12.5
6	1134	1344	15.63
7	1288	1568	17.86
8	1442	1792	19.53
9	1596	2016	20.83
10	1750	2240	21.86
20	3290	4480	26.56
40	6370	8960	28.91
80	12530	17920	30.08
100	15610	22400	30.31

The improvement rate is computed using $(224m - (154m + 210)) \cdot 100/224m$. As the number of messages increases, this rate converges to 31.5%. Variation of the number of multiplications required according to the number of messages in Batch Dilithium 5 and Classical Dilithium 5 is shown in Fig. 8. Variation of the improvement rate (%) provided by the batch algorithm compared to the classical version depending on the number of messages can be observed in Fig. 9.

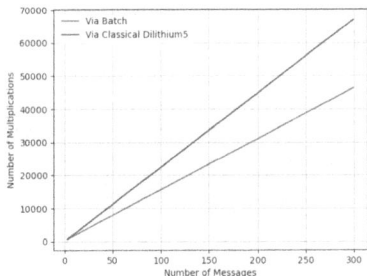

Fig. 8. Variation of the number of multiplications required according to the number of messages for Batch Dilithium 5 and Classical Dilithium 5

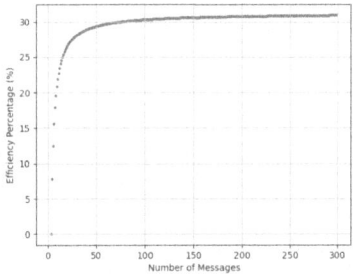

Fig. 9. Variation of the improvement rate (%) provided by the batch algorithm compared to the classical version depending on the number of messages for Dilithium 5

4.2 The Improvements Observed via the Implementations

Twenty random messages are generated to compare the batch implementations and the reference. Each message is signed individually using the reference implementation for CPU cycle counting. After that, the same messages are signed via the batch technique, and similar speed tests are performed. In this implementation, cycle counts are obtained utilizing a single core of the Intel Core i7-8700 processor.

Table 12 gives the cycle counts of only the multiplication process of signing 20 random messages with Classical Dilithium and Batch Dilithium. Using Batch Dilithium, 17 messages are signed, and CPU cycles are counted using the batch process. For the remaining three messages, classical Dilithium is used. The same 20 messages were also signed with Classical Dilithium. In order to make an accurate comparison, the cycle counts of signing 17 messages for Classical Dilithium are measured.

Table 12. CPU Cycle counts of the multiplication stage obtained by signing 17 random messages ($m = 20 - 3$) with reference and batch implementations of Dilithium 2, Dilithium 3, and Dilithium 5. Cycle counts are obtained on one core of Intel Core i7-8700.

Dilithium Signature Generation (Only the Multiplication Stage)				
Algorithm	Batch Size	Reference	Batch	Improvement (%)
Dilithium 2	4	2555202	1532297 (via [15])	40.03
Dilithium 2	4	2555202	1809040 (via [18])	29.20
Dilithium 3	5	4723348	3331311 (via [15])	29.47
Dilithium 5	4	5715468	4914677 (via [15])	14.01

Table 13 shows the tests' results for Dilithium's three security levels. Dilithium 2 is implemented using Strassen's algorithm and Rosowski's commutative algorithm. Since the latter has slightly better multiplicative complexity, it yields faster results. Dilithium 3 and Dilithium 5 are implemented only using Rosowski's method.

Table 13. CPU Cycle counts obtained by signing 20 random messages ($m = 20$) with reference and batch implementations of Dilithium 2, Dilithium 3, and Dilithium 5. Cycle counts are obtained on one core of Intel Core i7-8700.

Dilithium Signature Generation				
Algorithm	Batch Size	Reference	Batch	Improvement (%)
Dilithium 2	4	31382991	20645253 (via [15])	34.22
Dilithium 2	4	31382991	20991482 (via [18])	33.11
Dilithium 3	5	50128969	41407391 (via [15])	17.40
Dilithium 5	4	46078440	41833217 (via [15])	10.15

The fact that the method we have presented uses matrix-matrix multiplication instead of matrix-vector multiplication means that the entire matrix is represented instead of a single vector. Calculating and keeping the linear combinations contained in the structure of multiplication methods and obtaining entries of the matrix $\hat{W}$ by addition and subtraction operations increases stack memory consumption that can be reduced with appropriate implementation techniques (using loops to calculate linear combinations or parallelization techniques, etc.). Using matrices for the multiplication operations instead of vectors requires dynamic memory allocations for the operations to be appropriately calculated in the algorithm represented by the entire function. Thus, these create a time-memory trade-off, as expected. However, our computations show that it is not significant since the matrix is constructed by adding a few more vectors. It should also be noted that the proposed method can be implemented using SIMD instructions such as AVX2. This approach can significantly enhance performance and could be considered for future work.

5 Conclusion

This study proposes efficient batch signature generation with the Dilithium algorithm. The batch Dilithium signing algorithm is designed to enable many messages to be signed simultaneously. According to the repetition numbers of Dilithium 2, Dilithium 3, and Dilithium 5 signing algorithms, the column sizes of the first matrix (i.e., batch numbers) are determined as 4, 5, and 4, respectively. Dilithium's matrix-vector multiplication has been converted to matrix-matrix multiplication. This transformation allows us to employ efficient matrix-matrix multiplications for many signature generations. The matrix dimensions that provide improvements are determined, and efficient multiplication methods, such as commutative matrix multiplication by Rosowski and Strassen's multiplication algorithms, are integrated into Dilithium. As a result of those multiplications, the arithmetic complexities of generating many signatures are enhanced up to 28.1% for Dilithium 2, 33.3% for Dilithium 3, and 31.5% for Dilithium 5. Moreover, we implement the proposed batch signature generation by signing 20 messages using the efficient matrix-matrix multiplication algorithms for three security levels and obtain improvements in terms of CPU cycle counts, which are 34.22% for Dilithium 2, 17.40% for Dilithium 3, and 10.15% for Dilithium 5.

Acknowledgments. We would like to extend our heartfelt gratitude to Sevim Seda Odacıoğlu for her invaluable assistance in the implementation phase of this paper.

Availability of the software. All source code is available at https://github.com/denizzzture/Efficient-Batch-Dilithium.

References

1. Aguilar-Melchor, C., et al.: Batch signatures, revisited. Cryptology ePrint Archive (2023)
2. Alagic, G., et al.: Status report on the third round of the NIST post-quantum cryptography standardization process. US Department of Commerce, NIST (2022)
3. Alkim, E., et al.: Frodokem learning with errors key encapsulation. NIST PQC Stand.: Round **3** (2020)
4. Benjamin, D.: Batch signing for TLS. Internet Engineering Task Force, Internet-Draft Draft-Davidben-TLS-Batch-Signing-02 (2019)
5. Bos, J.W., Ofner, M., Renes, J., Schneider, T., van Vredendaal, C.: The matrix reloaded: Multiplication strategies in frodokem. In: Cryptology and Network Security: 20th International Conference, CANS 2021, Vienna, Austria, 13–15 December 2021, Proceedings 20, pp. 72–91. Springer (2021)
6. Cenk, M., Hasan, M.A.: On the arithmetic complexity of Strassen-like matrix multiplications. J. Symb. Comput. **80**, 484–501 (2017)
7. Chang, Y.S., Wu, T.C., Huang, S.C.: ElGamal-like digital signature and multisignature schemes using self-certified public keys. J. Syst. Softw. **50**(2), 99–105 (2000)
8. Ducas, L., et al.: Crystals-dilithium algorithm specifications and supporting documentation (version 3.1) (2021)
9. Elgamal, T.: A public key cryptosystem and a signature scheme based on discrete logarithms. IEEE Trans. Inf. Theory **31**(4), 469–472 (1985). https://doi.org/10.1109/TIT.1985.1057074
10. Fiat, A.: Batch RSA. In: Advances in Cryptology-CRYPTO 1989 Proceedings 9, pp. 175–185. Springer (1990)
11. Hwang, S.J., Lee, Y.H.: Repairing ElGamal-like multi-signature schemes using self-certified public keys. Appl. Math. Comput. **156**(1), 73–83 (2004)
12. Moriarty, K., Kaliski, B., Jonsson, J., Rusch, A.: PKCS #1: RSA Cryptography Specifications Version 2.2. RFC 8017 (2016). https://doi.org/10.17487/RFC8017
13. Paquin, C., Stebila, D., Tamvada, G.: Benchmarking post-quantum cryptography in TLS. In: Ding, J., Tillich, J.-P. (eds.) PQCrypto 2020. LNCS, vol. 12100, pp. 72–91. Springer, Cham (2020). https://doi.org/10.1007/978-3-030-44223-1_5
14. Pavlovski, C., Boyd, C.: Efficient batch signature generation using tree structures. In: International workshop on cryptographic techniques and E-commerce, CrypTEC. vol. 99, pp. 70–77. Citeseer (1999)
15. Rosowski, A.: Fast commutative matrix algorithms. J. Symbolic Comput. **114**, 302–321 (2023). https://doi.org/10.1016/j.jsc.2022.05.002, https://www.sciencedirect.com/science/article/pii/S0747717122000499
16. Shor, P.W.: Polynomial-time algorithms for prime factorization and discrete logarithms on a quantum computer. SIAM Rev. **41**(2), 303–332 (1999)
17. Sikeridis, D., Kampanakis, P., Devetsikiotis, M.: Post-quantum authentication in TLS 1.3: a performance study. Cryptology ePrint Archive (2020)
18. Strassen, V., et al.: Gaussian elimination is not optimal. Numer. Math. **13**(4), 354–356 (1969)
19. Tanwar, S., Kumar, A.: An efficient and secure identity based multiple signatures scheme based on rsa. Journal of Discrete Mathematical Sciences and Cryptography **22**(6), 953–971 (2019)
20. Winograd, S.: On multiplication of 2×2 matrices. Linear Algebra Appl. **4**(4), 381–388 (1971)

Ursa Minor: The Implementation Framework for Polaris

Mohammadtaghi Badakhshan[✉], Guiwen Luo, Tanmayi Jandhyala, and Guang Gong

University of Waterloo, Waterloo, Canada
{mbadakhshan,guiwen.luo,tjandhya,ggong}@uwaterloo.ca

Abstract. This paper conducts an analysis of algorithms within Polaris, a plausibly post-quantum zero-knowledge succinct non-interactive argument of knowledge (zkSNARK) protocol, by decomposing it into its construction components for detailed investigation. Recognizing the need for fast implementation in real-world applications, we introduce the Ursa Minor, an implementation framework tailored to evaluate Polaris's efficiency. Our contribution in this framework are twofold: Firstly, we proposed a concrete GKR arithmetic circuit to be integrated in Polaris. Secondly, we optimized the efficiency of FRI protocol employed in Polaris, by eliminating the field inversion operations.

Keywords: Zero-Knowledge Proof · Post-Quantum · Privacy-Preserving Application

1 Introduction

Zero-Knowledge Succinct Non-Interactive Arguments of Knowledge (zkSNARKs) have gained significant interest in academia for their ability to enable efficient and privacy-preserving verifiable computations. With the potential threat of quantum computers posing to the security of classical zkSNARKs, researchers have been exploring post-quantum secure zkSNARKs to ensure their resilience against quantum adversaries.

Polaris [9] is such a novel zkSNARK protocol that achieves plausibly post-quantum security, leveraging several construction components that are widely used in academia. Polaris is built upon an Rank-1 Constraint System (R1CS) like arithmetician over binary field, and integrates with sumcheck, GKR and FRI procotols [3,11,16]. The sumcheck protocol enables efficient zero-knowledge proofs for polynomial evaluations, and the GKR protocol extends this capability to handle general arithmetic circuits. Additionally, the FRI protocol is used to do succinct proximity proof to low-degree polynomials. Because the security of those protocols does not rely on the hardness assumption of classical math

---

Part of the research of Mohammadtaghi Badakhshan on this project is supported by the Ripple Graduate Fellows award from May 1st, 2023 to April 30th, 2024.

problems, Polaris is plausibly quantum secure. The landscape of post-quantum secure zkSNARKs continues to evolve, with ongoing research efforts focused on enhancing transparency and efficiency. Several notable constructions include Ligero [2], STARK [4], Aurora [5], Fractal [7], Spartan-RO [15].

Classic zkSNARKs, such as Groth16 [12] and Halo2 [1] built upon elliptic curve pairings, would usually show the efficiency advantages against those plausibly post-quantum secure protocols built upon GKR and FRI. However, efficient implementation of these protocols is crucial for enabling practical deployment in real-world applications. To address concerns regarding efficiency, this paper proposes the fast implementation of Polaris by emphasizing on the optimization of GKR and FRI protocols.

Contributions. Our main contributions are summarized as follows:

- In Sect. 4, we present an instantiation of the FRI protocol. This instantiation eliminates the field inversion operations in both the Commit phase and Query phase, expecting to show better efficiency.
- In Sect. 5, we present an instantiation of the GKR circuit tailored for the Polaris implementation. By designing the circuit as a satisfiability circuit, we ensure the verifiable computation of values essential for the Polaris protocol while minimizing the number of gates. This would reduce the communication overhead, and the verifier and the prover complexities.

2 Preliminaries

2.1 Algebraic Foundations

Here we first introduce the algebraic foundations of the paper. They are summarized from the existing literature.

Finite Fields. In this paper, $\mathbb{F}$ is used to denote a general finite field and $\mathbb{F}_2$ denotes the binary field $\{0, 1\}$, while $\mathbb{F}_{2^{64}}$ and $\mathbb{F}_{2^{192}}$ specifically refer to the following 2 finite fields,

$$\mathbb{F}_{2^{64}} := \mathbb{F}_2[X]/(X^{64} + X^4 + X^3 + X + 1), \tag{1}$$

the field over which we do the arithmetization for circuits. And

$$\mathbb{F}_{2^{192}} := \mathbb{F}_{2^{64}}[Y]/(Y^3 + Y + 1), \tag{2}$$

from which the FRI protocol's random challenges are picked up. Those two fields are adopted in Preon [6], a SNARK-based post-quantum secure signature scheme.

Interpolations. Given n points $\{(x_i, y_i) \in \mathbb{F}^2 \mid 0 \leq i < n\}$, we define the unique polynomial

$$P(X) := \sum_{j=0}^{n-1} a_j X^j \tag{3}$$

of degree less than n, such that

$$P(x_i) = y_i,\ 0 \leq i < n.$$

Vanishing Polynomial. For a set $H \subset \mathbb{F}$, we define $Z_H(X)$ as the following polynomial

$$\mathbb{Z}_H(X) := \prod_{x \in H}(X - x) \qquad (4)$$

that vanishes on H.

2.2 FRI

FRI protocol is a low-degree test for polynomials. It is used to determine whether a polynomial f is low-degree with respect to the size of the evaluation domain, without actually knowing f itself [3,13].

FRI protocol begins with a polynomial $f_0(X)$ and its evaluation domain L_0, which is an affine subspace in $\mathbb{F}$. The polynomial $f_0(X)$ and domain L_0 undergo a stepwise reduction process using a random folding procedure, resulting in a sequence of polynomials

$$f_0(X), f_1(X), \cdots, f_r(X) \in \mathbb{F}[X], \qquad (5)$$

and a sequence of domains

$$L_0 \supseteq L_1 \supseteq \cdots \supseteq L_r. \qquad (6)$$

Suppose d_k is the upper bound of polynomial's degree, i.e., $\deg f_k(X) < d_k$. We assume the degrees decrease with the same ratio as the domains, which means the quotients

$$\frac{d_k}{d_{k+1}} = \frac{|L_k|}{|L_{k+1}|} \qquad (7)$$

are the same, called *reduction factors*, and in this papar we always assume the reduction factors to be 2.

Let $f^{(k)}$ ($0 \leq k \leq r$) be a Reed Solomn codeword defined as follows,

$$\begin{aligned} f^{(k)} : L_k &\to \mathbb{F}, \\ x &\mapsto f_k(x). \end{aligned} \qquad (8)$$

The notation caveat should be noted here that $f_k(X)$ represents a polynomial, while $f^{(k)}$ is an array of points representing a Reed Solomn codeword. The interpolant of $f^{(k)}$ would construct the polynomial $f_k(x)$. The rate of Reed Solomn codeword is defined as

$$\rho = \frac{d_k}{|L_k|}. \qquad (9)$$

FRI is an interactive protocol containing a Commit phase and a Query phase, running for r rounds.

Commit Phase. During the k-th round of the Commit phase ($0 \leq k \leq r-1$), the prover commits to $f^{(k)}$ and the verifier has oracle access to $f^{(k)}$. In this paper, Merkle tree commitment is employed.

During the last round $k = r$, prover sends the $f^{(r)}$ to the verifier. By this point, the degree of $f_r(X)$, which is the interpolant of $f^{(r)}$, should be no more than $\rho \cdot |L_r| - 1$.

Query Phase. The verifier will validate that the prover adheres to the prescribed procedures.

Specifically, the verifier will randomly select $s^{(0)}$ from L_0, and iteratively computes a sequence of points $s^{(0)}, s^{(1)}, \cdots, s^{(r)}$. The verifier will query the value $f_{k+1}(s^{(k+1)})$, and two other points from $f^{(k)}$, and check the round consistency among those three points ($0 \leq k \leq r-1$). The verifier will repeat this check for ℓ times, and accept only when all checks pass.

If verifier accepts, it means that the degree of the original polynomial $f_0(X)$ should be no more than $\rho \cdot |L_0| - 1$.

2.3 GKR

GKR [11] is a public coin interactive proof protocol for any language computable by a log-space uniform layered (fan-in 2) arithmetic circuit $\mathcal{C}$, in which a prover $\mathcal{P}$ can run the computation and interactively prove the correctness of the result (output gate(s)) to a verifier $\mathcal{V}$. Consider a circuit with depth denoted by d and size represented by S, where the size is defined as the total number of gates. For any layer ℓ within the circuit, S_ℓ indicates the number of gates at that layer. Specifically, $\ell = 0$ corresponds to the output layer, while $\ell = d$ represents the input layer. Additionally, ν is the size of the input to the circuit $\mathcal{C}$. The communication cost is $O(S_0 + d\log(S))$, the cost of $\mathcal{V}$ is $O(\nu + d\log(S))$, and the runtime of the $\mathcal{P}$ is bounded by $O(S^3)$. In the following, we briefly describe the GKR protocol following Thaler's presentation of the protocol in [17].

Circuit Encoding. At each layer ℓ, the gates are numerically labeled in binary from 0 to $S_\ell - 1$, assuming S_ℓ is a power of two (expressed as $S_\ell = 2^{k_\ell}$). The functions $\mathsf{in1}_\ell$ and $\mathsf{in2}_\ell$, each defined as $\mathsf{in1}_\ell, \mathsf{in2}_\ell : \{0,1\}^{k_\ell} \to \{0,1\}^{k_{\ell+1}}$, map a binary gate label at layer ℓ to its input gates at layer $\ell+1$. This mapping explicitly encodes the wiring-how outputs from gates at layer $\ell + 1$ serve as inputs to a gate at layer ℓ. Accordingly, the functions add_ℓ and mult_ℓ, representing addition and multiplication gates at layer ℓ, are defined as

$$\mathsf{add}_\ell, \mathsf{mult}_\ell : \{0,1\}^{k_\ell} \times \{0,1\}^{k_{\ell+1}} \times \{0,1\}^{k_{\ell+1}} \to \{0,1\}.$$

For a gate labeled a at layer ℓ, these functions take as input the labels of three gates (a, b, c), and return 1 if and only if $(b, c) = (\mathsf{in1}_\ell(a), \mathsf{in2}_\ell(a))$. $\widetilde{\mathsf{add}}_\ell$ and $\widetilde{\mathsf{mult}}_\ell$ denote the Multilinear Extension (MLE) of add_ℓ and mult_ℓ. Additionally, the function $W_\ell : \{0,1\}^{k_\ell} \to \mathbb{F}$, maps gate at layer ℓ to the outputted value of the gate. Accordingly, $\widetilde{W}_\ell$ denote the MLE of W_ℓ. The following equation describes how $\widetilde{W}_\ell$ can be derived form $\widetilde{W}_{\ell+1}$, $\widetilde{\mathsf{add}}_\ell$, and $\widetilde{\mathsf{mult}}_\ell$:

$$\widetilde{W}_\ell(z) = \sum_{b,c \in \{0,1\}^{k_{\ell+1}}} \left(\widetilde{\mathsf{add}}_\ell(z,b,c) \left(\widetilde{W}_{\ell+1}(b) + \widetilde{W}_{\ell+1}(c) \right) \right.$$
$$\left. + \widetilde{\mathsf{mult}}_\ell(z,b,c) \left(\widetilde{W}_{\ell+1}(b) \cdot \widetilde{W}_{\ell+1}(c) \right) \right).$$

Multivariate Sum-Check. The GKR protocol consists of an iteration for each layer. In the iteration corresponding to layer $\ell < d$ of the circuit, $\mathcal{P}$ claims a specific value for $\widetilde{W}_\ell(r_\ell)$, with $r_\ell \in \mathbb{F}^{k_\ell}$ being a randomly selected point. Note that r_ℓ may have non-Boolean entries. In order to check the claim, $\mathcal{P}$ and $\mathcal{V}$ cooperate in a multivariate sum-check protocol [14] to the polynomial $f_{r_\ell}^{(\ell)}$ defined as

$$f_{r_\ell}^{(\ell)}(b,c) = \widetilde{\mathsf{add}}_\ell(r_\ell,b,c) \left(\widetilde{W}_{\ell+1}(b) + \widetilde{W}_{\ell+1}(c) \right)$$
$$+ \widetilde{\mathsf{mult}}_\ell(r_\ell,b,c) \left(\widetilde{W}_{\ell+1}(b) \cdot \widetilde{W}_{\ell+1}(c) \right).$$

Given that $\mathcal{V}$ does not know the polynomial $f_{r_\ell}^{(\ell)}$, to evaluate $f_{r_\ell}^{(\ell)}(b^*,c^*)$ in the final round of the sum-check protocol at a randomly chosen point $(b^*,c^*) \in \mathbb{F}^{k_{\ell+1}} \times \mathbb{F}^{k_{\ell+1}}$, $\mathcal{V}$ asks $\mathcal{P}$ to provide $z_1 = \widetilde{W}_{\ell+1}(b^*)$ and $z_2 = \widetilde{W}_{\ell+1}(c^*)$, which are then verified in the subsequent iteration $(i+1)$ through the *round consistency check* process. However, $\mathcal{V}$ can independently evaluate $\widetilde{\mathsf{add}}_\ell(r_\ell,b^*,c^*)$ and $\widetilde{\mathsf{mult}}_\ell(r_\ell,b^*,c^*)$, according to the circuit's structure.

Round Consistency Check. To verify z_1 and z_2, $\mathcal{V}$ aims to *reduce* these verifications into a single task: validating the $\mathcal{P}$'s claim of $\widetilde{W}_{\ell+1}(r_{\ell+1})$. This claim represents the summation outcome at the next layer denoted as

$$\widetilde{W}_{\ell+1}(r_{\ell+1}) = \sum_{b,c \in \{0,1\}^{k_{\ell+2}}} f_{r_{\ell+1}}^{(\ell+1)}(b,c).$$

To do so, let define the unique line $\lambda : \mathbb{F} \to \mathbb{F}^{k_{\ell+1}}$, such that $\lambda(0) = b^*$ and $\lambda(1) = c^*$. $\mathcal{P}$ then sends the polynomial $q = \widetilde{W}_{\ell+1}\big|_\lambda$ to $\mathcal{V}$, representing the restriction of $\widetilde{W}_{\ell+1}$ to the line λ. Upon receiving the polynomial, $\mathcal{V}$ first verifies that $q(0) = z_1$ and $q(1) = z_2$. Subsequently, $\mathcal{V}$ selects a random $r^* \in \mathbb{F}$ and sets $r_{\ell+1} = \lambda(r^*)$, then checks if $q(r^*) = \widetilde{W}_{\ell+1}(r_{\ell+1})$, which is the claim made by $\mathcal{P}$ for the next round.

Final Round Check. In the final round, $\mathcal{V}$ independently checks $\widetilde{W}_d(r_d)$.

3 Polaris

Polaris [9] is a zkSNARK protocol without a trusted setup that has quasi-linear time complexity for the prover and polylogarithmic proof size. Its verification time is relative to the size of the arithmetic circuit representing the statement to be proven. It achieves this efficiency by encoding the R1CS instance as a univariate polynomial in a quadratic arithmetic program (QAP) [10]. The main source of Polaris' efficiency is an arithmetic layered circuit design that allows the verifier to delegate query computations to the prover and verify the results using the GKR protocol [11]. Polaris combines univariate polynomial encoding described in Sect. 3.2 with univariate sumcheck protocol in Aurora [5]. By accomplishing this, the protocol constructs an interactive proof that is complete and sound, and then extends it to incorporate zero-knowledge and non-interativeness using Fiat-Shamir protocol [8]. Figure 1 below shows the building blocks of the Polaris protocol.

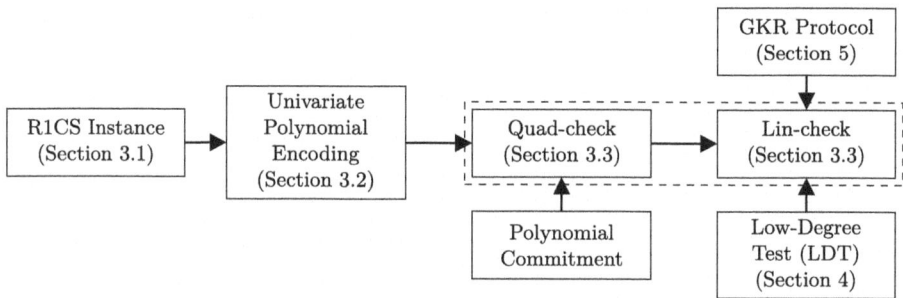

Fig. 1. Structure of the Polaris protocol with key sub-protocols.

3.1 R1CS Instance

The R1CS instance can be represented as a tuple $x = (\mathbb{F}, A, B, C, \mathbf{v}, m, n)$, where $\mathbb{F}$ is the finite field, A, B are input matrices and C is the output matrix of degree $m \times m$ from the R1CS construction. $\mathbf{v}$ represents the vector containing the instance's public parameter. There are at most n non-zero entries in each matrix.

An R1CS relation $\mathcal{R}_{\text{R1CS}}$ is said to be satisfiable if there exists a witness $\mathbf{w} \in \mathbb{F}^{m-|v|-1}$ consisting of the circuit's private input and wire values such that

$$(A\mathbf{z}) \circ (B\mathbf{z}) = C\mathbf{z},$$

where $\mathbf{z} := (1, \mathbf{w}, \mathbf{v})$ and "$\circ$" denotes the Hadamard product of the two vectors $A\mathbf{z}$ and $B\mathbf{z}$. This relation $\mathcal{R}_{\text{R1CS}}$ represents the input and output vectors of the gates in the arithmetic circuit.

3.2 Univariate Polynomial Encoding

Let H be an s-dimensional affine space of $\mathbb{F}$ such that $|H| = m$. The vector $\mathbf{z} = (1, \mathbf{v}, \mathbf{w}) \in \mathbb{F}^H$ is interpreted as a univariate function $Z : H \to \mathbb{F}$. This allows the accessing of any element of vector $\mathbf{z}$ using an index in H. The function $F_{\mathbf{w}}(\cdot)$ encodes $\mathbf{z}$, which contains the private witness $\mathbf{w}$, into a polynomial form for use in the protocol. It is defined as

$$F_{\mathbf{w}}(X) = \left(\sum_{y \in H} A(X, y) \cdot Z(y)\right) \cdot \left(\sum_{y \in H} B(X, y) \cdot Z(y)\right) - \left(\sum_{y \in H} C(X, y) \cdot Z(y)\right).$$

A witness-instance pair $(\mathbf{x}, \mathbf{w})$ is deemed valid, i.e., $(\mathbf{x}, \mathbf{w}) \in \mathcal{R}_{\text{R1CS}}$ if and only if $F_{\mathbf{w}}(x) = 0$ for any $x \in H$. Polaris utilises the polynomial extension of $F_{\mathbf{w}}(\cdot)$, for its arithmetisation in the protocol. We assign

$$\bar{A}(X) = \sum_{y \in H} A(X, y) \cdot Z(y),$$

$$\bar{B}(X) = \sum_{y \in H} B(X, y) \cdot Z(y),$$

$$\bar{C}(X) = \sum_{y \in H} C(X, y) \cdot Z(y).$$

To find the coefficients of the three polynomials, we can compute the vector products $A\mathbf{z}, B\mathbf{z}$ and $C\mathbf{z}$, and then interpolate those vector results over H. Given that A, B and C are sparse matrices, the prover employs *sparse encoding* approach for efficiently finding the coefficient of each $\bar{A}$, $\bar{B}$, and $\bar{C}$. To do so, we define three functions for each matrix M, where $M \in \{A, B, C\}$. These functions, row, col : $[n] \to H$, and val : $[n] \to \mathbb{F}$, respectively map from the set of indices $[n]$ to the row and column indices, and to the value of the non-zero entries within the matrix M. Here $[n] := \{1, 2, \cdots, n\}$. Thus, for every $x \in H$,

$$\bar{M}(x) = \sum_{\substack{i \in [n] \text{ s.t. row}(i) = x}} \text{val}(i) \cdot Z(\text{col}(i)).$$

We can represent any sequence of length N over $\mathbb{F}$ as a univariate polynomial by using Lagrange interpolation. For a sequence $\{f_i\}$, applying Lagrange interpolation would result in a polynomial f, where $f(\rho(i)) = f_i$. Here, $\rho(i) = i_0\alpha_0 + \cdots + i_{s-1}\alpha_{s-1}$ with $(i_0, \cdots, i_{s-1})$ as the binary representation of i. The points used in the Lagrange interpolation are $(\rho(i), f_i), 0 \leq i < 2^s - 1$. In other words, we have

$$f(x) = \sum_{i=0}^{2^s-1} f_i \sigma_i(x), \tag{10}$$

where $\{\sigma_i \mid 0 \leq i < 2^s\}$ is the Lagrange basis, given by

$$\sigma_i(x) = \frac{\prod_{j \neq i}(x - \rho(j))}{\prod_{j \neq i}(\rho(i) - \rho(j))}. \tag{11}$$

Since $f(x)$ is a polynomial with coefficients in $\mathbb{F}$, we can naturally extend f from a function mapping $N_{2^s} \to \mathbb{F}$ to a function over $\mathbb{F}$.

Given the definition of the vanishing polynomial in Equation (4), we can further represent $\mathbb{Z}_H(x)$, an affine linearized polynomial of $\mathbb{F}$ with dimension s, as below,

$$\mathbb{Z}_H(x) = x^{2^k} + \sum_{i=1}^{s} c_i x^{2^{i-1}} + c_0, c_i \in \mathbb{F}.$$

If H is linear, then $c_0 = 0$. Note that $c_1 \neq 0$, since $\mathbb{Z}_H(x)$ has no repeated roots. This also means that $\mathbb{Z}_H$ has degree $|H|$. To perform univariate encoding, Polaris makes use of the following bivariate polynomial:

$$\Delta_H(x, y) := \frac{\mathbb{Z}_H(x) - \mathbb{Z}_H(y)}{x - y}, \tag{12}$$

which is a polynomial of degree $|H| - 1$ because $\mathbb{Z}_H(x) - \mathbb{Z}_H(y)$ is divisible by $x - y$. With the above notation, the Lagrange basis in 11 is given as

$$\sigma_i(x) = \frac{\Delta_H(x, \rho(i))}{c_1} = \frac{1}{c_1} \frac{\mathbb{Z}_H(x)}{x - \rho(i)}, 0 \leq i < 2^s,$$

and (10) now becomes

$$f(x) = \frac{1}{c_1} \sum_{i \in N_{2^s}} f_i \Delta_H(x, \rho(i)) = \frac{1}{c_1} \sum_{i \in N_{2^s}} f_i \frac{\mathbb{Z}_H(x)}{x - \rho(i)}. \tag{13}$$

Equation (13) represents a univariate polynomial leveraged from bivariate interpolation.

3.3 Quadratic and Linear Checks

In the Polaris protocol, the prover $\mathcal{P}$ wants to convince the verifier $\mathcal{V}$ that $(\mathbf{x}, \mathbf{w}) \in \mathcal{R}_{\text{R1CS}}$ which is true if and only if the univariate $\mathbb{F}_{\mathbf{w}}(x) = 0$ at all points within the affine subspace H of $\mathbb{F}$. According to the factor theorem (Theorem 1 in [9]), this verification is the same as determining if there's a polynomial $G(X)$ where $\deg(F_{\mathbf{w}}) - |H| \leq |H| - 2$, such that

$$F_{\mathbf{w}}(X) = \mathbb{Z}_H(X) \cdot G(X). \tag{14}$$

This is verified through two distinct checks, namely, the Quad-Check and the Lin-Check, as presented in the following sections.

Quad-Check. In the Quad-Check protocol, $\mathcal{V}$ conducts a probabilistic check of Equation (14) at a randomly selected point. To realize this, $\mathcal{P}$ commits to the polynomial $G(X)$ using FRI univariate polynomial commitment and sends the commitment to $\mathcal{V}$. Section 4 presents the instantiation of the FRI commitment scheme in our protocol. Then, $\mathcal{V}$ sends the randomly selected point $r_x \in F \setminus H$

to $\mathcal{P}$. Then, the prover evaluates $\eta = G(r_x)$ and sends the result to $\mathcal{V}$. Finally, $\mathcal{V}$ query the commitment to verify the evaluation.

To verify $F_w(r_x) = \mathbb{Z}_H(r_x) \cdot G(r_x)$, the verifier also needs to compute $F_w(r_x)$. Recall that
$$F_w(r_x) = \bar{A}(r_x) \cdot \bar{B}(r_x) - \bar{C}(r_x).$$
The prover makes three separate claims to $\mathcal{V}$, say that $\bar{A}(r_x) = v_A$, $\bar{B}(r_x) = v_B$, and $\bar{C}(r_x) = v_C$. Then the verifier $\mathcal{V}$ can check if
$$v_A \cdot v_B - v_C = G(r_x) \cdot \mathbb{Z}_H(r_x).$$
This equation is represents the quadratic-check of the R1CS instance, or *quad-check*, owing to it checking the R1CS verifiability in $\rho(i)$, where i is within the range of the matrix degree m. But the verifier must now additionally verify three new claims from the prover $\mathcal{P}$:
$$\bar{A}(r_x) = v_A, \bar{B}(r_x) = v_B, \text{ and } \bar{C}(r_x) = v_C. \tag{15}$$
To do so, $\mathcal{P}$ and $\mathcal{V}$ engage in the Lin-check protocol.

Lin-Check. In the Lin-Check protocol, the three claims presented in Equation (15) are combined into one to be evaluated in a single claim. Accordingly, the verifier chooses $r_A, r_B, r_C \in \mathbb{F}$ uniformly at random and sends them to the prover, which can test if
$$c = r_A \cdot \bar{A}(r_x) + r_B \cdot \bar{B}(r_x) + r_C \cdot \bar{C}(r_x). \tag{16}$$
We can rewrite c as follows.
$$c = r_A \cdot \sum_{y \in H} A(r_x, y) \cdot Z(y) + r_B \cdot \sum_{y \in H} B(r_x, y) \cdot Z(y) + r_C \cdot \sum_{y \in H} C(r_x, y) \cdot Z(y)$$
$$= \sum_{y \in H} \left(r_A \cdot A(r_x, y) + r_B \cdot B(r_x, y) + r_C \cdot C(r_x, y) \right) \cdot Z(y).$$
where $y \in H$. We denote
$$Q_{r_x}(Y) := \left(r_A \cdot A(r_x, Y) + r_B \cdot B(r_x, Y) + r_C \cdot C(r_x, Y) \right) \cdot Z(Y). \tag{17}$$
Consequently, to verify $c = \sum_{y \in H} Q_{r_x}(y)$, $\mathcal{P}$ and $\mathcal{V}$ engage in the univariate sumcheck protocol realized by the FRI low degree test (LDT). Section 4 presents our instantiation of the FRI LDT protocol. At the end of the FRI protocol, $\mathcal{V}$ needs oracle access to the evaluations of $Q_{r_x}(\cdot)$ at κ points $r_y \in L$, where L is an affine subspace $L \subseteq \mathbb{F}$ such that $|L| > 2|H|$ and $L \cap H = \varnothing$.

We denote the evaluations of $A(r_x, r_y)$, $B(r_x, r_y)$, and $C(r_x, r_y)$ in $Q_{r_x}(r_y)$ as $M(r_x, r_y)$ where $M \in \{A, B, C\}$ using row(i), col(i), and val(i) functions as the following equation:
$$M(r_x, r_y) = \frac{\mathbb{Z}_H(r_x) \cdot \mathbb{Z}_H(r_y)}{c_1^2} \cdot \sum_{i \in [n]} \frac{\text{val}(i)}{(r_x - \text{row}(i))(r_y - \text{col}(i))}, \tag{18}$$

where we denote the second part as $C_M(r_x, r_y)$, such that

$$C_M(r_x, r_y) := \sum_{i \in [n]} \frac{\mathsf{val}(i)}{(r_x - \mathsf{row}(i))(r_y - \mathsf{col}(i))}. \tag{19}$$

Given that $\mathsf{row}(i), \mathsf{col}(i) \in H$, $r_x \in \mathbb{F} \backslash H$ and $r_y \in L$, such that $L \cap H = \emptyset$, the denominators in $C_M(r_x, r_y)$ are non-zero. To reduce the computational burden on the verifier, the task of evaluating $C_M(r_x, r_y)$ at κ points $r_y \in L$ is not performed locally. Instead, this computation is outsourced to the prover. The verifier then uses the GKR protocol to validate the results provided by the prover. In Sect. 5, we explain how we utilized the GKR protocol and designed the corresponding circuit for $C_M(r_x, r_y)$.

3.4 Adding Zero-Knowledge

The interactive protocol reveals information about the witness $\mathbf{w}$ when $\mathcal{P}$ sends evaluations of $G(r_x), \bar{A}(r_x), \bar{B}(r_x), \bar{C}(r_x)$ and $Z(\cdot)$, and invokes the low degree test on related polynomials of $((Q_{r_x}(Y))$. To prevent these "leakages" and achieve zero-knowledge, three main modifications are employed:

1. **Eliminating leakage of queries on $Z(\cdot)$.** The prover chooses a random polynomial $R_Z(\cdot)$ of degree κ and computes a $\widetilde{Z}(Y) := Z(Y) + \mathbb{Z}_H(Y) \cdot R_Z(Y)$. Even though $\widetilde{Z}(y) = Z(y)$ for $y \in H$, $\widetilde{Z}(\cdot)$ polynomial evaluations outside H preserve zero knowledge as $R_Z(\cdot)$ masks the information of polynomial $Z(\cdot)$.
2. **Modifications to the evaluations of $\bar{A}(\cdot), \bar{B}(\cdot), \bar{C}(\cdot)$.** The prover $\mathcal{P}$ samples some random polynomials $R_A(\cdot), R_B(\cdot), R_C(\cdot)$ of degree $|H| - 1$ and provides that

$$\widetilde{A}(X) := \sum_{y \in H} A(X, y) \cdot \widetilde{Z}(y) + \mathbb{Z}_H(X) \cdot \sum_{y \in H} R_A(y),$$

$$\widetilde{B}(X) := \sum_{y \in H} B(X, y) \cdot \widetilde{Z}(y) + \mathbb{Z}_H(X) \cdot \sum_{y \in H} R_B(y),$$

$$\widetilde{C}(X) := \sum_{y \in H} C(X, y) \cdot \widetilde{Z}(y) + \mathbb{Z}_H(X) \cdot \sum_{y \in H} R_C(y).$$

As the $R(\cdot)$ are random, revealing the evaluations of $\widetilde{A}(\cdot), \widetilde{B}(\cdot), \widetilde{C}(\cdot)$ outside H does not leak information about the values in the witness w.
3. **Modifications to the polynomial $Q(r_x)(\cdot)$.** To uphold the zero-knowledge property of the $Q_{r_x}(\cdot)$ polynomial from Equation (17) in the univariate sumcheck phase, the prover $\mathcal{P}$ picks a random polynomial $S_Q(\cdot)$ of degree $2|H| + \kappa - 1$, and sends an $s_1 = \sum_{y \in H} S_Q(y)$ to $\mathcal{V}$. To this, $\mathcal{V}$ responds with a random challenge $\alpha_1 \in \mathbb{F}$. $\mathcal{P}$ and $\mathcal{V}$ then run a sumcheck on the following linearised representation:

$$\alpha_1 \cdot \widetilde{c} + s_1 = \sum_{y \in H} (\alpha_1 \cdot \widetilde{Q}r_x(y) + S_Q(y)).$$

where $\tilde{c} = r_A \cdot \widetilde{A}(r_x) + r_B \cdot \widetilde{B}(r_x) + r_C \cdot \widetilde{C}(r_x)$, referenced from Equation (16). This ensures $\tilde{c}$ and s_1 can be correctly computed due to the random linear combination of the sumcheck, while revealing no information about $\widetilde{Q}r_x(\cdot)$ as it is masked by the random polynomial $S_Q(\cdot)$ [5] [19].

To obtain the full zero-knowledge protocol, we replace relevant components with their zero-knowledge versions, with the additional need for the prover $\mathcal{P}$ to commit to the random polynomials using Merkle tree commitments at the beginning, to be later opened at κ points by $\mathcal{V}$. A key advantage is that the GKR protocol remains unchanged, avoiding expensive cryptographic computations.

4 FRI Instantiation

In the FRI protocol, let r be the number of rounds, let $\{\beta_0 = 1, \beta_1, \beta_2, \cdots, \beta_{191}\}$ be one of the $\mathbb{F}_2$-basis of $\mathbb{F}_{2^{192}}$ defined in Sect. 2.1.

Evaluation Domains. We assume that the verifier and prover have agreed upon the evaluation domains L_k ($0 \leq k \leq r$), whose sizes are the powers of 2, and $m = \log_2(|L_0|)$. Those affine subspaces are adopted in Preon [6]. The evaluation domains are recursively defined as follows. First,

$$L_0 = <\beta_0, \beta_1, \cdots, \beta_{m-1}> + \beta_m. \tag{20}$$

For an integer i ($0 \leq i \leq 2^m - 1$), its binary expression is

$$i = (i_{m-1} i_{m-2} \cdots i_1 i_0)_2 = i_0 + i_1 \cdot 2 + \cdots + i_{m-1} \cdot 2^{m-1}, i_j \in \{0, 1\}.$$

Then we can define the i-th element in L_0 as

$$L_0[i] = (i_0 \cdot \beta_0 + i_1 \cdot \beta_1 + \cdots + i_{m-1} \cdot \beta_{m-1}) + \beta_m, \ 0 \leq i < 2^m. \tag{21}$$

Let us define the polynomial

$$q_0(X) = X(X - \beta_0) \tag{22}$$

and

$$\beta_j^{(1)} = q_0(\beta_{j+1}), \ 0 \leq j \leq m-1,$$

then L_1 is defined as

$$L_1 = q_0(L_0) = <\beta_0^{(1)}, \beta_1^{(1)}, \cdots, \beta_{m-2}^{(1)}> + \beta_{m-1}^{(1)}, \tag{23}$$

and the i-th element in L_1 is

$$L_1[i] = (i_0 \cdot \beta_0^{(1)} + i_1 \cdot \beta_1^{(1)} + \cdots + i_{m-2} \cdot \beta_{m-2}^{(1)}) + \beta_{m-1}^{(1)}, \ 0 \leq i < 2^{m-1}.$$

We can thus recursively define that, for $1 \leq k \leq m-1$,

$$q_k(X) = X(X - \beta_0^{(k)}),$$
$$\beta_j^{(k+1)} = q_k(\beta_{j+1}^{(k)}), \text{ for } 0 \leq j \leq m-k-1, \tag{24}$$
$$L_{k+1} = q_k(L_k) = <\beta_0^{(k+1)}, \beta_1^{(k+1)}, \cdots, \beta_{m-k-2}^{(k+1)}> + \beta_{m-k-1}^{(k+1)},$$

and for $0 \leq i < 2^{m-k-1}$, the i-th element of L_{k+1} is defined as

$$L_{k+1}[i] = (i_0 \cdot \beta_0^{(k+1)} + i_1 \cdot \beta_1^{(k+1)} + \cdots + i_{m-1} \cdot \beta_{m-k-2}^{(k+1)}) + \beta_{m-k-1}^{(k+1)}.$$

FRI Commit Phase.
Prover's input: $f^{(0)} : L_0 \rightarrow \mathbb{F}$, a purported Reed Solomn codeword corresponding to polynomial $f_0(X)$, with rate ρ.

Loop for $0 \leq k \leq r - 1$:

1. Prover commits to all codewords.
   - $f^{(k)} : L_k \rightarrow \mathbb{F}$ is recursively defined in Step 3.
   - Prover computes a Merkle commitment to $f^{(k)}$ and sends out the Merkle root.
2. Verifier sends a uniformly random $\alpha^{(k)} \in \mathbb{F}$.
3. Prover defines the codeword $f^{(k+1)}$ with domain L_{k+1}, such that for each $0 \leq i < |L_{k+1}|$,
   - It is easy to check that

   $$q_k(L_k[2i]) = q_k(L_k[2i+1]) = L_{k+1}[i].$$

   - The values in codeword $f^{(k+1)}$ is derived from the previous codeword,

   $$f_{k+1}(L_{k+1}[i]) = \frac{f_k(L_k[2i]) - f_k(L_k[2i+1])}{L_k[2i] - L_k[2i+1]}(\alpha^{(k)} - L_k[2i]) + f_k(L_k[2i]).$$

   Here the denominator $L_k[2i] - L_k[2i+1] = \beta_0^{(k)}$, whose inverse can be precomputed.

For $k = r$:

- $f^{(r)} : L_r \rightarrow \mathbb{F}$ is defined in Step 2.
- prover sends out the last codeword $f^{(r)}$.

FRI Query Phase.

1. Verifier extracts all Merkle roots, all challenges $\alpha^{(0)}, \alpha^{(1)}, \cdots, \alpha^{(r-1)}$, and the last codeword $f^{(r)}$. Verifier has oracle access to $f^{(0)}, f^{(1)}, \cdots, f^{(r-1)}$.
2. Verifier computes the interpolant $f_r(X)$ from points $f^{(r)}$, then check if the degree of $f_r(X)$ is no more than $\rho \cdot |L^{(r)}| - 1$. If not, reject.
3. Verifier does the consistency check between two neighboring codewords. Repeat ℓ times:
   - Sample random index $s^{(0)} = i$ from $0 \leq i < |L_0|$, and for $0 \leq k \leq r - 1$, compute $s^{(k+1)} = \lfloor s^{(k)}/2 \rfloor$.
   - If $s^{(r)}$ is repeated, resample $s^{(0)}$.

– **Round consistency check:**
Denote
$$x_1^{(k)} = L_k[2\,s^{(k+1)}], \ x_2^{(k)} = L_k[2\,s^{(k+1)} + 1],$$
the verifier first queries
$$f_{k+1}(L_{k+1}[s^{(k+1)}]), f_k(x_1^{(k)}), f_k(x_2^{(k)}),$$
and checks the Merkle commit paths for those three points, then checks that for every $k \in \{0, 1, 2, \cdots, r-1\}$,
$$f_{k+1}(L_{k+1}[s^{(k+1)}]) = \frac{f_k(x_1^{(k)}) - f_k(x_2^{(k)})}{x_1^{(k)} - x_2^{(k)}}(\alpha^{(k)} - x_1^{(k)}) + f_k(x_1^{(k)}). \quad (25)$$

If any one equation of the consistency check fails, reject.

Notice that $x_1^{(k)} - x_2^{(k)} = \beta_0^{(k)}$, whose inverse can be precomputed.

4. Accept if all checks pass. This implies that the degree of the original polynomial $f_0(X)$ is no more than $\rho \cdot |L_0| - 1$.

5 GKR Circuit

In this section we are going to present the circuit $\mathfrak{C}$ to provide a verifiable computation for $C_M(r_x, r_y)$. By considering that the arithmetic circuit can only include addition and multiplication operations, computing the multiplicative inverse is expensive. Therefore, we turn the straightline computation, in which we should compute the multiplicative inverse, into an *satisfiability* circuit instance [17]. Therefore, the circuit $\mathfrak{C}$ receives a set of inputs that includes $\overline{\mathsf{c}^{(d)}} = \{\mathsf{c}^{(d)}(i) \mid i \in [n]\}$, alongside r_x, r_y, $\overline{\mathsf{row}} = \{\mathsf{row}(i) \mid i \in [n]\}$, $\overline{\mathsf{col}} = \{\mathsf{col}(i) \mid i \in [n]\}$, and $\overline{\mathsf{val}} = \{\mathsf{val}(i) \mid i \in [n]\}$, where d denotes the depth of the circuit and $\mathsf{c}^{(\ell)}(i)$ is defined as follows:

$$\mathsf{c}^{(\ell)}(i) := \begin{cases} \frac{\mathsf{val}(i)}{(r_x - \mathsf{row}(i))(r_y - \mathsf{col}(i))}, & \ell = d,\ i \in [n] \%1 \leq i \leq n \\ \mathsf{c}^{(\ell+1)}(2i-1) + \mathsf{c}^{(\ell+1)}(2i), & \ell \in [0,d),\ i \in [2^{\ell-d}n] \\ 0 & \text{otherwise.} \end{cases} \quad (26)$$

This definition specifies that $\mathfrak{C}$ encompasses all terms included in the summation outlined in Eq. (19), utilizing these terms as inputs (where $\ell = d$). To enable verifiable computation of this summation, $\mathfrak{C}$ is equipped with a *summation* component structured as a binary tree. Within this structure, for any level $\ell < d$, the function $\mathsf{c}^{(\ell)}(i)$ calculates the sum of two preceding terms from the immediately lower layer (i.e., layer $\ell + 1$), using addition gates. In Eq. (26), we simplify our notation by assuming a balanced binary tree structure for ease of definition. This assumption entails that the input layer, denoted by $\mathsf{c}^{(d)}(i)$, comprises a power of two elements. Consequently, the number of layers are $d = \log n$. The output layer, serving as the tree's root, is designated by $\mathsf{c}^{(0)}(1) = C_M(r_x, r_y)$. However, this binary structure is not a strict requirement for the actual implementation. In practice, given that each addition gate necessitates two inputs, layers featuring an odd number of elements incorporate a zero as the supplemental input for the subsequent layer. This zero is consistently available at every layer of the circuit.

For $\mathfrak{C}$ to qualify as a satisfiability circuit, it must verify that each asserted term $\mathsf{c}^{(d)}(i)$ aligns with the parameters r_x, r_y, $\mathsf{row}(i)$, $\mathsf{col}(i)$, and $\mathsf{val}(i)$. Consequently, we need to design a *consistency* component embedded to the circuit. The number of layers $d = \log n$ is determined by the summation component presented above.

Layer $\ell = d-1$: (The immediate layer beyond the input layer) $\alpha(i) = r_x - \mathsf{row}(i)$, $\beta(i) = r_y - \mathsf{col}(i)$, where $\overline{\alpha} = \{\alpha(i) \mid i \in [n]\}$ and $\overline{\beta} = \{\beta(i) \mid i \in [n]\}$ are realized by addition gates.

Layer $\ell = d-2$: $\gamma(i) = \alpha(i)\beta(i)$ for $i \in [n]$, where $\overline{\gamma} = \{\gamma(i) \mid i \in [n]\}$ is realized by multiplication gates

Layer $\ell = d-3$: $\mathsf{val}'(i) = \gamma(i)\mathsf{c}^{(d)}(i)$, where $\mathsf{c}^{(d)}(i)$ is copied to this layer from the input layer by being added to zero in previous two layers. This zero is consistently available at every layer of the circuit. $\overline{\mathsf{val}'} = \{\mathsf{val}'(i) \mid i \in [n]\}$ is realized by multiplication gates.

Layer $\ell = d-4$: $\zeta(i) = \mathsf{val}'(i) + \mathsf{val}(i)$, where $\mathsf{val}(i)$ is also copied from the input layer to this layer by being added to zero. This layer performs an addition of $\mathsf{val}'(i)$ to $\mathsf{val}(i)$, such that, when $\mathsf{val}'(i)$ equals $\mathsf{val}(i)$, $\zeta(i) = 0$, since the addition is equivalent to the XOR operation in this field. This property ensures $\mathsf{val}'(i)$ and $\mathsf{val}(i)$ are equal. $\overline{\zeta} = \{\zeta(i) \mid i \in [n]\}$ is realized by addition gates.

Layers $\ell < d-4$: $\overline{\zeta}$ is copied to the upper layer till the output layer.

Figure 2 illustrates the GKR circuit used in Polaris. For clarity, only connections from the input layer are shown, as the other connections follow a similar pattern and have been omitted from this visual representation.

The summation component of the presented GKR circuit consists of $2n - 1$ gates, under the assumption that it forms a balanced binary tree (i.e., n is a power of two). If this condition is not met, the circuit will have a number of gates that is close to this figure. The consistency component of the circuit has $(n + 1) \log n + 7n + 2$ gates. Consequently the size of the circuit is $S = (n + 1) \log n + 9n + 1$ ($2n$ of them are multiplication and the rest are addition gates). The prover only needs to send $\mathsf{c}^{(0)}(1)$ to the verifier as the output of the circuit because the verifier assumes that the other $n + 1$ gate values should be zero; therefore, $S_0 = 1$ (S_0 is the size of the output layer). Note that the depth of the circuit is $d = \log(n)$. According to Sect. 2.3, the communication cost is $O(S_0 + d\log(S)) = O(\log(n) \cdot \log((n+1)\log n + 9n))$ which simplifies to $O(\log(n) \cdot \log(n \log n)) = O((\log(n))^2)$.

The input size to the circuit is $\nu = 4n + 3$. According to Sect. 2.3, the verifier's computation cost should be $O(\nu + d\log(S)) = O(n + \log(n) \cdot (n \log(n))) = O(n(\log(n))^2)$. However, in the implementation of the GKR protocol, this computation can be delegated to the prover via verifiable polynomial delega-

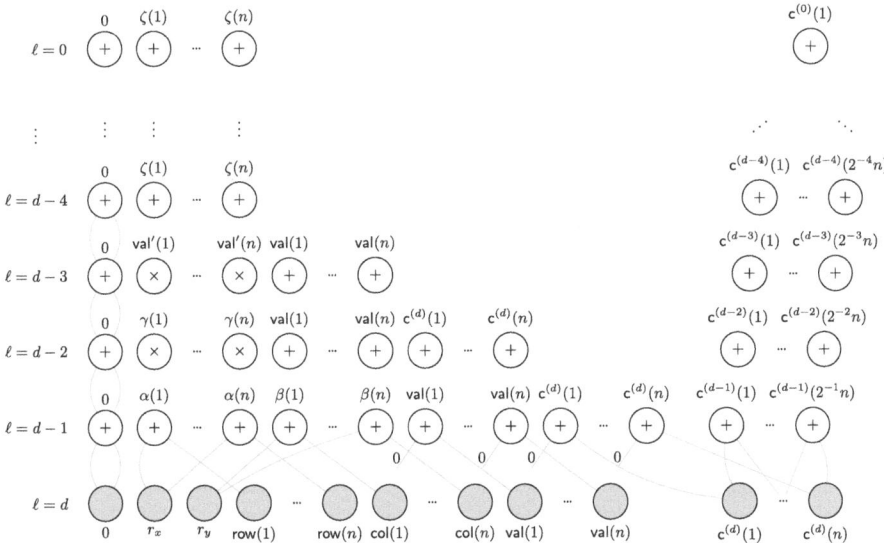

Fig. 2. The GKR satisfiability circuit for calculating $C_M(r_x, r_y)$ (Equation (19)). Sample connections are shown for clarity. $\forall i, \zeta(i) = 0$, and $\mathsf{c}^{(0)}(1) = C_M(r_x, r_y)$.

tion (VPD) schemes such as Virgo [18]. Finally, the prover's computation is $O(S^3) = O((n \log(n))^3)$.

6 Conclusion

In this paper, we introduced the Polaris protocol and its construction components. We presented an instantiation of the GKR circuit to minimize the number of gates, with the aim to reduce communication overhead, as well as the verifier and prover's complexities. We also explained an instantiation of the FRI protocol that eliminates the field inversion operations in both the Commit phase and Query phase, while would help to achieve better efficiency. The complete performance benchmark and the expected improvements will be demonstrated in the upcoming full implementation of Polaris.

References

1. The halo2 Book. https://zcash.github.io/halo2/. Accessed 19 Mar 2024
2. Ames, S., Hazay, C., Ishai, Y., Venkitasubramaniam, M.: Ligero: lightweight sublinear arguments without a trusted setup. In: Proceedings of the 2017 ACM SIGSAC Conference on Computer and Communications Security, pp. 2087–2104 (2017)
3. Ben-Sasson, E., Bentov, I., Horesh, Y., Riabzev, M.: Fast reed-solomon interactive oracle proofs of proximity. In: 45th International Colloquium on Automata, Languages, and Programming (ICALP 2018). Schloss Dagstuhl-Leibniz-Zentrum fuer Informatik (2018)

4. Ben-Sasson, E., Bentov, I., Horesh, Y., Riabzev, M.: Scalable zero knowledge with no trusted setup. In: Boldyreva, A., Micciancio, D. (eds.) CRYPTO 2019. LNCS, vol. 11694, pp. 701–732. Springer, Cham (2019). https://doi.org/10.1007/978-3-030-26954-8_23
5. Ben-Sasson, E., Chiesa, A., Riabzev, M., Spooner, N., Virza, M., Ward, N.P.: Aurora: transparent succinct arguments for R1CS. In: Ishai, Y., Rijmen, V. (eds.) EUROCRYPT 2019. LNCS, vol. 11476, pp. 103–128. Springer, Cham (2019). https://doi.org/10.1007/978-3-030-17653-2_4
6. Chen, M.S., et al.: Preon: zk-SNARK based signature scheme (2023)
7. Chiesa, A., Ojha, D., Spooner, N.: FRACTAL: post-quantum and transparent recursive proofs from holography. In: Canteaut, A., Ishai, Y. (eds.) EUROCRYPT 2020. LNCS, vol. 12105, pp. 769–793. Springer, Cham (2020). https://doi.org/10.1007/978-3-030-45721-1_27
8. Fiat, A., Shamir, A.: How to prove yourself: practical solutions to identification and signature problems. In: Conference on the Theory and Application of Cryptographic Techniques, pp. 186–194. Springer (1986)
9. Fu, S., Gong, G.: Polaris: transparent succinct zero-knowledge arguments for R1CS with efficient verifier. Proc. Priv. Enhancing Technol., 544–564 (2022). https://doi.org/10.2478/popets-2022-0027
10. Gennaro, R., Gentry, C., Parno, B., Raykova, M.: Quadratic span programs and Succinct NIZKs without PCPs. In: Johansson, T., Nguyen, P.Q. (eds.) EUROCRYPT 2013. LNCS, vol. 7881, pp. 626–645. Springer, Heidelberg (2013). https://doi.org/10.1007/978-3-642-38348-9_37
11. Goldwasser, S., Kalai, Y.T., Rothblum, G.N.: Delegating computation: interactive proofs for muggles. In: Proceedings of the Fortieth Annual ACM Symposium on Theory of Computing, STOC 2008, pp. 113–122. Association for Computing Machinery, New York (2008). https://doi.org/10.1145/1374376.1374396
12. Groth, J.: On the size of pairing-based non-interactive arguments. In: Fischlin, M., Coron, J.-S. (eds.) EUROCRYPT 2016. LNCS, vol. 9666, pp. 305–326. Springer, Heidelberg (2016). https://doi.org/10.1007/978-3-662-49896-5_11
13. Haböck, U.: A summary on the FRI low degree test. Cryptology ePrint Archive (2022)
14. Lund, C., Fortnow, L., Karloff, H., Nisan, N.: Algebraic methods for interactive proof systems. J. ACM **39**(4), 859–868 (1992)
15. Setty, S.: Spartan: efficient and general-purpose zkSNARKs without trusted setup. In: Annual International Cryptology Conference, pp. 704–737. Springer (2020)
16. Thaler, J.: A note on the GKR protocol (2015)
17. Thaler, J.: Proofs, arguments, and zero-knowledge. Now Found. Trends (2022). https://doi.org/10.1561/978163828125
18. Zhang, J., Xie, T.: Virgo: zero knowledge proofs system without trusted setup (2019). https://api.semanticscholar.org/CorpusID:221606763
19. Zhang, J., Xie, T., Zhang, Y., Song, D.: Transparent polynomial delegation and its applications to zero knowledge proof. In: 2020 IEEE Symposium on Security and Privacy (SP), pp. 859–876. IEEE (2020)

SMALL: Scalable Matrix OriginAted Large Integer PoLynomial Multiplication Accelerator for Lattice-Based Post-Quantum Cryptography

Jiafeng Xie[1](✉), Pengzhou He[1], Samira Carolina Oliva Madrigal[2], and Çetin Kaya Koç[3]

[1] Department of Electrical and Computer Engineering, Villanova University, Villanova, PA, USA
{jiafeng.xie,phe}@villanova.edu
[2] San José State University, San Jose, USA
[3] NUAA, Iğdir University, UCSB, Iğdir, Turkey
cetinkoc@ucsb.edu

Abstract. Along with the rapid development in quantum computing, more attention has been switched to post-quantum cryptography (PQC) and related research including their hardware implementations. Following this trend, this paper presents a novel strategy to implement a special type of polynomial multiplication used in lattice-based PQC, where the coefficients of two input polynomials are unequal, and modulus and polynomial size are power-of-two numbers (not in favor of deploying number theoretic transform). In particular, we have proposed a Scalable Matrix originAted Large integer poLynomial multiplication Accelerator (SMALL) for flexible and compact implementation of the targeted polynomial multiplication that is constant-time. In total, our efforts include: (i) we have formulated and derived a scalable matrix originated computation strategy for the targeted polynomial multiplication in a general format; (ii) we have then presented the detailed internal structures for the proposed polynomial multiplication accelerator based on novel algorithm-to-architecture design techniques; (iii) we have implemented the proposed accelerator based on two case study PQC schemes to demonstrate the superior efficiency of the proposed design over the state-of-the-art solutions. We hope the outcome of this work will be useful for further PQC development.

Keywords: Hardware design · lattice-based PQC · polynomial multiplication accelerator · scalable matrix originated computation

1 Introduction

In light of the fast advances in quantum technology, the post-quantum cryptography (PQC) related research and development have reached an all-time high

[23]. Quite a number of cryptographic algorithms have been proposed for possible PQC candidates, and lattice-based cryptography is regarded as one of the most important categories of algorithms due to their strong security proof and relatively easy implementation complexity [15,18]. Along with this direction of exploration, advances in the field have gradually switched to the hardware implementation side [14,23].

As polynomial multiplication over ring $\mathbb{Z}_q[x]/(x^n + 1)$ is one of the most important components for lattice-based PQC, a number of works have been released on related hardware implementation [5,20]. In particular, we notice a special type of polynomial multiplication where the coefficients of the two inputs are unequal, and the corresponding polynomial length and modulus q are numbers of power-of-two. This polynomial multiplication can be observed in one of the National Institute of Standards and Technology (NIST) PQC standardization third-round candidates Saber [1] and ring binary learning-with-errors (RBLWE)-based encryption scheme (RBLWE-ENC, a promising lightweight PQC suitable for lightweight applications) [2,4].

Motivation. Unlike the other large integer polynomial multiplications used in typical lattice-based PQC that can be implemented with popular fast algorithm number theoretic transform (NTT) [17], the targeted polynomial multiplication here is not in favor of deploying of NTT unless field extension is executed (which is out of the scope of this research). Meanwhile, due to its specific parameter sets that the two input polynomials' coefficients do not have the same size, deploying traditional fast algorithms such as Karatsuba or Toeplitz Matrix-Vector Product (TMVP) may cause the small coefficients involved operations to become larger-sized computations (more resource usage) because of the pre-addition related operations (may eventually offset the gain) [2]. As a result, the recent works based on traditional fast algorithms all have relatively large resource usage [2,11,21]. In this case, a compact design (also with flexibility) for this polynomial multiplication is highly desirable.

Proposal. Following these considerations, in this paper, we propose a novel polynomial multiplication algorithm to design the targeted accelerator, i.e., a Scalable Matrix originAted Large integer poLynomial multiplication Accelerator (SMALL). In particular, we have made three layers of innovative efforts:

(i) We present a detailed mathematical formulation process to propose a novel scalable matrix originated computation strategy for the targeted polynomial multiplication in a general format.
(ii) We then provide the architecture design process and related component details to obtain the proposed SMALL with the help of novel algorithm-to-architecture design techniques.
(iii) Finally, we give a thorough evaluation to showcase the superior performance of the proposed accelerator over the state-of-the-art solutions (based on two case study examples).

Overall Layout. The rest of the paper is arranged as follows. The proposed polynomial multiplication algorithm is formulated in Sect. 2. Details of the proposed accelerator are provided in Sect. 3. The evaluation is conducted in Sect. 4. And the conclusion is given in Sect. 5.

2 Related Work

This work follows a rather simple method from numerical linear algebra or a technique from polynomial arithmetic transformations (i.e., polynomial matrix multiplication). Specifically, the process of transforming the multiplication of two polynomials into a Matrix-Vector Product, which some works may describe as block-matrix multiplication. This type of transformation and strategies with similar decomposition is observed in numerous cryptographic works and the sciences in general and can be described as follows [7,8,12,13,16,19,22].

Consider an $(n \times n)$ matrix decomposed into v^2 total $(u \times u)$ sub-matrices and a vector decomposed into v total $(1 \times u)$ subvectors. In a similar manner, we can express the multiplication of two univariate polynomials $A(x)$ and $B(x)$ consisting of n coefficients each. First, each of these can be expressed as a finite sum of terms where each term consists of a coefficient and power according to the respective position (e.g., $A(x) = a_0 + a_1 x + a_2 x^2 + ... + a_{n-1} x^{n-1}$). Second, the product of two such polynomials can be decomposed in different ways according to the computational goal. For example, if our goal is a small Matrix-Vector Product decomposition, we can decompose, say, $A(x)$ into a sum of v smaller polynomials each of degree $u - 1$. For instance, $A(x) = (a_0 + a_1 x + ... + a_{u-1} x^{u-1}) + (a_u + a_{u+1} x + ... + a_{2u-1} x^{u-1}) + ... + (a_{uv-u} + a_{uv-u+1} x + ... + a_{uv-1} x^{u-1}) = (A_0 + A_1 + ... + A_{v-1})$. Clearly, we can multiply out with $B(x)$ to have $B(x)(A_0 + A_1 + ... + A_{v-1})$. Then, for each such sub-product (e.g., $B(x)(A_0)$) can we can decompose $B(x)$ and multiply out to have a sum of v^2 small products (e.g., $(B_0 A_0 + B_0 A_1 + ... + B_0 A_{v-1}) + (B_1 A_0 + B_1 A_1 + ... + B_1 A_{v-1}) + ... + (B_{v-1} A_0 + B_{v-1} A_1 + ... + B_{v-1} A_{v-1})$. Each such small product (e.g., $B_0 A_0$) can be expressed in Matrix-Vector Product form where B_0 accounts for multiplication with respective power terms of A_0 and becomes a $(u \times u)$ matrix and A_0 is left as a vector of coefficients or a $(u \times 1)$ vector. Ideally, we would like to explore further optimizations in addition to small component decomposition and parallelism in hardware.

3 SMALL: Proposed Algorithm

Mathematical Formulation. Without loss of generality, we can just define the targeted polynomial multiplication over ring $\mathbb{Z}_q/(x^n + 1)$ as

$$W = GD \bmod f(x), \tag{1}$$

where $f(x) = x^n + 1$, $W = \sum_{i=0}^{n-1} w_i x^i$, $G = \sum_{i=0}^{n-1} g_i x^i$, and $D = \sum_{i=0}^{n-1} d_i x^i$ (w_i (t-bit), g_i (t-bit), and d_i (h-bit) are integers over ring such that $t = \log_2 q$ and $h < t$, and the actual h and t are determined by the specific PQC scheme).

Proposed Mathematical Derivation Strategy. For efficient implementation, it will be ideal that the original polynomial multiplication can be transformed into a number of small-size sub-components, where these sub-components can be realized through the form of serial accumulation (desirable for low-complexity implementation). Following this principle, we set our mathematical formulation and derivation strategy with the following goals: (i) deriving the polynomial multiplication into the equivalent form of the additions of small-size sub-polynomial-multiplications (where each sub-polynomial-multiplication retains certain degree of similarity and modularity); (ii) looking for unique/common features from these sub-components such that they can be easily processed by a format of low-resource usage.

Following the above strategy, let us rewrite (1) as

$$W = (Gd_0 + Gd_1 x + \cdots + Gd_{n-1} x^{n-1}) \bmod f(x)$$
$$= G^{(0)} d_0 + G^{(1)} d_1 + \cdots + G^{(n-1)} d_{n-1}, \quad (2)$$

where $G \bmod f(x) = G^{(0)} = G$, $Gx \bmod f(x) = G^{(1)}, \ldots, Gx^{n-1} \bmod f(x) = G^{(n-1)}$. We can then substitute x^n with $x^n \equiv -1$ to have

$$G^{(1)} = -g_{n-1} + g_0 x + \cdots + g_{n-2} x^{n-1},$$
$$\cdots \cdots \cdots \quad (3)$$
$$G^{(n-1)} = -g_1 - g_2 x - \cdots + g_0 x^{n-1}.$$

Let $n = u \times v$ (u, v are integers). We can then define that

$$D = D_0 + D_1 x^u + D_2 x^{2u} + \cdots + D_{v-1} x^{(v-1)u}, \quad (4)$$

where $D_0 = d_0 + d_1 x + d_2 x^2 + \cdots + d_{u-1} x^{u-1}$, $D_1 = d_u + d_{u+1} x + d_{u+2} x^2 + \cdots + d_{2u-1} x^{u-1}$, $\cdots$, $D_{v-1} = d_{uv-u} + d_{uv-u+1} x + \cdots + d_{uv-1} x^{u-1}$.

Similarly, we can have $G = G_0 + \cdots + G_{v-1} x^{(v-1)u}$, where (the similar decomposition strategy applies to other $G^{(i)}$ for $1 \leq i \leq n-1$) $G_0 = g_0 + g_1 x + g_2 x^2 + \cdots + g_{u-1} x^{u-1}$, $G_1 = g_u + g_{u+1} x + g_{u+2} x^2 + \cdots + g_{2u-1} x^{u-1}$, $\cdots$, $G_{v-1} = g_{uv-u} + g_{uv-u+1} x + \cdots + g_{uv-1} x^{u-1}$. Then, we can rewrite (2) into

$$W = G(D_0 + D_1 x^u + \cdots + D_{v-1} x^{(v-1)u}) \bmod f(x)$$
$$= GD_0 + G^{(u)} D_1 + \cdots + G^{(uv-u)} D_{v-1}, \quad (5)$$

where the original polynomial multiplication has been decomposed into the addition of several sub-polynomial-multiplications. For further decomposition, we just cover GD_0 of (5) first (without loss of generality)

$$GD_0 = (G_0 + G_1 x^u + G_2 x^{2u} + \cdots + G_{v-1} x^{(v-1)u}) D_0$$
$$= G_0 D_0 + G_1 x^u D_0 + \cdots + G_{v-1} x^{(v-1)u} D_0, \quad (6)$$

where we define $T_0^{(0)} = G_0 D_0, \cdots, T_{v-1}^{(0)} = G_{v-1} x^{(v-1)u} D_0$. We can then substitute them into (6) to have $GD_0 = T_0^{(0)} + T_1^{(0)} + \cdots + T_{v-1}^{(0)}$, where the sub-polynomial-multiplication is further decomposed into the addition of smaller-size components (which satisfies the first aspect of the proposed derivation strategy).

It is clear that (consider $T_0^{(0)}$ first)

$$\begin{aligned}T_0^{(0)} &= G_0(d_0 + d_1 x + d_2 x^2 + \cdots + d_{u-1} x^{u-1}) \\ &= G_0 d_0 + G_0^{(1)} d_1 + G_0^{(2)} d_2 + \cdots + G_0^{(u-1)} d_{u-1},\end{aligned} \quad (7)$$

which can be transformed into a matrix-vector product form of (connecting (3))

$$[T_0^{(0)}] = \begin{bmatrix} g_0 & -g_{n-1} & \cdots & -g_{N-u+1} \\ g_1 & g_0 & \cdots & -g_{N-u+2} \\ \vdots & \vdots & \ddots & \vdots \\ g_{u-1} & g_{u-2} & \cdots & g_0 \end{bmatrix} \begin{bmatrix} d_0 \\ d_1 \\ \vdots \\ d_{u-1} \end{bmatrix} = [G_0][D_0], \quad (8)$$

where $[G_0]$ is an $u \times u$ matrix. Then, from $[G_0]$ we observe that: (i) the elements in the main diagonal are identical (say g_0); (ii) the rest of elements are regularly distributed in two regions (the upper-right and the lower-left ones) and meanwhile the values in the specific region are symmetrically identical along with the direction of the main diagonal of the matrix; (iii) the subscripts of the values of each row/column within each region are following a pattern of decreasing sequence (e.g., from g_{u-1} to g_0 and then to $-g_{n-u+1}$); (iv) there are actually in total $(2u - 1)$ values contained in the $[G_0]$ (counting the related signs), namely $g_{u-1}, \cdots, g_0, \cdots, -g_{N-u+1}$, which is the values in the far left column and the first top row. These unique features indicate that all the elements within the matrix $[G_0]$ can be obtained through the circularly shifting of the coefficients of polynomial G, which facilitates the actual implementation (see Sect. 3).

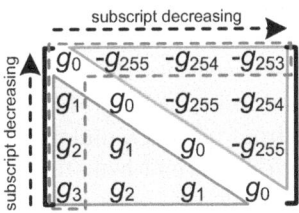

Fig. 1. Example of $n = 256$ and $u = 4$ ($[G_0]$), where the values are regularly distributed in the regions (colored areas). (Color figure online)

For a clear demonstration and clarification, we have used a case study example of $n = 256$ and $u = 4$ and have shown $[G_0]$ in Fig. 1, where the mentioned two regions are highlighted as the blue and green areas, respectively. One can see that the actual values contained in the dotted red area are in total $(2u - 1 = 7)$ numbers (where the subscripts are decreasing). Besides, the values in the respective region are symmetrically identical along with the line of the main diagonal.

In summary, one can conclude that these unique properties are very much related to the elements in the matrix ($[G_0]$) main diagonal and the other elements are distributed following a specific order. Besides that, the matrix size u is not a fixed number, i.e., <u>scalable matrix originated computation</u>.

For a more general conclusion, one can find that these observed unique features do not apply to $[G_0]$ only. In fact, these properties apply also to other sub-products of (6). For instance, we can have $T_1^{(0)}$ as

$$T_1^{(0)} = \begin{bmatrix} g_u & g_{u-1} & \cdots & g_1 \\ g_{u+1} & g_u & \cdots & g_2 \\ \vdots & \vdots & \ddots & \vdots \\ g_{2u-1} & g_{2u-2} & \cdots & g_u \end{bmatrix} \begin{bmatrix} d_0 \\ d_1 \\ \vdots \\ d_{u-1} \end{bmatrix} = [G_1][D_0], \tag{9}$$

where all the elements within $[G_1]$ follow the same pattern (as specified above). Likewise, $T_2^{(0)}, \ldots, T_{v-1}^{(0)}$ can be transformed into similar matrix-vector products, and the involved matrices share the same features.

Similarly, $G^{(u)}D_1$ can be composed as

$$G^{(u)}D_1 = T_0^{(1)} + T_1^{(1)} + \cdots + T_{v-1}^{(1)}, \tag{10}$$

where $T_0^{(1)} = G_0^{(u)}D_1$, $T_1^{(1)} = G_1^{(u)}x^u D_1, \cdots, T_{v-1}^{(1)} = G_{v-1}^{(u)}x^{(v-1)u}D_1$. The same strategy can be extended to $G^{(2u)}D_2, \cdots, G^{(uv-u)}D_{v-1}$, as

$$\begin{aligned} G^{(2u)}D_2 &= T_0^{(2)} + T_1^{(2)} + \cdots + T_{v-1}^{(2)}, \\ &\cdots \cdots \cdots \\ G^{(uv-u)}D_{v-1} &= T_0^{(v-1)} + T_1^{(v-1)} + \cdots + T_{v-1}^{(v-1)}, \end{aligned} \tag{11}$$

where we can find that each sub-polynomial-multiplication of (5) has now been further decomposed into v number of sub-components. Besides that, all these sub-components can be transformed into the matrix-vector product forms, following the examples presented in (8), (9), and Fig. 1.

The above steps, mainly from (4)–(11), undoubtedly have fully satisfied the mentioned two goals of the proposed mathematical derivation strategy. Hence, we can summarize the proposed decomposition strategy as follows:

Proposed Strategy. For a general polynomial multiplication over ring $\mathbb{Z}_q/(x^n+1)$, we propose a constant-time solution in which we can follow the above steps of (4)–(11) to decompose the polynomial multiplication into the addition of a total v^2 number of regular sub-components, where each sub-component is equivalent to a matrix-vector product involved with a main matrix sharing the pattern of scalable matrix based processing.

Overall, the whole polynomial multiplication can be computed as follows. Let us again decompose W into v sub-polynomials as

$$W = W_0 + W_1 x^u + W_2 x^{2u} + \cdots + W_{v-1} x^{(v-1)u}, \tag{12}$$

where $W_0 = w_0 + w_1 x + w_2 x^2 + \cdots + w_{u-1} x^{u-1}$, $W_1 = w_u + w_{u+1} x + w_{u+2} x^2 + \cdots + w_{2u-1} x^{u-1}, \cdots, W_{v-1} = w_{uv-u-1} + w_{uv-u} x + \cdots + w_{uv-1} x^{u-1}$.

From (5), one can further have $W_0 = T_0^{(0)} + T_0^{(1)} + \cdots + T_0^{(v-1)} = \sum_{j=0}^{v-1} T_0^{(j)}$, $W_1 = T_1^{(0)} + T_1^{(1)} + \cdots + T_1^{(v-1)} = \sum_{j=0}^{v-1} T_1^{(j)}, \cdots, W_{v-1} = T_{v-1}^{(0)} + T_{v-1}^{(1)} + \cdots + T_{v-1}^{(v-1)} = \sum_{j=0}^{v-1} T_{v-1}^{(j)}$, where each output sub-polynomial becomes the accumulation of v number of $T_k^{(j)}$ (for $W_k = \sum_{j=0}^{v-1} T_k^{(j)}$). We can thus have:

Algorithm 1. Proposed polynomial multiplication algorithm (general form)

Inputs: G and D are polynomials (the actual bit-width of the coefficients of G and D follows the specific PQC scheme).
Output: $W = GD \bmod f(x)$ (where $f(x) = x^n + 1$).
1. Initialization (preparation) step
1.1. make ready input polynomials G and D.
1.2. $\overline{W} = 0$.
2. Main step
2.1. decompose D into $\{D_0, D_1, \cdots, D_{v-1}\}$. // see (4)
2.2. obtain $G^{(1)}, G^{(2)}, \cdots, G^{(n-1)}$ from G, respectively. // (3)
2.3. decompose G into $\{G_0, G_1, \cdots, G_{v-1}\}$.
   2.4. for $k = 0$ to $v - 1$.
     2.5. for $j = 0$ to $v - 1$.
     2.6. obtain all the corresponding $G_k^{(ju)}$. // (13)
     2.7. $\overline{W} = \overline{W} + T_k^{(j)}$. // scalable matrix based processing strategy (14)
     2.8. end for.
     2.9. $W_k = \overline{W}$.
   2.10. end for.
3. Final step
3.1. obtain the output W from serially delivered W_k.

Details of the Algorithm. Overall, the procedures presented in Algorithm 1 are clearly expressed (see the above-detailed derivation processes) except the computation of each $T_k^{(j)}$ as well as the obtaining of related $G_k^{(ju)}$ during the actual implementation process. Here we present the details of them as below.

(a) Obtaining of Related $G_k^{(ju)}$ Sequentially. As the related T_k^j are serially accumulated, the obtaining of corresponding $G_k^{(ju)}$ also needs to be carried out in a sequential format. For instance, we can have $[G_0^{(0)}]$ as

$$[T_0^{(u)}] = \begin{bmatrix} -g_{n-u} & -g_{n-u-1} & \cdots & -g_{n-2u+1} \\ -g_{n-u+1} & -g_{n-u} & \cdots & -g_{n-2u+2} \\ \vdots & \vdots & \ddots & \vdots \\ -g_{n-1} & -g_{n-2} & \cdots & -g_{n-u} \end{bmatrix}, \quad (13)$$

where there are actually $(2u-1)$ number of values involved within, i.e., $\{-g_{n-1}, \cdots, -g_{n-u}, \cdots, -g_{n-2u+1}\}$. Comparing with the actual $(2u-1)$ values contained in $[G_0]$, namely $\{g_{u-1}, \cdots, g_0, \cdots, -g_{n-u+1}\}$, these values (subscripts) are circularly related to one another and there also exist an overlap of $(u-1)$ values (i.e., $\{-g_{n-1}, \cdots, -g_{n-u+1}\}$). This property facilitates the use of circular shift-register (CSR) to deliver out the desired outputs per every cycle for the construction of proper $G_k^{(ju)}$ (the detailed hardware structure is presented in Sect. 3). Similar strategy applies to the following obtaining $G_0^{(2u)}$ from $G_0^{(u)}$, which can be extended to the obtaining of other $G_k^{(ju)}$ in a sequential order.

Another aspect of obtaining $G_k^{(ju)}$ in Algorithm 1 also involves the assigning of correct signs to the corresponding coefficient within a certain $[G_k^{(ju)}]$ since the original coefficients of the polynomial G are assumed to have positive values (no additional sign inverting). Again, here we combine the proposed computational strategy with the feature of the sign distributions within two regions of the matrix $[G_k^{(ju)}]$ to obtain the accurate sign assignment and the detailed implementation process can also be seen in the next section.

(b) Computation of Each $T_k^{(j)}$. The computation of each $T_k^{(j)}$ follows the regular calculation process, i.e., transform each $T_k^{(j)}$ into the equivalent matrix-vector product and then obtain the related output (u number) in parallel through point-wise multiplication-and-addition operations (Step 2.7 of Algorithm 1 is the serial accumulation of $T_k^{(j)}$). For example, (8) can be calculated as

$$[T_0^{(0)}] = \begin{bmatrix} g_0 d_0 - g_{n-1} d_1 - \cdots - g_{N-u+1} d_{u-1} \\ g_1 d_0 + g_0 d_1 - \cdots - g_{N-u+2} d_{u-1} \\ \cdots \cdots \cdots \\ g_{u-1} d_0 + g_{u-2} d_1 + \cdots + g_0 d_{u-1} \end{bmatrix}, \quad (14)$$

which applies to other $T_k^{(j)}$ of Algorithm 1.

4   SMALL: Proposed Accelerator

The overall structure of the proposed accelerator (SMALL) is shown in Fig. 2, where it consists of five major components, namely the input processing component, the sign processing component, the main computation component, the control generating component, and the output delivering component. In terms of the constitution of each component, there are: (i) two CSRs in the input processing component; (ii) a sign block in the sign processing component; (iii) one multiplication-and-addition (MAA) cell and one accumulation (AC) cell in the main computation component; (iv) a control unit in the control generating component; and (v) a parallel-in serial-out (PISO) shift-register (SR) in the output delivering component. Note that the bit-width of the each data path depends on the setup in the specific PQC scheme.

The input processing component (two CSRs) is firstly loaded with the necessary coefficients from the two inputs (which takes N cycles) and then in the

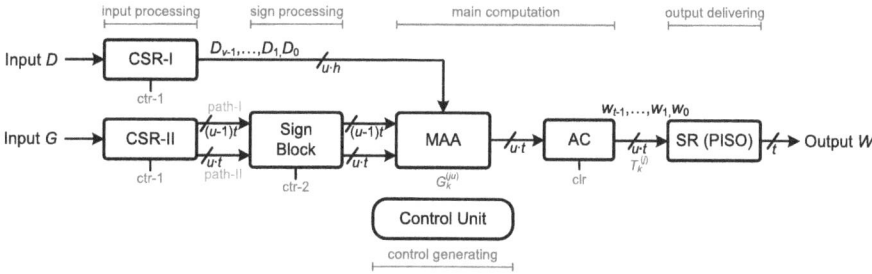

Fig. 2. The proposed polynomial multiplication accelerator (SMALL).

following cycles it produces the correct outputs to the following components. Connecting with Algorithm 1, while the CSR-I is producing D_j ($j = 0$ to $v-1$) in a sequential format, the CSR-II is responsible for generating the necessary values (in total $2u - 1$ values) to construct the corresponding $G_k^{(ju)}$. Of course, the sign processing component (sign block) assists with the sign assigning to all the delivered $2u - 1$ values (in two paths) to form the accurate $G_k^{(ju)}$. When all the necessary values have been fed to the main computation component, the MAA cell functions to execute the computation of $T_k^{(j)}$ and the following AC cell executes the related accumulation to deliver the desired W_k ($k = 0$ to $v-1$). As the output of the AC cell has u parallel output coefficients, the final output delivering component transfers the parallel output into serial style to be stored in the external memory or for other usage. The overall operation, of course, is carried out through different types of control signals generated from the control unit (see control generating component for more details). The whole process, including the input loading and output delivering time, requires $(n + v^2 + u)$ cycles of operations. The detailed internal structures as well as related functions are presented below.

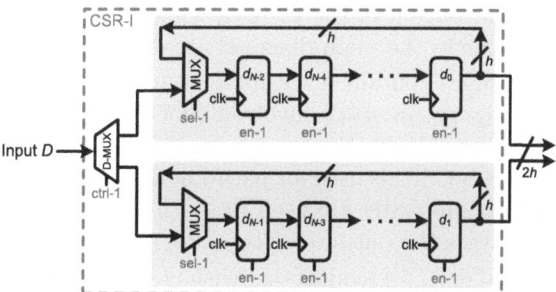

Fig. 3. The CSR-I (for $u = 2$), where the values in the registers are initial values.

The Input Processing Component. As seen from Fig. 2, there are two CSRs contained in this component. The CSR-I is responsible for generating the proper D_j ($j = 0$ to $v-1$) to the MAA cell in a repeated format (repeats every v cycles after all the input coefficients are loaded in the CSR-I). To realize this specific function, we have used a multi-path based CSR, as shown in Fig. 3, where we have presented a case study example when $u = 2$. This multi-path based CSR consists of two sub-CSRs (each sub-CSR has $n/2$ registers), where the input to each sub-CSR is directly from a De-MUX (D-MUX) and two connected MUXes attached to the input of the sub-CSR. During the loading time, the control signal to the D-MUX operates according to the sequence of "0101···0101", which splits the coefficients of the polynomial D into two groups (each group corresponds to the specific sub-CSR), i.e., group of $\{d_{n-2}, d_{n-4}, \cdots, d_0\}$ and another group of $\{d_{n-1}, d_{n-3}, \cdots, d_1\}$. When all the necessary values are loaded into the corresponding registers, the two MUXes in the sub-CSRs then switch to close the loop such that in the following cycles, the output of the CSR-1 produces D_j ($j = 0$ to $v - 1$) correctly. The design of Fig. 3 can be extended to other u.

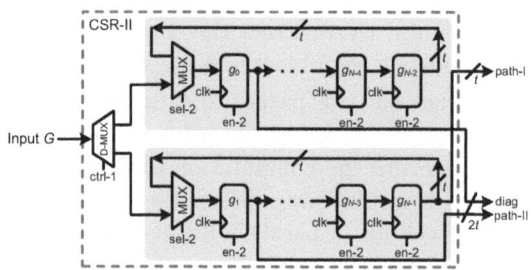

Fig. 4. The CSR-II ($u = 2$), where the values in the registers are initial values. Note "diag" refers to the value in the diagonal direction of the scalable matrix, and is part of the path-II output.

The CSR-II has a similar design structure as that in Fig. 3 except the output setup. When connecting with the actual values contained in the two regions of Fig. 1 (applies to other matrices also), the values in the upper-right region are generated by the path-I output of the CSR-II, while the values in the lower-left region as well as the one in the main diagonal are delivered out by the path-II.

Considering the values contained within each $[G_k^{(ju)}]$, e.g., $[G_0^{(0)}]$ (for $[G_0]$, see (8)), there are only $\{g_0, g_1, g_{255}\}$ involved (not counting the sign, which is done by the following sign block). Hence, as seen in Fig. 4, the far right register's output (only bottom sub-CSR) is used for path-I delivering while both the far left registers' outputs (two sub-CSR) are used for path-II delivering. In the second cycle, the CSR-II delivers the outputs of $\{g_{255}, g_{254}, g_{253}\}$, which is exactly the actual values contained in $[G_0^{(2)}]$ (connecting with (13)). When the desired output for $[G_0^{(254)}]$ (for the example here, at the $n/2$th cycle) is delivered (g_2, g_1, g_3), all the registers in the CSRs will be disabled for one cycle, i.e., the same output values are delivered out for the next cycle, which matches the actual values contained within $[G_0^{(2)}]$ (see (13)). Then, the registers in the CSR-II will be enabled again in the following cycles (the disabling of registers in the CSR-II repeats every $n/2$ cycles until all the proper outputs are produced).

In a more general sense u is selected as other values. All the outputs of the far-right registers in all the sub-CSRs (not including the top one) are used to deliver the values required for path-I, while all the far-left registers (in all the sub-CSRs) are used to form the path-II output (can be extended to other u).

The Sign Processing Component. The sign block in the sign processing component functions to assign the delivered outputs from the CSR-II with proper signs according to the distribution within each $G_k^{(ju)}$. As shown in Fig. 5, there are basically two inverter cells (marked as $x = -x$) attached correspondingly to two MUXes. The inverter cell contains $(u-1)$ (or u) sign inverters (SIs) according to the two's complement representation requirement that all the bits of a certain value are all inverted and then pass through the same number of half-adder (HD) (with one carry-in set as '1'). Note s-0 and s-1 are generated by the control unit.

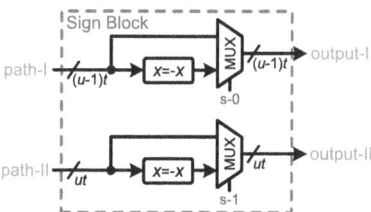

Fig. 5. The internal structure of the sign block.

The Main Computation Component. In this component, the MAA cell is responsible for the calculation of corresponding $T_k^{(j)}$ in Step 2.7 of Algorithm 1, while the AC cell executes the following accumulation in the same step. As specified in (14), the MAA directly obtains the output from standard matrix-vector based calculation, which applies to other values of u (see Fig. 6).

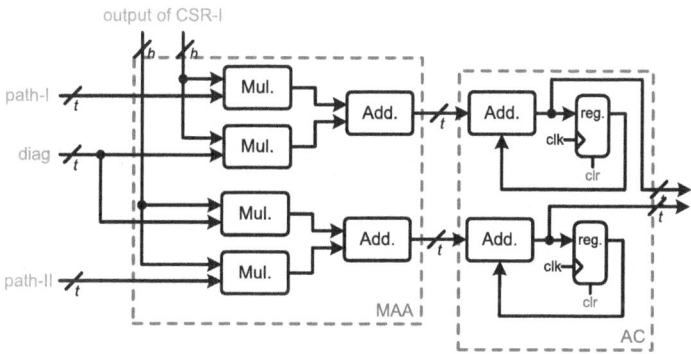

Fig. 6. The main computation component (for $u = 2$), where Mul. and Add. denote the multiplier and adder, respectively (reg. is the register).

As seen from Fig. 6, the MAA cell mainly consists of necessary multipliers and adders to perform the matrix-vector product (connecting (14)). In the case study example of $u = 2$, there is only one value contained in the upper-right region of the main matrix of $[T_k^{(j)}]$ (path-I) as well as the one element in the lower-left region (there are two values from path-II as the one in the main diagonal is also included). Following this setup, we can have the arrangement of multipliers and adders in the MAA cell, as shown in Fig. 6, where one input value (the element in the main diagonal) from path-II is reused twice as the input to the multiplier while the other input values (including the ones delivered from the CSR-I) are connected to the corresponding multipliers, respectively, following the principle of matrix-vector product (size of 2×2). The produced two outputs, namely the outputs of $[T_k^{(j)}]$, are then accumulated in the following AC cell through parallel processing to produce two outputs. Note that the outputs of the adders in the AC cell are directly connected to the outside as outputs of the main computation component, for the sake of saving one extra clock cycle spent on the registers. The structure shown in Fig. 6 can easily extended to the design of other u.

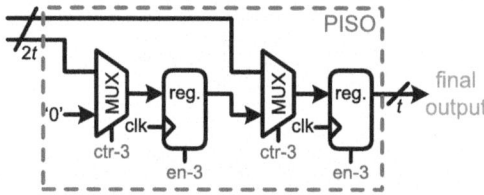

Fig. 7. The internal structure of the PISO SR for $u = 2$.

The Output Delivering Component. This component is relatively simple, and only a PISO SR is used to transfer the parallel output from the AC cell into a serial format for further processing (which is very important in practical applications). Figure 7 gives an example for such SR when $u = 2$, which can be extended to other u. One can see that the two MUXes function to load the output from the AC cell into the registers in the SR for serial output delivery.

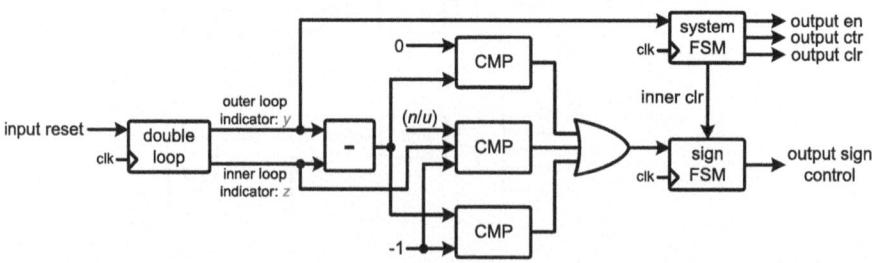

Fig. 8. The control unit, where y (horizontal) and z (vertical) are the indices for the elements within matrix $[G_k^{ju}]$. CMP: comparator. FSM: finite state machine.

The Control Generating Component. This component plays a key role in the whole accelerator, i.e., generating the sign controlling signals (sign block), clearing signals (mainly for the registers), enabling signals (mainly registers), and selecting signals (for MUXes/D-MUXes), etc. All the necessary control signals can be easily added/generated since we have used a double loop component centered control unit, as shown in Fig. 8, where the entire work status for the entire computation of $W = GD \bmod f(x)$ is divided into two stages, namely the loading and calculating (including the final output delivering) stages.

During the load stage, the control unit takes n cycles to serially receive all the coefficients of G and D into the corresponding registers in the CSRs. Note the control signal (ctr-1, see Fig. 2) is set as '0' throughout this stage such that the two CSRs are all working in the loading mode. Once all the values are initiated in the related registers, the control unit switches to the calculating stage.

The overall calculating stage takes $(n/u)^2$ cycles to produce (n/u) batches of desired results, namely $[w_0, ..., w_{u-1}], ..., [w_{n-u}, w_{n-1}]$ (from W_0 to W_{v-1}). One part of the work for the control unit during this stage is to generate the necessary sign control signals for the sign block in Figs. 2 and 5. To achieve the accurate sign assigning to the correct coefficients, we have used a novel sign control generating strategy here: (i) first of all, we observe that the signs for all the element within a certain matrix $[G_k^{ju}]$ (connecting Algorithm 1) can be categorized into three conditions, namely a) all the values in the whole matrix have positive signs, b) the values in the lower-left region have positive signs but all negative signs in the upper-right region of the matrix, c) all the values in the whole matrix have negative signs; (ii) secondly, we hence propose to use indices y (horizontal) and z (vertical) to represent every element in the matrix $[G_k^{ju}]$ (e.g., $y = 0$ and $z = 0$ represent the element in the left-top corner of the matrix) such that we only need to consider three conditions of a) $y - z = 0$, b) $y - z = -1$, c) $z = (n/u) - 1$ (the transition between two states happens whenever one of three transition condition is satisfied), which can be realized by a three-state finite state machine (sign FSM). As shown in Fig. 8, the sign FSM produces the correct sign control signals, where the first bit of the sign signal determines the signs in the lower-left region (including the main diagonal), and the second bit of sign control signal determines the signs in the upper-right region of the matrix.

The control unit also sets the enable signal for the CSR-II to '0' when transition condition c) is satisfied because the main matrix in the last computation block of $T_k^{(v-1)}$ and the main matrix in the first computation block of $T_{k+1}^{(0)}$ consists of exactly the same values (with only different signs). Furthermore, the checking of the transition condition c) also enables the control unit to generate the clear signal for the AC cell as well as the loading signal for the final PISO component. The proposed control unit fully utilizes the reusability of checking transition of condition c) to reduce the area usage of the entire control unit.

5 Evaluation: Implementation and Comparison

This section focuses on the thorough complexity analysis and comparison of the proposed accelerator along with the state-of-the-art designs.

Table 1. Comparison of the Complexities of The Proposed Polynomial Multiplication Accelerator for Saber (FPGA Platform)

design	ALMs	Fmax (MHz)	latency	T[1] (μs)	ADP[2]
Stratix V device $n = 256$ (Proposed Design)					
$u = 1$	151	353	65,536	186	28,025
$u = 2$	340	240	16,384	68	23,211
$u = 4$	969	181	4,096	23	21,937
$u = 8$	2,766	142	1,024	7	19,966
Cyclone V device $n = 256$ (Proposed Design)					
$u = 1$	149	137	65,536	478	71,245
$u = 2$	403	109	16,384	151	60,732
$u = 4$	929	83	4,096	49	45,812
$u = 8$	2,908	64	1,024	16	46,311

[1]: T = total execution time ((latency cycle) $\times$ (1/Fmax)) . [2]: ADP=#ALM $\times$ T.

Complexity Analysis. In general, the proposed polynomial multiplication accelerator (SMALL), as shown in Fig. 2, uses two CSRs (where each CSR has n registers and $(v+1)$ MUXes/D-MUXes), one sign block $((2u-1)$ MUXes and $(2u-1)$ SIs), one MAA cell (u^2 number of multipliers and $(u-1)u$ adders), one AC cell (u number of adders and the same number of registers), one PISO SR (u number of MUXes and the same number of registers), and one control unit. The polynomial multiplication structure requires n cycles of input loading into the related CSRs, $(n/u)^2$ cycles of calculation, and additional u cycles of output delivering (the first $(v-1)$ groups of outputs, from W_0 to W_{v-2}, are delivered out during the computation time). The actual critical-path of the accelerator is mainly determined by the selection of u and n.

Case Study Examples. For a detailed evaluation, we have used the polynomial multiplication used in NIST PQC third-round standardization candidate Saber [1] and RBLWE-ENC [2,4], respectively. Specifically, we notice that: (i) For Saber, one polynomial has coefficients of either 13-bit or 10-bit (the 13-bit design covers the 10-bit one), while another polynomial has coefficients of [-4,4] (4-bit). The polynomial size n is fixed at 256. (ii) For RBLWE-ENC, one polynomial consists of integer coefficients of $\log_2 256 = 8$-bit. While another polynomial involves merely binary coefficients of $\in \{0,1\}$. The polynomial size can be $n = 256$ and $n = 512$,

Experimental Setup. The experimental setup of our evaluation is set as follows: (a) we have coded the proposed polynomial multiplication accelerator (SMALL), based on Saber and RBLWE-ENC's respective parameters, with VHDL and have verified its functionality through ModelSim ($t = 13$, $h = 4$, and $n = 256$ for Saber and $t = 8$, $h = 1$, $n = 256$ and $n = 512$ for RBLWE-ENC, respectively, as well as $u = 1$, $u = 2$, $u = 3$, and $u = 4$); (b) we have

Table 2. Comparison of the Complexities with the Existing Polynomial Multiplication Accelerators for Saber

design	LUT	FF	Fmax	latency	ADP[1] (μs)
Xilinx Artix-7 XC7A12TLCSG325-2L ($n = 256$)					
[3]	541	301	100	19,471	105,338
[25][2]	561	302	130	16,384	270,336
TW ($u = 1$)	202	605	178	65,536	74,372
TW ($u = 2$)	318	670	125	16,384	41,681
TW ($u = 4$)	664	798	109	4,096	24,952
TW ($u = 8$)	2,424	1,069	102	1,024	24,335

TW: This Work. Unit for Fmax: MHz.
[1]: ADP=#LUT × latency time.
[2]: This design also uses [25] uses 2 DSPs and 2 BRAMs (which need to be transferred into equivalent LUT usage, i.e., 1 BRAM (8k) equals 70 slices, 1 DSP equals 128 slices, and 1 slice has 4 LUTs).

synthesized and implemented the coded designs on the FPGA devices (both AMD-Xilinx and Intel-Altera FPGAs) to obtain their detailed area-time complexities along with the competing designs; (c) more detailed, we used the Intel Quartus Prime 17.0 to obtain the performance for all the coded designs on the Stratix V 5SGXMABN1F45C2 and/or Cyclone V 5CSXFC6D6F31I7ES devices; (d) we have also obtained the corresponding implementation results for Saber on the Artix-7 XC7A12TLCSG325-2L device through Xilinx Vivado 2019.2; (e) for the accelerator deployed with Saber's parameter, the coefficients of D are represented in the sign magnitude binary numbers (following [3]), while the coefficients in the design for RBLWE-ENC are denoted by the 2's complement.

FPGA Based Implementation Results and Comparison. The obtained area-time complexities, in terms of the number of adaptive logic modules (ALMs) (or slice LUT/FF), maximum frequency (Fmax), latency cycles, total execution time T = ((latency cycle)×(1/Fmax)), and area-delay product (ADP), of the proposed designs with different parameter settings are shown in Tables 2, and 3 along with those of the existing designs of [3,25], respectively. Note in Table 2, we have used the equivalent LUT calculation method to obtain the overall ADP for [25], for a fair comparison. Meanwhile, [3] indicated that external memory is used during the computation process, yet we don't count it here.

As seen from Tables 1 and 2, the proposed accelerator for Saber has superior performance on both Intel-Altera and AMD-Xilinx FPGA devices, not only limited to its scalability (but also the overall complexity). For instance, the proposed design ($u = 4$) has at least 76.3% less ADP than the recent one of [3] (also at least 90.8% less ADP than another recent report of [25]), not even counting the conditions that the existing design of [3] very much relies on the memory access

Table 3. Comparison of the Complexities of The Polynomial Multiplication Designs for RBLWE-ENC

design	ALMs	Fmax (MHz)	latency	T (μs)	ADP[1]
Stratix V device $n=256$ (Existing Design)					
[10]	1,793	318.47	65,536	206	369,358
Stratix V device $n=256$ (Proposed Design)					
$u=1$	98	553	65,536	119	11,618
$u=2$	179	421	16,384	39	6,959
$u=4$	357	315	4,096	13	4,644
$u=8$	914	202	1,024	5	4,624
Stratix V device $n=512$ (Existing Design)					
[10]	3,491	288.77	262,144	908	3,169,828
Stratix V device $n=512$ (Proposed Design)					
$u=1$	104	562	262,144	466	48,501
$u=2$	192	395	65,536	166	31,822
$u=4$	430	290	16,384	56	24,294
$u=8$	996	207	4,096	20	19,749

[1]: ADP=#ALM $\times$ T (total execution time).

(this part of resource is not included in Table 2). Meanwhile, one can notice that the proposed design overall maintains very low complexity, especially the ones of $u=1,2,4$, which are desirable for lightweight applications.

For completeness, we highlight relevant NTT-based solutions. For instance, Saber work from [24] requires ~1,680 clock cycles with 2,247 LUTs (with higher clock frequency) while our work has a latency of 1,024 clock cycles and 2,424 LUTs for $u=8$. For RBLWE-ENC ($n=512$) the work from [9] using t=8 and t=32 parallel processing units, has similar latency as this work ($u=2$ and $u=4$) but the ALMs are much higher, 5,073 and 6,076, respectively.

Meanwhile, as seen from Table 3, one can find that the proposed polynomial multiplication structure significantly outperforms the recent one in [10]. For example, the proposed design of $u=4$ involves at least 98.7% less ADP than the one in [10] on the Stratix V device when $n=256$ (the similar situation applies to nearly every case presented in Table 3). Meanwhile, considering that the existing one also requires external resource assistance and has a limited processing style, the proposed design has completely outperformed [10].

Discussion. The proposed algorithmic computation and accelerator design strategies overall are highly efficient: (i) the proposed scalable matrix originated strategy brings both flexibility and compactness to the polynomial multiplication's scalable processing and low complexity computation; (ii) the proposed novel algorithm-to-architecture design techniques have produced an exceptionally optimized accelerator (with resource usage significantly minimized).

The proposed design strategy can be further developed into other applications. For instance, we notice that the polynomial multiplication used in the homomorphic encryption scheme BFV [6] also involves a similar coefficients setup. Other research directions can also be further extending the proposed strategy into polynomial multiplication used for other PQC schemes.

6 Conclusion

In this paper, we propose a novel constant-time scalable matrix originated computation strategy for the efficient implementation of the targeted polynomial multiplication in important PQC schemes. In total, we have: (i) proposed a new polynomial multiplication algorithm; (ii) presented the details of the polynomial multiplication accelerator (SMALL); (iv) demonstrated the superior efficiency of the proposed polynomial multiplication accelerator through two case study examples. We hope the outcome of this work will produce significant impact on the PQC development and related computer arithmetic technique research.

Acknowledgement. J. Xie was supported by NIST-60NANB20D203. Ç. Koç was supported by TUBITAK Project 1001-121F348.

References

1. Saber. https://www.esat.kuleuven.be/cosic/pqcrypto/saber/
2. Bao, T., He, P., Bai, S., Xie, J.: TINA: TMVP-initiated novel accelerator for lightweight ring-LWE-based PQC. IEEE Trans. Very Large Scale Integr. (VLSI) Syst. (01), 1–13 (2023). https://doi.org/10.1109/TVLSI.2023.3341037
3. Basso, A., Roy, S.S.: Optimized polynomial multiplier architectures for post-quantum KEM saber. In: 2021 58th ACM/IEEE Design Automation Conference (DAC), pp. 1285–1290. IEEE (2021). https://doi.org/10.1109/DAC18074.2021.9586219
4. Buchmann, J., Göpfert, F., Güneysu, T., Oder, T., Pöppelmann, T.: High-performance and lightweight lattice-based public-key encryption. In: Proceedings of the 2nd ACM international workshop on IoT privacy, trust, and security, pp. 2–9 (2016). https://doi.org/10.1145/2899007.2899011
5. Choi, P., Kim, D.K.: Lightweight polynomial multiplication accelerator for NTRU using shared SRAM. IEEE Trans. Circuits Syst. II Express Briefs **70**(12), 4574–4578 (2023). https://doi.org/10.1109/TCSII.2023.3290192
6. Fan, J., Vercauteren, F.: Somewhat practical fully homomorphic encryption. Cryptology ePrint Archive (2012)
7. Fawzi, A., et al.: Discovering faster matrix multiplication algorithms with reinforcement learning. Nature **610**, 47–53 (2022). https://doi.org/10.1038/s41586-022-05172-4
8. Han, J., Fan, H.: Toeplitz matrix-vector product based $GF(2^n)$ shifted polynomial basis multipliers for all irreducible pentanomials. Cryptology ePrint Archive, Paper 2013/427 (2013). https://eprint.iacr.org/2013/427

9. He, P., Bao, T., Xie, J., Amin, M.: FPGA implementation of compact hardware accelerators for ring-binary-LWE-based post-quantum cryptography. ACM Trans. Reconfigurable Technol. Syst. **16**(3) (2023). https://doi.org/10.1145/3569457
10. He, P., Guin, U., Xie, J.: Novel low-complexity polynomial multiplication over hybrid fields for efficient implementation of binary ring-LWE post-quantum cryptography. IEEE J. Emerg. Sel. Top. Circuits Syst. **11**(2), 383–394 (2021). https://doi.org/10.1109/JETCAS.2021.3075456
11. He, P., Tu, Y., Xie, J., Jacinto, H.: KINA: karatsuba initiated novel accelerator for ring-binary-LWE (RBLWE)-based post-quantum cryptography. IEEE Trans. Very Large Scale Integr. (VLSI) Syst. **31**(10), 1551–1564 (2023). https://doi.org/10.1109/TVLSI.2023.3302289
12. Hu, J., Wang, W., Cheung, R.C., Wang, H.: Optimized polynomial multiplier over commutative rings on FPGAS: a case study on bike. In: 2019 International Conference on Field-Programmable Technology (ICFPT), pp. 231–234 (2019). https://doi.org/10.1109/ICFPT47387.2019.00035
13. Khayyat, A., Manjikian, N.: Analysis of blocking and scheduling for FPGA-based floating-point matrix multiplication analyse du blocage et de l'ordonnancement d'une multiplication matricielle à virgule flottante sur un FPGA. Can. J. Electr. Comput. Eng. **37**(2), 65–75 (2014). https://doi.org/10.1109/CJECE.2014.2317983
14. Lucas, B.J., et al.: Lightweight hardware implementation of binary ring-LWE PQC accelerator. IEEE Comput. Archit. Lett. **21**(1), 17–20 (2022). https://doi.org/10.1109/LCA.2022.3160394
15. Lyubashevsky, V., Peikert, C., Regev, O.: On ideal lattices and learning with errors over rings. In: Gilbert, H. (ed.) EUROCRYPT 2010. LNCS, vol. 6110, pp. 1–23. Springer, Heidelberg (2010). https://doi.org/10.1007/978-3-642-13190-5_1
16. Mouilleron, C.: Efficient computation with structured matrices and arithmetic expressions (2011)
17. Pollard, J.M.: The fast fourier transform in a finite field. Math. Comput. **25**(114), 365–374 (1971)
18. Regev, O.: On lattices, learning with errors, random linear codes, and cryptography. J. ACM (JACM) **56**(6), 1–40 (2009). https://doi.org/10.1145/1568318.1568324
19. Samsi, S., Helfer, B., Kepner, J., Reuther, A., Ricke, D.O.: A linear algebra approach to fast DNA mixture analysis using GPUS. In: 2017 IEEE High Performance Extreme Computing Conference (HPEC), pp. 1–6 (2017). https://doi.org/10.1109/HPEC.2017.8091027
20. Tan, W., Wang, A., Zhang, X., Lao, Y., Parhi, K.K.: High-speed VLSI architectures for modular polynomial multiplication via fast filtering and applications to lattice-based cryptography. IEEE Trans. Comput. **72**(09), 2454–2466 (2023). https://doi.org/10.1109/TC.2023.3251847
21. Wong, Z.Y., Wong, D.C.K., Lee, W.K., Mok, K.M., Yap, W.S., Khalid, A.: KaratSaber: new speed records for saber polynomial multiplication using efficient karatsuba FPGA architecture. IEEE Trans. Comput. **72**(07), 1830–1842 (2023). https://doi.org/10.1109/TC.2023.3238129
22. Xie, J., He, P., Lee, C.Y.: Crop: FPGA implementation of high-performance polynomial multiplication in saber KEM based on novel cyclic-row oriented processing strategy. In: 2021 IEEE 39th International Conference on Computer Design (ICCD), pp. 130–137 (2021). https://doi.org/10.1109/ICCD53106.2021.00031
23. Xie, J., Zhao, W., Lee, H., Roy, D.B., Zhang, X.: Hardware circuits and systems design for post-quantum cryptography-a tutorial brief. IEEE Trans. Circuits Syst.

II Express Briefs **71**(3), 1670–1676 (2024). https://doi.org/10.1109/TCSII.2024.3357836
24. Xu, T., Cui, Y., Liu, D., Wang, C., Liu, W.: Lightweight and efficient hardware implementation for saber using NTT multiplication. In: 2022 IEEE Asia Pacific Conference on Circuits and Systems (APCCAS), pp. 601–605 (2022). https://doi.org/10.1109/APCCAS55924.2022.10090310
25. Zhang, Y., et al.: A lightweight and efficient schoolbook polynomial multiplier for saber. In: 2022 IEEE International Symposium on Circuits and Systems (ISCAS), pp. 2251–2255. IEEE (2022). https://doi.org/10.1109/ISCAS48785.2022.9937496

Author Index

A
Arpin, Sarah 171
Askeland, Amund 127

B
Badakhshan, Mohammadtaghi 258
Bartoli, Daniele 191

C
Camps-Moreno, Eduardo 37
Carlet, Claude 213
Castryck, Wouter 171
Cenk, Murat 237
Chan, Chin Hei 19
Crespo Bofill, Pedro 70

D
Dastbasteh, Reza 70
deMarti iOlius, Antonio 70

E
Eriksen, Jonathan Komada 171
Etxezarreta Martinez, Josu 70

G
Gao, Zhicheng 154
García-Marco, Ignacio 37
Gómez, Ana I. 117
Gómez-Pérez, Domingo 117
Gong, Guang 258

H
He, Pengzhou 274
Hernández, Félix 3

J
Jandhyala, Tanmayi 258

K
Koç, Çetin Kaya 274
Kölsch, Lukas 191
Kuttner, Simon 154

L
López, Hiram H. 37
Lorenzon, Gioella 171
Luo, Dongxia 53
Luo, Guiwen 258

M
Madrigal, Samira Carolina Oliva 274
Márquez-Corbella, Irene 37
Martínez-Moro, Edgar 37
Micheli, Giacomo 191
Moura, Lucia 53

O
Oliva del Moral, Javier 70

P
Pal, Mohit 99
Pastor-Díaz, Ulises 213
Polverino, Olga 139

R
Reis, Lucas 91

S
Santonastaso, Paolo 139
Sanz Larrarte, Olatz 70
Sarmiento, Eliseo 37
Soto, Francisco-Javier 117

T
Tornero, José M. 213
Türe, Nazlı Deniz 237

V
Vega, Gerardo 3
Vercauteren, Frederik 171

W
Wang, Qiang 154

X
Xie, Jiafeng 274
Xiong, Maosheng 19

Z
Zullo, Ferdinando 139

The manufacturer's authorised representative in the EU is Springer Nature Customer Service Centre GmbH, Europaplatz 3, 69115 Heidelberg, Germany. If you have any concerns regarding our products, please contact ProductSafety@springernature.com

Printed and bound by CPI Group (UK) Ltd, Croydon, CR0 4YY

26/03/2026

02078952-0007